T0257535

# X-Ray Spectroscopy

# X-Ray Spectroscopy

Edited by **Hugo Kaye**

New York

Published by NY Research Press,
23 West, 55th Street, Suite 816,
New York, NY 10019, USA
www.nyresearchpress.com

**X-Ray Spectroscopy**
Edited by Hugo Kaye

© 2015 NY Research Press

International Standard Book Number: 978-1-63238-465-2 (Hardback)

Printed in the United States of America.

# Contents

# Preface

In my initial years as a student, I used to run to the library at every possible instance to grab a book and learn something new. Books were my primary source of knowledge and I would not have come such a long way without all that I learnt from them. Thus, when I was approached to edit this book; I became understandably nostalgic. It was an absolute honor to be considered worthy of guiding the current generation as well as those to come. I put all my knowledge and hard work into making this book most beneficial for its readers.

Advanced information regarding the topic of x-ray spectroscopy has been described in this book. X-ray is the only invention that became a routine diagnostic tool in hospitals within a week of its first observation by Roentgen in 1895. Till date, x-ray technology serves as an excellent characterization tool for scientists engaged in nearly all fields, including physics, space science, medicine, archeology, medicine, chemistry, metallurgy, and material science. With an already extensive range of existing applications of x-rays, it comes as a surprise that every day people are discovering novel applications of x-rays or developing the existing methods. The book has been compiled with selective information regarding the current applications of x-ray spectroscopy which are of significant interest to the engineers and scientists engaged in the fields of astrophysics, material science, astrochemistry, chemistry, instrumentation, physics, and methods of x-ray based characterization. The book elucidates fundamental principles of satellite x-rays as characterization devices for the physics of detectors, chemical properties and x-ray spectrometer. It also elucidates techniques like EPMA, EDXRF, satellites, WDXRF, particle induced XRF, matrix effects, and micro-beam analysis. The characterization of ceramic materials and thin films with the help of x-rays has also been described.

I wish to thank my publisher for supporting me at every step. I would also like to thank all the authors who have contributed their researches in this book. I hope this book will be a valuable contribution to the progress of the field.

**Editor**

# Part 1

# XRF Processes and Techniques

# X-Ray Spectroscopy Tools for the Characterization of Nanoparticles

Murid Hussain, Guido Saracco and Nunzio Russo
*Politecnico di Torino*
*Italy*

## 1. Introduction

Photocatalysts are solids that can promote reactions in the presence of light without being consumed in the overall reaction (Bhatkhande et al., 2001), and they are invariably semiconductors. A good photocatalyst should be photoactive, able to utilize visible and/or near UV light, biologically and chemically inert, photostable, inexpensive and non-toxic. For a semiconductor to be photochemically active as a sensitizer for the aforementioned reaction, the redox potential of the photogenerated valence band hole should be sufficiently positive to generate OH$^\bullet$ radicals that can subsequently oxidize the organic pollutant. The redox potential of the photogenerated conductance band electron must be sufficiently negative to be able to reduce the adsorbed $O_2$ to a superoxide. $TiO_2$, $ZnO$, $WO_3$, $CdS$, $ZnS$, $SrTiO_3$, $SnO_2$ and $Fe_2O_3$ can be used as photocatalysts.

$TiO_2$ is an ideal photocatalyst for several reasons (Bhatkhande et al., 2001; Fujishima et al., 2000, 2006; Mills & Hunte, 1997; Park et al., 1999; Peral et al., 1997; Periyat et al., 2008). It is relatively cheap, highly stable from a chemical point of view and easily available. Moreover, its photogenerated holes are highly oxidizing, and the photogenerated electrons are sufficiently reducing to produce superoxides from dioxygen groups. $TiO_2$ promotes ambient temperature oxidation of most indoor air pollutants and does not need any chemical additives. It has also been widely accepted and exploited as an efficient technology to kill bacteria.

Volatile organic compounds (VOCs) are considered to be as some of the most important anthropogenic pollutants generated in urban and industrial areas (Avila et al., 1998). VOCs are widely used in (and produced by) both industrial and domestic activities since they are ubiquitous chemicals that are used as industrial cleaning and degreasing solvents (Wang et al., 2007). VOCs come from many well-known indoor sources, including cooking and tobacco smoke, building materials, furnishings, dry cleaning agents, paints, glues, cosmetics, textiles, plastics, polishes, disinfectants, household insecticides, and combustion sources (Jo et al., 2004; Wang et al., 2007; Witte et al., 2008).

Moreover, ethylene ($C_2H_4$) is an odorless and colorless gas which exists in nature and is generated by human activities as a petrochemical derivative, from transport engine exhausts, and from thermal power plants (Saltveit, 1999). However, it is produced naturally by plant tissues and biomass fermentation and occurs along the food chain, in packages, in storage chambers, and in large commercial refrigerators (Martinez-Romero et al., 2007). The effect of ethylene on fruit ripening and vegetable senescence is of significant interest for the scientific community. During the postharvest storage of fruit and vegetables, ethylene can induce negative effects, such as senescence, overripening, accelerated quality loss, increased fruit pathogen susceptibility, and physiological disorders. Fruit, vegetables, and flowers have ethylene receptors on their surface. Their actuation promotes ethylene production by the fruit itself and accelerates its ripening and aging (Kartheuser & Boonaert, 2007). Thus, the prevention of postharvest ethylene action is an important goal. Conventional as well as commercial techniques and technologies are used to control the action of ethylene, e. g. ethylene scavengers, especially the potassium permanganate ($KMnO_4$) oxidizer. However, $KMnO_4$ cannot be used in contact with food products due to its high toxicity. Ozone ($O_3$) is also an alternative oxidant, but it is highly unstable and decomposes into $O_2$ in a very short time. Carbons and zeolites are used as ethylene adsorbers and they play a key role in the control of ethylene. This technique only transfers the ethylene to another phase rather than destroying it. Hence, additional disposal or handling steps are needed.

New, safe and clean chemical technologies and processes for VOC and ethylene (generated by fruit) abatement are currently being developed (Hussain et al., 2010, 2011a, 2011b, Toma et al., 2006). Conventionally, VOC pollutants are removed by air purifiers that employ filters to remove particulate matter or use sorption materials (e.g. granular activated carbon) to adsorb the VOC molecules. These techniques also transfer the contaminants to another phase instead of destroying them and hence, additional disposal or handling steps are again needed. Moreover, all these sequestration techniques have inherent limitations, and none of them is decisively cost effective. Therefore, there is great demand for a more cost effective and environmentally friendly process that is capable of eliminating VOCs from gas streams, for example photochemical degradation, UV photolysis and photo-oxidation in the presence of some oxidants such as ozone. The photocatalytic oxidation (PCO) of VOCs is a very attractive and promising alternative technology for air purification. It has been demonstrated that organics can be oxidized to carbon dioxide, water and simple mineral acids at low temperatures on $TiO_2$ catalysts in the presence of UV or near-UV illumination. PCO requires a low temperature and pressure, employs inexpensive semiconducting catalysts, and is suitable for the oxidation of a wide range of organics. Some researchers (Augugliaro et al., 1999; Kumar et al., 2005) have already focused on this promising technique, and a great deal of beneficial advancement has been made in the field of VOC abatement. The performance of semiconducting photocatalyst depends above all on its nature and morphology.

Most of the studies have shown that the photocatalytic activity of titanium dioxide is influenced to a great extent by the crystalline form, although controversial results have also been reported in the literature. Some authors have stated that anatase works better than rutile (Zuo et al., 2006), others have found the best photocatalytic activity for rutile (Watson et al., 2003), and some others have detected synergistic effects in the photocatalytic activity for anatase–rutile mixed phases (Bacsa & Kiwi, 1998). It has recently been demonstrated that photo-activity towards organic degradation depends on the phase composition and on the oxidizing agent; for example, when the performance of different crystalline forms was compared, it was discovered that rutile shows the highest photocatalytic activity with $H_2O_2$, whereas anatase shows the highest with $O_2$ (Testino et al., 2007). It has also been found that photoformed OH species, as well as $O^{2-}$ and $O^{3-}$ anion radicals, play a significant role as a key active species in the complete photocatalytic oxidation of ethylene with oxygen into carbon dioxide and water (Kumar et al., 2005).

Therefore, in this chapter we have focused in particular on the synthesis of titania nano-particles (TNP) at a large scale by controlling the optimized operating conditions and using a special passive mixer or vortex reactor (VR) to achieve TNPs with a high surface area and a mixed crystalline phase with more anatase and small amounts of rutile in order to obtain the synergistic effect that occurs between anatase and rutile. These TNPs were characterized and compared with $TiO_2$, synthesized by the solution combustion (TSC) method, and commercially available $TiO_2$ by Degussa P-25 and Aldrich. X-ray diffraction (XRD), energy dispersive X-ray (EDX) spectroscopy and X-ray photoelectron spectroscopy (XPS) techniques were used to screen the best candidate with the best characteristics for the above mentioned catalytic applications.

## 2. Titania synthesis and optimization by means of the XRD technique

Many processes can be employed to produce titanium dioxide particles, e. g. flame aerosol synthesis, hydrothermal synthesis, and sol–gel synthesis (Hussain et al., 2010). Flame aerosol synthesis offers the main advantage of being easily scalable to the industrial level, but also suffers from all the disadvantages of high temperature synthesis. Hydrothermal synthesis is instead particularly interesting as it directly produces a crystalline powder, without the need of a final calcination step, which is necessary in the sol–gel process. However, a lack of knowledge on the chemical equilibria of the species in solution and on the kinetics of the nucleation and growth of the different phases makes it difficult to control the overall process. Therefore, at the moment, the sol–gel process is the most common and promising at a lab scale. Although the sol–gel process has been known for almost a century and some of the most important aspects have been clarified, there is still room for improvement as far as individuating the synthesis conditions that result in a powder with improved properties, compared with the commercial products that are currently available, is concerned. Furthermore, upscaling the process from the laboratory to the industrial scale is still a complex and difficult to solve problem. Mixing plays an important role, but its effects are usually underestimated, as can be seen by the qualitative statements (e.g. add drop wise or mix vigorously) that are generally used to define ideal mixing conditions.

In our previous studies, TNPs were synthesized at a large scale (2 L gel), and the optimized operating parameters were controlled using the vortex reactor (VR) (Hussain et al., 2010, 2011a). Titanium tetra-isopropoxide (TTIP: Sigma–Aldrich) was used as a precursor in these studies, because of its very rapid hydrolysis kinetics. Two solutions of TTIP in isopropyl alcohol and of water (Milli-Q) in isopropyl alcohol were prepared separately under a nitrogen flux to control the alkoxide reactivity with humidity. Hydrochloric acid (HCl: Sigma–Aldrich) was added to the second solution as a hydrolysis catalyst and deagglomeration agent. The TTIP/isopropanol concentration was taken equal to 1 M to obtain the maximum $TiO_2$ (1 M), whereas the water and hydrochloric acid concentrations were chosen in order to result in a water-to-precursor ratio, $W = [H_2O]/[TTIP]$, equal to four, and an acid-to-precursor ratio, $H = [H^+]/[TTIP]$, equal to 0.5. The two TTIP and water solutions in isopropyl alcohol were stored in two identical vessels, then pressurized at 2 bars with analytical grade nitrogen, and eventually fed and mixed in the VR. The inlet flow rates were kept equal to 100 mL/min by using two rotameters. This inlet flow rate guarantees very fast mixing, and induces the formation of very fine particle. Equal volumes of reactant solutions (i.e. 1 L) were mixed at equal flow rates at 28 °C and then, for both configurations, the solutions exiting the VR were collected in a beaker thermostated at 28 °C and gently stirred. The TTIP conversion into $TiO_2$ through hydrolysis and condensation can be summarized in the following overall chemical reaction:

$$Ti(OC_3H_7)_4 + 4H_2O \rightarrow TiO_2 + 2H_2O + 4C_3H_7OH$$

It is well known that a very fast chemical reaction is characterized by an equilibrium that is completely shifted towards the products, and that $TiO_2$ is a thermodynamically very stable substance which results in an almost 100% yield. The reaction product (i.e. gel) was dried in three different ways; dried by a rotavapor, directly in an oven, and in an oven after filtration. The resulting dried powders were eventually calcined at 400 °C for 3 h. TSC was synthesized by following the procedure reported in (Sivalingam et al., 2003), but with modified precursors and ratios. TTIP was used as the precursor, glycine/urea as the fuel, at stoichiometric as well as non-stoichiometric ratios, and 400/500 °C was adopted as the combustion temperature. After the combustion reaction, the samples were calcined at 400 °C for 3 h. Different commercial $TiO_2$ were purchased from Sigma–Aldrich and Aerosil for comparison purposes.

The TNPs were dried according to three different commercial processes in order to find the best method. After drying and before calcination, the powder is mainly amorphous and no distinct peak can be observed, as shown in Fig. 1 (Hussain et al., 2010). However, after calcination at 400 °C for 3 h, the main crystalline form was anatase (denoted as "A") and rutile (denoted as "R") was present to a lower extent (Fig. 1). The optimal drying condition was found by drying in a rotavapor, this resulted in an anatase-to-rutile ratio of 80:20. Details and a comparison with the other drying conditions are reported in Table 1. Compared to TNP, TSC showed a greater rutile phase in all the cases shown in Fig. 2. However, TSC (glycine, 500 °C, 1:1) and TSC (urea, 500 °C, 1:3) were comparatively better in this category, as shown in Table 1. Fig. 3 shows the XRD patterns of three different commercial $TiO_2$. The $TiO_2$ by Aldrich (anatase) showed a pure anatase phase whereas the $TiO_2$ by Aldrich (technical) had a mixed phase. Degussa P 25 also showed mixed anatase and rutile phases.

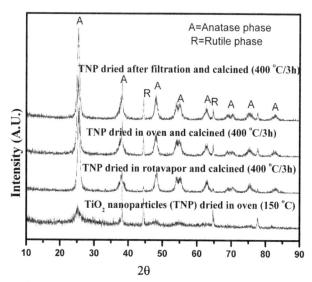

Fig. 1. XRD patterns of different dried TNPs showing the anatase and rutile phases

In order to determine the different polymorphs, XRD patterns were recorded on an X'Pert Phillips diffractomer using Cu Kα radiation, in the following conditions: range = (10–90°) 2θ; step size 2θ = 0.02. Moreover, quantification of the anatase:rutile phases was performed on the basis of the X'Pert database library (Hussain et al., 2010).

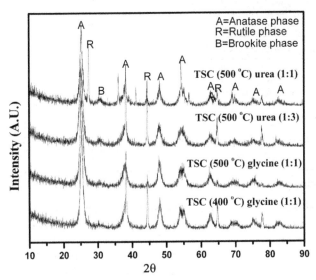

Fig. 2. XRD patterns of different titania synthesized by the solution combustion method (TSC) showing the anatase and rutile phases

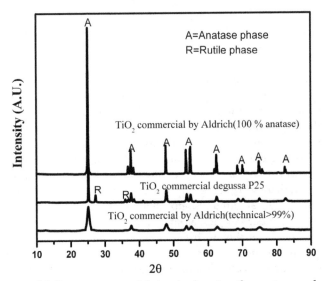

Fig. 3. XRD patterns of different commercial titania showing the anatase and rutile phases

| Sample | Anatase:Rutile (%) |
|---|---|
| TNP (rotavapor dried and calcined) | 80:20 |
| TNP (filtered, dried and calcined) | 71:29 |
| TNP (oven dried and calcined) | 69:31 |
| TSC (glycine, 400 °C, 1:1) | 55:45 |
| TSC (glycine, 500 °C, 1:1) | 60:40 |
| TSC (urea, 500 °C, 1:3) | 61:39 |
| TSC (urea, 500 °C, 1:1) | 58:42 |
| $TiO_2$ commercial (aldrich, technical) | 80:20 |
| $TiO_2$ commercial (aldrich, anatase) | 100:0 |
| $TiO_2$ commercial (degussa P 25) | 70:30 |

Table 1. Crystalline phases of different $TiO_2$ obtained by means of XRD

It is generally accepted that anatase demonstrates a higher activity than rutile, for most photocatalytic reaction systems, and this enhancement in photoactivity has been ascribed to the fact that the Fermi level of anatase is higher than that of rutile (Porkodi & Arokiamary, 2007). The precursors and the preparation method both affect the physicochemical properties of the specimen. In recent years, Degussa P 25 $TiO_2$ has set the standard for photoreactivity in environmental VOC applications. Degussa P 25 is a non-porous 70:30% (anatase to rutile) material. Despite the presence of the rutile phase, this material has proved to be even more reactive than pure anatase (Bhatkhande et al., 2001). Therefore, a mixed anatase–rutile phase seems to be preferable to enable some synergistic effects for photocatalytic reactions since the conduction band electron of the anatase partly jumps to the less positive rutile part, thus reducing the recombination rate of the electrons and the positive holes in the anatase part. The synthesized TNP is characterized by a similar anatase–rutile mixture.

Fig. 4(a) (Hussain et al., 2011a) shows the effect of calcination temperatures at a specific moderate calcination time (3 h) on the TNP by XRD patterns in order to establish the optimized calcination temperature. It was found that when the calcination temperature was below 500 °C, the TNP sample dominantly displayed the anatase phase with just small amounts of rutile. The synthesized TNP was dried in a rotary evaporator and this process was followed by complete water evaporation at 150 °C in an oven before calcination. Just after drying, the powder was mainly amorphous and no distinct peak was found, as shown in Fig. 4(b), although there were some very low intensity peaks at $2\theta$ = 38.47, 44.7, 65.0 and 78.2, which were due to the aluminum sample holder. These aluminum sample holder peaks can also be observed in other samples, as shown in Fig. 4. The effect of calcination times on the characteristics of TNP is shown in Fig. 4(b); these data indicate that even the longer calcination time (7 h) has no significant effect on the TNP and the main phase remains anatase with low amounts of rutile. Therefore, the effect of calcination times at 400 °C is not so severe.

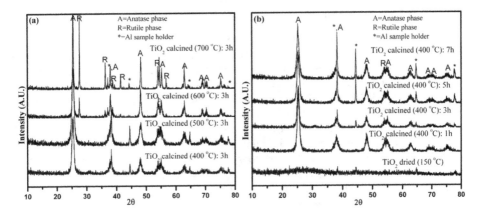

Fig. 4. XRD patterns of TNP at different calcinations (a) temperatures and (b) times

## 3. Synthesis confirmation of optimized TNP with EDX analysis

The elemental composition of TNPs was checked by EDX analysis equipping a high-resolution FE-SEM instrument (LEO 1525). Figure 5 shows the EDX analysis of the optimized TNP (dried by rotavapor and calcined at 400 ºC for 3 h). This figure demonstrates that the main components are O and Ti with small amounts of Cl impurity. This Cl impurity originated from the HCl that was added during the synthesis and it is usually favorable for the photocatalytic reaction (Hussain et al., 2011b).

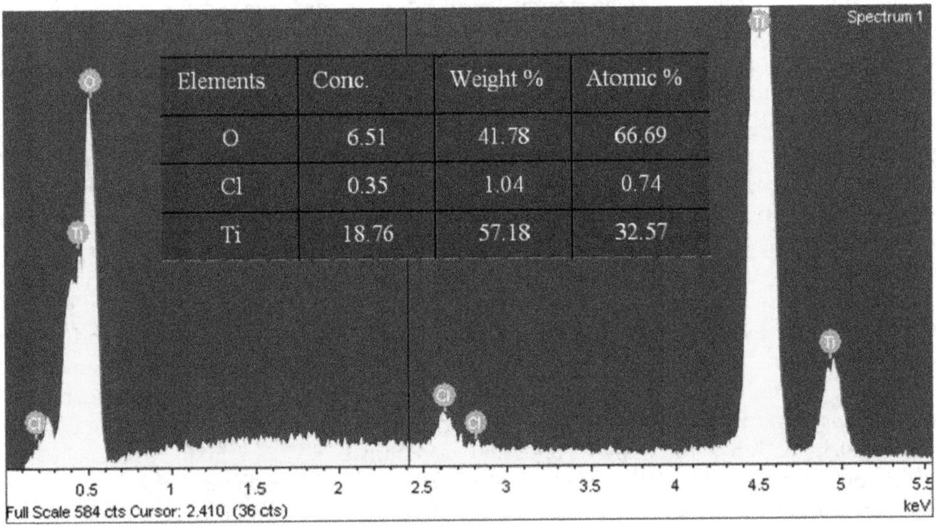

| Elements | Conc. | Weight % | Atomic % |
|----------|-------|----------|----------|
| O | 6.51 | 41.78 | 66.69 |
| Cl | 0.35 | 1.04 | 0.74 |
| Ti | 18.76 | 57.18 | 32.57 |

Spectrum 1

Full Scale 584 cts Cursor: 2.410 (36 cts)

Fig. 5. EDX analysis of TNP

# 4. Photocatalytic reaction

All the ethylene, propylene, and toluene photocatalytic degradation tests were performed in a Pyrex glass reactor with a total volume of 2 L. The experimental setup of the photocatalysis reaction includes a Pyrex glass reactor (transparent to UV light), connectors, mass flow controllers (MFC, Bronkhorst high tech), and a UV lamp (Osram ULTRA-VITALUX 300W. This lamp has a mixture of UVA light ranging from 320–400 nm and UVB light with a 290–320 nm wavelength which produces 13.6 and 3.0 W radiations, respectively; it is ozone-free and the radiations are produced by a mixture of quartz burner and a tungsten wire filament, as mentioned in the manufacturer's indications). The set up also has gas cylinders (1000ppm ethylene/propylene/toluene), a gas chromatograph (Varian CP-3800) equipped with a capillary column (CP7381) and a flame ionization detector (FID) with a patented ceramic flame tip for ultimate peak shape and sensitivity, which was used for the gas analysis of the products (Hussain et al., 2010, 2011a, 2011b).

## 4.1 Screening of the best photocatalyst for ethylene photodegradation

The calcined $TiO_2$ photocatalyst sample was spread homogeneously, by hand, on a support placed inside the Pyrex glass reactor. An initial humidity of 60% was supplied to the photocatalyst to initiate the photocatalytic reaction. The VOC (ethylene, propylene or toluene) was continuously flushed in the reactor, with the help of the MFC, at a constant flow rate of 100 mL/min. After achieving equilibrium in the peak intensity, the UV light was turned on, the reaction products were analyzed by GC, and the conversion was calculated. The reaction experiments were repeated twice and the results showed reproducibility.

PCO of the ethylene over TNP was performed at ambient temperature and compared with different TSC and commercial $TiO_2$ photocatalysts. The important feature of this reaction is the use of air instead of conventional oxygen. In this situation, the required oxygen for the photocatalytic reaction is obtained from the air, leading towards the commercialization step. Fig. 6 shows the percentage conversion of ethylene as a function of time (Hussain et al., 2010). The TNP showed significantly higher conversion than all the other samples. Degussa P 25 showed comparable results. Even the 100% anatase commercial $TiO_2$ showed very low conversion in this reaction. Obviously, TSC synthesized in different ways using urea and glycine were also not suitable for this application. The TNP was active for 6 h of reaction time, unlike degussa P 25, which started to deactivate at this time. This deactivation of degussa P 25 is due to its inferior properties. Moreover, the TNP showed higher activity and better stability because of its superior properties. The main superior characteristic of TNP in ethylene photodegradation is that it has a main anatase phase with limited rutile (Table 1). The photocatalyst become active when photons of a certain wavelength hit the surface, which promotes electrons from the valence band and transfers them to the conductance band (Bhatkhande et al., 2001). This leaves positive holes in the valence band, and these react with the hydroxylated surface to produce OH· radicals, which are the most active oxidizing agents. In the absence of suitable electron and hole scavengers, the stored energy is dissipated, within a few nanoseconds, by recombination. If a suitable scavenger or a surface defect state is available to trap the electron or hole, their recombination is prevented and a subsequent redox reaction may occur. In TNP, which is similar to degussa P 25, the conduction band electron of the anatase part jumps to the less positive rutile part, thus reducing the rate of recombination of the electrons and positive holes in the anatase part.

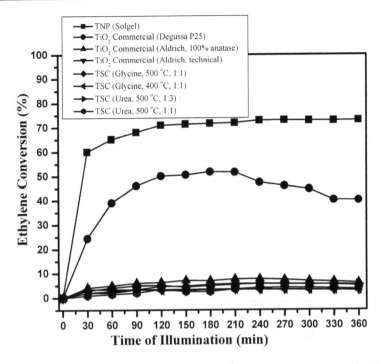

Fig. 6. Ethylene photodegradation over different titania photocatalysts with the illumination time

## 4.2 Optimization of photocatalyst for ethylene photodegradation

The PCO of ethylene was performed at ambient temperature over TNPs calcined at different calcination temperatures and times (Fig. 7) in order to check the catalytic performance of the developed TNP material (Hussain et al., 2011a). Air was again used instead of conventional oxygen in order to obtain more representative data for practical application conditions, in view of commercialization. After a preliminary saturation of the sample under an ethylene flow, conversion did not occur in the dark in any of the experiments, even in the presence of a catalyst or in the presence of UV light and the absence of a catalyst. Therefore, it can be concluded that the reaction results reported hereafter are only induced photocatalytically. Figs. 7(a) and 7(b) show the percentage conversion of ethylene as a function of illumination time. A steady-state conversion is reached after approximately 6 h of illumination for all the samples. This rather long time is necessary because of the type of experimental apparatus that has been employed; on the one hand because of fluid-dynamic reasons and on the other hand to make the surface of the sample reach a steady, equilibrium coverage value. The CO and $CO_2$ measurements of the outlet gases demonstrated that ethylene oxidizes completely to $CO_2$, and only traces of CO are observable. The TNP sample at the highest calcination temperature (700 °C) showed the worst performance, as can be observed in Fig. 7(a). However, the highest conversion was obtained for TNP calcined at 400 °C for 3 h. As expected, all the other sample preparation conditions resulted in a lower catalytic activity. In other words, the TNP sample at the highest calcination time (7 h), also showed the lowest activity, as shown in Fig. 7(b). These reaction results of TNP calcined at different calcination temperatures and times are in perfect agreement with the characterization results of their XRD analysis. The TNP sample calcined at 400 °C for 3 h proved to be the best performing of all the samples. As previously mentioned, it possesses superior characteristics of the mixed anatase (80%)–rutile(20%) phase, a confined band gap energy of 3.17 eV, and the highest BET specific surface area, of 151 m²/g, compared to all the others and therefore showed the best catalytic performance.

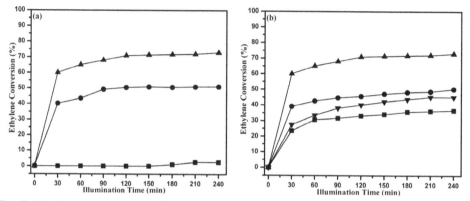

Fig. 7. Ethylene oxidation over TNP photocatalysts at different calcination (a) temperatures: (▲) 400 °C/3 h; (●) 600 °C/3 h; (■) 700 °C/3 h and (b) times: (▼) 400 °C/1 h; (▲) 400 °C/3 h; (●) 400 °C/5 h; (■) 400 °C/7 h. Operating conditions: feed concentration = 200 ppm; flow rate = 100 mL/min, room temperature; 1 g of TNP catalyst

### 4.3 Optimized TNP vs Degussa P 25 titania for VOC photodegradation

Fig. 8 shows a comparison of the optimized TNP and the Degussa P 25 catalysts for ethylene, propylene, and toluene at room temperature (Hussain et al., 2011a). In all three cases of VOC photodegradation for mineralization, the optimized TNP has shown a better activity and stability than that of Degussa P 25. TNP has small nanoparticles with a higher surface area and porosity than the non-porous Degussa P 25 (Hussain et al., 2010). It is also possible that the TNP material has a more amenable anataseto-rutile ratio (80:20) compared to Degussa P 25. Anatase phase based TiO₂ is usually better than rutile for photocatalytic reactions, due to its better adsorption affinity (Periyat et al., 2008). This difference is due to the structural difference of anatase and rutile. Both anatase and rutile have tetragonal structures with $[TiO_6]^{2-}$ octahedra, which share edges and corners in different ways, but maintain the overall TiO₂ stoichiometry. Four edges of the $[TiO_6]^{2-}$ octahedra are shared in anatase and a zigzag chain of octahedra that are linked to each other through shared edges is thus formed, but as far as rutile is concerned, two opposite edges of each $[TiO_6]^{2-}$ octahedra are shared to form the corner oxygen atoms. For this reason, the surface properties of anatase and rutile show considerable differences. Rutile is characterized by a surface on which the dissociation of the adsorbed organic molecules takes place much more easily than on anatase. These essential differences in the surface chemistry of the two TiO₂ phases result in differences in photocatalytic properties since the photocatalysis reactions mainly take place on the surface of the catalyst rather than in the bulk. Rutile titania has a much lower specific surface area than that of anatase. As the specific surface area of the catalyst increases, it can adsorb more VOCs. Moreover, anatase exhibits lower electron–hole recombination rates than rutile due to its 10-fold greater electron trapping rate. Therefore, the mixed optimized TNP phase with a high surface area is the main characteristic which makes it better than Degussa P 25. The XRD data shown in Figs. 1-4 and Table 1 support this affirmation.

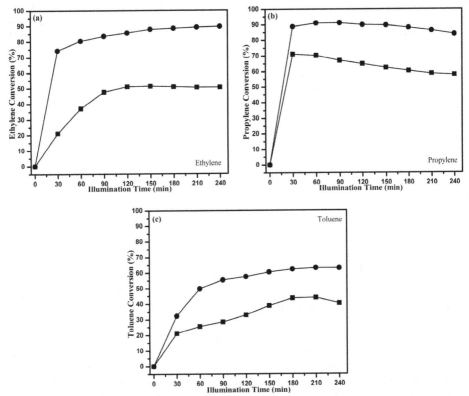

Fig. 8. Comparison of the optimized TNP vs. Degussa P 25 for VOC abatement (a) ethylene, (b) propylene, (c) toluene, fed at 100 ppm, in a 100 mL/min flow rate, over 1 g of catalyst at room temperature: (●) TNP (400 °C/3 h); (■) Degussa P 25

## 4.4 Optimized TNP vs Degussa P 25 titania for VOC photodegradation

The photocatalytic degradation of ethylene was performed in the reaction system explained in (Hussain et al., 2011b), at 3 °C, using ice, an artificial temperature atmosphere that is very similar to that commonly used for the cold storage of fruit. Air was also used here instead of conventional oxygen for the photocatalytic reaction to obtain more representative data of practical application conditions, for commercialization purposes.

Fig. 9(a) shows the effect of surface hydroxyl groups on ethylene degradation (Hussain et al., 2011b). The ethylene degradation reduced very significantly as the surface hydroxyl groups decreased due to increasing calcination temperature.

It has been observed that water has a significant effect on the photocatalytic degradation of ethylene, as shown in Fig. 9(b). After complete drying of the titania, the ethylene degradation reduced significantly. It became very low at the initial illumination time, due to a lack of water, which is necessary for the reaction. However, there was a slight improvement as illumination time increased, which might be due to the water produced during the reaction. This was confirmed when the fully dried titania was kept in a closed vessel with water for 12 h. After 12 h of contact time, the titania showed much higher activity than the fully dried samples. However, there was a slight improvement in ethylene degradation after keeping the normal titania in contact with water. In all of these cases, the TNPs showed better ethylene degradation than Degussa P 25, which might be due to the higher surface area of TNPs available for the adsorption of water and ethylene.

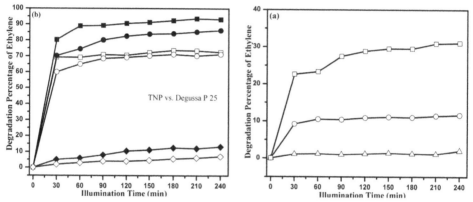

Fig. 9. (a) Effect of OH groups on ethylene degradation at 100 ppm, 100 mL/min flow rate, 3 °C using ice, and 1 g of photocatalyst: (□) TNP (400 °C/(3h))/UV lamp turned down to 75 cm; (○) TNP (600 °C/(3h))/UV lamp turned down to 75 cm; (Δ) TNP (800 °C/(3h))/UV lamp turned down to 75 cm, (b) Water effect on ethylene degradation over TNPs and Degussa P 25 photocatalysts at 100 ppm, 100 mL/min flow rate, 3 °C using ice, and 1 g of photocatalyst, UV lamp turned down to 12 cm/(short converging pipe + lens): (■) TNP kept with 1 g of water for 12 h before reaction; (□) Degussa P 25 kept with 1 g of water for 12 h before reaction; (●) TNP fully dried in oven at 150 °C for 12 h and then kept with water for 12 h before reaction; (○) Degussa P 25 fully dried in oven at 150 °C for 12 h and then kept with water for 12 h before reaction; (♦) TNP fully dried in oven at 150 °C for 12 h and immediate reaction; (◊) Degussa P 25 fully dried in oven at 150 °C for 12 h and immediate reaction.

## 5. XPS analysis of the optimized TNP vs Degussa P 25 titania

The XPS spectra were recorded using a PHI 5000 Versa Probe with a scanning ESCA microscope fitted with an Al monochromatic (1486.6 eV, 25.6 W) X-ray source, a beam diameter of 100 μm, a neutralizer at 1.4 eV and 20 mA, and a FAT analyzer mode. All the binding energies were referenced to the C1s peak at 284.6 eV of the surface carbon. The individual components were obtained by curve fitting (Hussain et al., 2011b).

XPS measurements were conducted to evaluate the hydroxyl groups and the evolution of the valence state of titanium on the TiO$_2$ surfaces. Fig. 10 shows the oxygen O1s XPS spectra

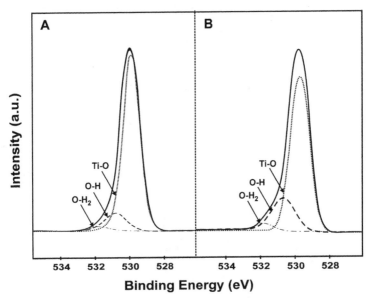

Fig. 10. XPS analysis showing the OH and O-H$_2$ comparison by O1s: (A) Degussa P 25; (B) TNPs

and the deconvolution results of the TNPs and Degussa P 25 from a quantitative point of view (Hussain et al., 2011b). The O1s spectrum displayed peaks at 529.6 eV associated with Ti-O bonds in TiO$_2$, at 530.8 eV, which correspond to the hydroxyl Ti-OH, (Hou et al., 2008; Kumar et al., 2000), whereas, at 532 eV, it shows Ti-OH$_2$ (Toma et al., 2006), which can be observed in the XPS spectra in Fig. 10 (A, Degussa P 25; B, TNPs). The TNPs clearly show more OH groups and OH$_2$ on the surface than Degussa P 25. The quantitative results are given in Table 1. The mass fraction of O1s, the hydroxyl groups, and the water of the two samples were calculated from the results of the curve fitting of the XPS spectra for the O1s region. The O1s values for the TNP and Degussa P 25 were 70.57 and 69.87%, respectively, and are similar. However, the O-H species for TNPs (22.59%) and Degussa P 25 (11.10%) are different. The water attached to Ti for TNPs (5.38%) and Degussa P 25 (2.29%) is also comparable. The higher OH groups on the surface of the TNPs than on Degussa P 25 might be responsible for obtaining superior photocatalytic activity in ethylene photodegradation at low temperature.

A comparison of the Ti2p spectra for TNPs and Degussa P 25 shows a Ti2p$_{3/2}$ peak at 458.5 and Ti2p$_{1/2}$ at 464 eV, as can be observed in Fig. 11(a) (Hussain et al., 2011b). The Ti species peaks, which occur at binding energies of 456.7 (Ti$^{3+}$) and 458.5 eV (Ti$^{4+}$) (Kumar et al., 2000), are shown in Fig. 11 for Degussa P 25 (B) and TNPs (C). It is clear that the TNPs have more Ti$^{3+}$ species than Degussa P 25. After proper calculation through curve fitting, Table 2 shows that the TNP and Degussa P 25 catalysts have similar Ti2p values, but different Ti species. The TNP material has 17.77% Ti$^{3+}$, while Degussa P 25 only shows 8.93%. The Ti$^{3+}$ species are responsible for oxygen photoadsorption, which results in the formation of O-ads, and which, together with the OH radical, is essential for photocatalytic oxidation (Fang et al., 2007; Suriye et al., 2007; Xu et al., 1999). The presence of surface Ti$^{3+}$ causes distinct differences in the nature of the chemical bonding between the adsorbed molecule and the substrate surface. These results are also correlated to the titania photocatalytic mechanism equations (1) and (2) (Hussain et al., 2011a) shown below:

$$Ti(IV) + e^- \rightarrow Ti(III) \tag{1}$$

$$Ti(III) + O_2 \rightarrow Ti(IV) + O_2- \tag{2}$$

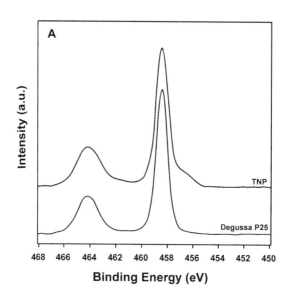

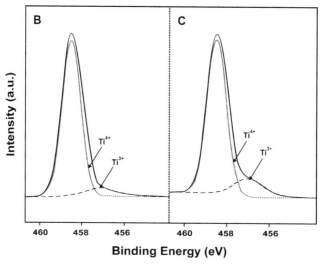

Fig. 11. XPS analysis showing the comparison between Ti2p$_{3/2}$ and Ti2p$_{1/2}$ (A), and the Ti species comparison: (B) Degussa P 25; (C) TNPs

| Catalyst | O1s | Ti2p$_{3/2,1/2}$ | Ti-O | O-H | O-H$_2$ | Ti$^{3+}$ | Ti$^{4+}$ |
|----------|-----|------------------|------|-----|---------|-----------|-----------|
| Degussa P 25 | 69.87 | 30.13 | 86.61 | 11.10 | 2.29 | 8.93 | 91.07 |
| TNP | 70.57 | 29.43 | 72.03 | 22.59 | 5.38 | 17.77 | 82.23 |

Table 2. Atomic concentrations (%) of TiO$_2$ using XPS

## 6. Conclusion

An attempt has been made to synthesize titania nano-particles (TNP) at a large scale by controlling the optimized operating conditions and using a special passive mixer or vortex reactor (VR) to achieve TNPs with a high surface area and a mixed crystalline phase with more anatase and smaller amounts of rutile in order to obtain a synergistic effect between the anatase and rutile. These TNPs were characterized and compared with TiO$_2$ synthesized by means of the solution combustion (TSC) method and commercially available TiO$_2$ by Degussa P-25 and Aldrich. XRD and EDX spectroscopy techniques were used to establish the best candidate with the best characteristics for the above catalytic applications. A higher photocatalytic ethylene conversion was observed for TNP than for TSC or commercial TiO$_2$. The superior TNP photocatalyst was then further optimized by conducting an effective control of the calcination temperatures (400-700 ºC) and times (1-7 h). The optimized TNP was achieved by calcining at 400 ºC for 3 h, which also resulted in rather pure crystalline anatase with small traces of rutile, relatively more Ti$^{3+}$ on the surface, and higher OH surface groups. This was confirmed by means of XRD and XPS investigations. The optimized TNP photocatalyst was then applied for photocatalytic degradation of different VOCs (ethylene, propylene and toluene) at near room temperature. Higher photocatalytic activity for VOC abatement was obtained for TNP than the Degussa P-25 TiO$_2$, due to the optimized mixed phase with a high surface area and the increased Ti$^{3+}$ species, which might induce the adsorption of VOCs and water and generated OH groups which act as oxidizing agents on the TNP surface, leading to higher photocatalytic activity characteristics. The TNPs optimized with the help of XRD and XPS were also applied and compared with Degussa P-25 for the photocatalytic degradation of ethylene (emitted by fruit) at 3 ºC to consider the possibility of its use for the cold storage of fruit. An efficient way of utilizing this optimized TNP photocatalyst for the target application has been developed. The role of the XRD, EDX and XPS characterization tools in the development of TNP for photocatalytic application seems to be very promising and encourages further research in this field.

## 7. Acknowledgment

M.H. is grateful to the Regione Piemonte and the Politecnico di Torino, Italy for his postdoctoral fellowship grant.

## 8. References

Augugliaro, V.; Coluccia, S.; Loddo, V.; Marchese, L.; Martra, G.; Palmisano, L. & Schiavello, M., (1999). Photocatalytic Oxidation of Gaseous Toluene on Anatase TiO$_2$ Catalyst: Mechanistic Aspects and FT-IR Investigation. *Appl. Catal. B: Environ.*, 20, 15–27, ISSN 0926-3373

Avila, P.; Bahamonde, A.; Blanco, J.; Sanchez, B.; Cardona, A. I. & Romero, M., (1998). Gas-Phase Photo-Assisted Mineralization of Volatile Organic Compounds by Monolithic Titania Catalysts. *Appl. Catal. B: Environ.*, 17, 75–88, ISSN 0926-3373

Bacsa, R. R. & Kiwi, J., (1998). Effect of Rutile Phase on the Photocatalytic Properties of Nanocrystalline Titania during the Degradation of p-Coumaric Acid. *Appl. Catal. B: Environ.*, 16, 19–29, ISSN 0926-3373

Bhatkhande, D. S.; Pangarkar, V. G. & Beenackers, A. A. C. M., (2001). Photocatalytic Degradation for Environmental Applications — a Review. *J. Chem. Technol. Biotechnol.*, 77, 102–116, ISSN 0268-2575

Fang, X.; Zhang, Z.; Chen, Q.; Ji, H. & Gao, X., (2007). Dependence of Nitrogen Doping on $TiO_2$ Precursor Annealed under $NH_3$ Flow. *J. Solid State Chem.*, 180, ISSN 1325-1332

Fujishima, A.; Rao, T. N. & Tryk, D. A., (2000). Titanium Dioxide Photocatalysis. *J. Photochem. Photobiol. C: Photochem. Rev.*, 1, 1–21, ISSN 1389-5567

Fujishima, A. & Zhang, X., (2006). Titanium Dioxide Photocatalysis: Present Situation and Future Approaches. *C. R. Chimie*, 9, 750–760, ISSN 1631-0748

Hou, Y. D.; Wang, X. C.; Wu, L.; Chen, Z. X.; Ding, X. X. & Fu, X. Z., (2008). N-Doped $SiO_2/TiO_2$ Mesoporous Nanoparticles with Enhanced Photocatalytic Activity under Visible-Light Irradiation. *Chemosphere*, 72, 414-421, ISSN 0045-6535

Hussain, M.; Ceccarelli, R., Marchisio, D. L.; Fino, D.; Russo, N. & Geobaldo, F., (2010). Synthesis, Characterization, and Photocatalytic Application of Novel $TiO_2$ Nanoparticles. *Chem. Eng. J.*, 157, 45–51, ISSN 1385-8947

Hussain, M.; Russo, N. & Saracco, G., (2011a). Photocatalytic Abatement of VOCs by Novel Optimized $TiO_2$ Nanoparticles. *Chem. Eng. J.*, 166, 138–149, ISSN 1385-8947

Hussain, M.; Bensaid, S.; Geobaldo, F.; Saracco, G. & Russo, N., (2011b). Photocatalytic Degradation of Ethylene Emitted by Fruits with $TiO_2$ Nanoparticles. *Ind. Eng. Chem. Res.*, 50, 2536-2543, ISSN 0888-5885

Jo, W.-K. & Park, K.-H., (2004). Heterogeneous Photocatalysis of Aromatic and Chlorinated Volatile Organic Compounds (VOCs) for Non-Occupational Indoor Air Application. *Chemosphere*, 57, 555–565, ISSN 0045-6535

Kartheuser, B. & Boonaert, C., (2007). Photocatalysis: A Powerful Technology for Cold Storage Applications. *J. Adv. Oxid. Technol.*, 10, 107-110, ISSN 1203-8407

Kumar, P. M.; Badrinarayanan, S. & Sastry, M., (2000). Nanocrystalline $TiO_2$ Studied by Optical, FT-IR and X-ray Photoelectron Spectroscopy: Correlation to Presence of Surface States. *Thin Solid Films*, 358, 122-130, ISSN 0040-6090

Kumar, S.; Fedorov, A. G. & Gole, J. L., (2005). Photodegradation of Ethylene using Visible Light Responsive Surfaces Prepared from Titania Nanoparticle Slurries. *Appl. Catal. B: Environ.*, 57, 93–107, ISSN 0926-3373

Martinez-Romero, D.; Bailen, G.; Serrano, M.; Guillen, F.; Valverde, J. M.; Zapata, P.; Castillo, S. & Valero, D., (2007). Tools to Maintain Postharvest Fruit and Vegetable Quality through the Inhibition of Ethylene Action: A Review. *Crit. Rev. Food Sci.*, 47, 543-560, ISSN 1040-8398

Mills, A. & Hunte, S. L., (1997). An Overview of Semiconductor Photocatalysis. *J. Photochem. Photobiol. A: Chem.*, 108, 1–35, ISSN 1010-6030

Park, D. R.; Zhang, J.; Ikeue, K.; Yamashita, H. & Anpo, M., (1999). Photocatalytic Oxidation of Ethylene to $CO_2$ and $H_2O$ on Ultrafine Powdered $TiO_2$ Photocatalysts in the Presence of $O_2$ and $H_2O$. *J. Catal.*, 185, 114–119, ISSN 0021-9517

Peral, J.; Domenech, X. & Ollis, D. F., (1997). Heterogeneous Photocatalysis for Purification, Decontamination and Deodorization of Air. *J. Chem. Technol. Biotechnol.*, 70, 117–140, ISSN 0268-2575

Periyat, P.; Baiju, K. V.; Mukundan, P.; Pillai, P. K. & Warrier, K. G. K., (2008). High Temperature Stable Mesoporous Anatase $TiO_2$ Photocatalyst Achieved by Silica Addition. *Appl. Catal. A: Gen.*, 349, 13–19, ISSN 0926-860X

Porkodi, K. & Arokiamary, S. D., (2007). Synthesis and Spectroscopic Characterization of Nanostructured Anatase Titania: a Photocatalyst. *Mater. Charact.*, 58, 495–503, ISSN 1044-5803

Saltveit, M. E., (1999). Effect of Ethylene on Quality of Fresh Fruits and Vegetables. *Postharvest Biol. Technol.*, 15, 279-292, ISSN 0925-5214

Sivalingam, G.; Nagaveni, K.; Hegde, M. S. & Madras, G., (2003). Photocatalytic Degradation of Various Dyes by Combustion Synthesized Nano Anatase $TiO_2$. *Appl. Catal. B: Environ.*, 45, 23–38, ISSN 0926-3373

Suriye, K.; Praserthdam, P. & Jongsomjit, B., (2007). Control of $Ti^{3+}$ Surface Defect on $TiO_2$ Nanocrystal using Various Calcination Atmospheres as the First Step for Surface Defect Creation and Its Application in Photocatalysis. *Appl. Surf. Sci.*, 253, 3849-3855, ISSN 0169-4332

Testino, A.; Bellobono, I. R.; Buscaglia, V.; Canevali, C.; D'Arienzo, M.; Polizzi, S.; Scotti, R. & Morazzoni, F., (2007). Optimizing the Photocatalytic Properties of Hydrothermal $TiO_2$ by the Control of Phase Composition and Particle Morphology. A systematic Approach. *J. Am. Chem. Soc.*, 129, 3564–3575, ISSN 0002-7863

Toma, F.-L.; Bertrand, G.; Begin, S.; Meunier, C.; Barres, O.; Klein, D. & Coddet, C., (2006) Microstructure and Environmental Functionalities of $TiO_2$-Supported Photocatalysts Obtained by Suspension Plasma Spraying. *Appl. Catal., B: Environ.*, 68, 74-84, ISSN 0926-3373

Wang, S.; Ang, H. M. & Tade, M. O., (2007). Volatile Organic Compounds in Indoor Environment and Photocatalytic Oxidation: State of the Art. *Environ. Int.*, 33, 694–705, ISSN 0160-4120

Watson, S. S.; Beydoun, D.; Scott, J. A. & Amal, R., (2003). The Effect of Preparation Method on the Photoactivity of Crystalline Titanium Dioxide Particles. *Chem. Eng. J.*, 95, 213–220, ISSN 1385-8947

Witte, K. De; Meynen, V.; Mertens, M.; Lebedev, O. I.; Tendeloo, G. Van; Sepulveda-Escribano, A.; Rodriguez-Reinoso, F.; Vansant, E. F. & Cool, P., (2008). Multi-step Loading of Titania on Mesoporous Silica: Influence of the Morphology and the Porosity on the Catalytic Degradation of Aqueous Pollutants and VOCs. *Appl. Catal. B: Environ.*, 84, 125–132, ISSN 0926-3373

Xu, Z.; Shang, J.; Liu, C.; Kang, C.; Guo, H. & Du, Y., (1999). The Preparation and Characterization of TiO₂ Ultrafine Particles. *Mater. Sci. Eng. B*, 56, 211-214, ISSN 0921-5107

Zuo, G.-M.; Cheng, Z.-X.; Chen, H.; Li, G.-W. & Miao, T., (2006). Study on Photocatalytic Degradation of Several Volatile Organic Compounds. *J. Hazard. Mater.*, B 128, 158–163, ISSN 0304-3894

# A Practical Application of X-Ray Spectroscopy in Ti-Al-N and Cr-Al-N Thin Films

Leonid Ipaz[1], William Aperador[2,3], Julio Caicedo[1],
Joan Esteve[4] and Gustavo Zambrano[1]
*[1]Thin Films Group, Physics Department, University of Valle*
*[2]Colombian School of Engineering Julio Garavito, Bogotá DC*
*[3]Universidad Militar Nueva Granada, Bogota DC*
*[4]Department of Applied Physics Optics, Universitat de Barcelona*
*[1,2,3]Colombia*
*[4]Spain*

## 1. Introduction

Binary and ternary transition metal nitrides coatings have been used in numerous applications to increase the hardness and improve the wear and corrosion resistance of structural materials, as well as in various high-tech areas, where their functional rather than tribological and mechanical properties are of prime importance (Münz, 1986; Chen & Duh, 1991; PalDey & Deevi, 2003; Ipaz et al., 2010). Up to now, Ti-Al-N and Cr-Al-N films have been synthesized by a variety of deposition techniques including cathodic arc evaporation (Cheng et al., 2001), ion plating (Setsuhara et al., 1997), chemical vapor deposition (CVD) or plasma-enhanced CVD (Shieh & Hon, 2001) and d.c. / r.f. reactive magnetron sputtering (Musil & Hruby, 2000; Sanchéz et al., 2010). Performance of these coatings is equally dependent on their chemical composition and long-range crystalline structure, as well as on the nature and amount of impurities and intergranular interactions. Significant improvement in the mechanical properties has recently been achieved with multi-component superlattice, multilayers and nanocomposite nitride coatings. In the case of such multilayers systems, not only is close control of the elemental composition (stoichiometry) and modulation period necessary to optimize the properties of the coatings, but the influence of chemical bond formation between the components is also of prime importance. Therefore, it is necessary to take special care when the conditions of preparation are non-equilibrium, activation of CVD and PVD by plasmas or energetic particle beams are applied, occasionally leading to unpredicted deviations, both in composition and structure. As is highlighted in this study, nitride coatings or nitrided surfaces based in Chromium and Aluminium materials can be analyzed in detail by X-ray photoelectron spectroscopy (XPS) due to its excellent element selectivity, quantitative character and high surface sensitivity. More importantly, XPS reflects the atomic scale chemical interactions, i.e. the bonds between neighboring atoms Cr-N, Al-N, Ti-N, Ti-Al-N and thus it also provides reliable structural characteristics for amorphous or nano-crystalline coatings of complex composition, for which application of diffraction techniques is not straightforward.

## 2. Experimental details and results

[Ti-Al/Ti-Al-N]$_n$ and [Cr-Al/Cr-Al-N]$_n$ multilayers films were deposited at a substrate temperature of 250 °C onto silicon (100) and AISI D3 steel substrates through magnetron co-sputtering pulsed dc method in (Ar/N2) gas mixture, from metallic binary target, 1:1 area ratio of Ti-Al and 0.25:0.75 area ratio of Cr-Al, See fig 1. The d.c. density power was 7.4 W/cm². The substrates were fixed to a holder 6.0 cm above the sputtering target. The deposition chamber was evacuated to a base pressure of 10⁻⁶ mbar before the introduction of the sputtering gas. Metallic single layer films (Ti-Al and Cr-Al) and metal nitride single layer films (Ti-Al-N and Cr-Al-N) were deposited by two different processes: the first one (Ti-Al and Cr-Al materials), was carried out by using a pure Ar gas at 1.5x10⁻³ mbar, the deposition time corresponding to the metallic layer ranged from 3 seconds and 300 seconds for n = 100 and n = 1, respectively (process A); the second one (TiAlN and Cr-Al-N materials), was carried out by using a 50:50 and 90:10 Ar-N₂ ratio in gas mixture for Ti-Al-N and Cr-Al-N, respectively, at 1.5x10⁻³ mbar, the deposition time corresponding to the nitride layer (Ti-Al-N) ranged from 15 seconds to 1500 seconds for n = 100 and n = 1, respectively (process B). So, the multilayers were obtained by alternatively changing the deposition conditions from process A to process B until desired number of bilayers was deposited. To ensure good quality of the interface includes a stabilization period of 20 seconds between process A and B during this period were evacuated gases used in the above process.

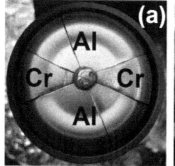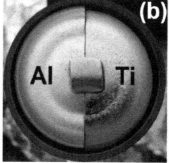

Fig. 1. Metallic binary targets (a) Cr-Al target with 0.25:0.75 area ratio; (b) Ti-Al target with 1:1 area ratio.

The total multilayer thickness was approximately of 2.5 μm for the multilayers deposited with different bilayer periods (Λ) and of bilayers number (n), which was varied between 2.5 μm and 25 nm and 1 and 100, respectively. The deposition parameters are showed in the Table 1. An exhaustive X-ray diffraction (XRD) study was carried out using a PANalytical X'Pert PRO diffractometer with Cu-Kα radiation (α = 1.5406 Å) at Bragg–Brentano configuration (θ/2θ) in high- and low-angle ranges.

The bilayer period and multilayer assembly modulation were observed via scanning electron microscopy (SEM) by using a Leika 360 Cambridge Instruments, equipped with a height sensitivity back-scattered electron detector. The total thickness and the ratio between thickness of Ti-Al and Ti-Al-N single layers was obtained by means of a (Dektak 3030)

Profilometer. Additionally, a detailed X-ray photoelectron spectroscopy (XPS) study of Ti-Al-N coatings, was carried out, in order to determine the [Ti-Al/Ti-Al-N]$_n$ multilayer coatings chemical composition and the bonding of aluminum, titanium, and nitrogen atoms by using ESCAPHI 5500 monochromatic Al-Kα radiation and a passing energy of 0.1 eV, taking in account that surface sensitivity of this technique is so high and any contamination can produce deviations from the real chemical composition and, therefore, the XPS analysis is typically performed under vacuum conditions with some sputter cleaning source to remove any undesired contaminants.

Raman spectroscopy measurements were carried out on a micro Raman system (HR LabRam II by Jobin Yvon), using the 632.8 nm He-Ne laser line, attenuated 100 times with a 2 microns spot. For each Raman measurement the sample was exposed to the laser light for 5 s and 100 acquisitions were performed in order to improve the signal to noise ratio.

| Process | A<br>Metallic (Cr-Al; Ti-Al) layer | B<br>Nitride (Cr-Al-N; Ti-Al-N) layer |
|---|---|---|
| Reactive gas sputtering | Pure Ar | 90/10; 50/50 (%Ar/ %N$_2$) Mixture |
| Substrate temperature | 250°C | |
| Sputtering gas pressure | 1.5x10$^{-3}$ mbar | |
| d.c. power | 7.4 W/cm$^2$ | |
| Target | (Cr-Al; Ti-Al) Binary | |
| Substrate | Silicon (100) | |
| Target-substrate distance | 6 cm | |

Table 1. Process parameters for deposition of the [Ti-Al/Ti-Al-N]$_n$ and [Cr-Al/Cr-Al-N]$_n$ multilayers.

## 2.1 Multilayer modulation analyzed by scanning electron microscopy

A first glimpse on [Ti-Al/Ti-Al-N]$_n$ and [Cr-Al/Cr-Al-N]$_n$ multilayers modulation and microstructure was accomplished by SEM micrographs. Fig. 2a presents the cross-sectional image of a [Ti-Al/Ti-Al-N]$_n$ coating with bilayer number (n = 10) and bilayer period (Λ = 250 nm). The darkest contrast of Ti-Al-N layers with respect to Ti-Al layers allowed a clear determination of the layer structure. These [Ti-Al/Ti-Al-N]$_{10}$ coatings presented a buffer layer of Ti-Al with thickness around 300 nm, the Ti-Al/Ti-Al-N multilayer with total thickness around 2.5 μm with well-defined and uniform periodicity. All the multilayer stacks were resolved by SEM and confirmed quite precisely the previously designed values of bilayer thickness, as well as the total thicknesses. The only slight deviation observed by SEM imaging was on relative thicknesses. Fig. 2b presents the cross-sectional image of a [Cr-Al/Cr-Al-N]$_n$ multilayer films with n = 10 (Λ = 200 nm). The darkest contrast of Cr-Al-N layers with respect to Cr-Al layers allowed a clear determination of the layer structure. These [Cr-Al/Cr-Al-N]$_{10}$ films presented a buffer layer of Cr-Al with thickness around 300 nm, the Ti-Al/Ti-Al-N multilayer with thickness around 2.0 μm with well-defined and uniform periodicity.

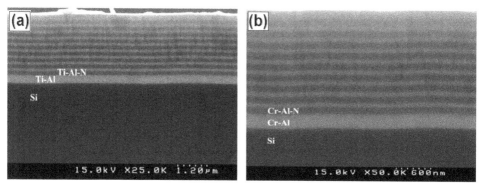

Fig. 2. SEM micrograph of (a) [Ti-Al/Ti-Al-N]$_{10}$ with n = 10, $\Lambda$ =200 nm and (b) [Cr-Al/Cr-Al-N]$_{10}$ films with n = 10, $\Lambda$ = 300 nm.

## 2.2 X-ray analyses of [Ti-Al/Ti-Al-N]$_n$ and [Cr-Al/Cr-Al-N]$_n$ multilayer coatings
### 2.2.1 [Ti-Al/Ti-Al-N]$_n$ multilayer

Fig. 3 shows a typical XRD diffraction patterns of [Ti-Al/Ti-Al-N]$_n$ multilayers deposited onto Si (100) substrate with a period of $\Lambda$ = 42 nm (n = 60), and $\Lambda$ = 25 nm (n = 100). Diffraction patterns show that the [Ti-Al/Ti-Al-N]$_n$ multilayer systems exhibit a polycrystalline structure with a (200) preferential orientation corresponding to TiAlN phase located at 43.72°. Other peaks present in the diffraction patterns can be seen in Table 2. (Schönjahn et al., 2000; Qu et al., 2002)

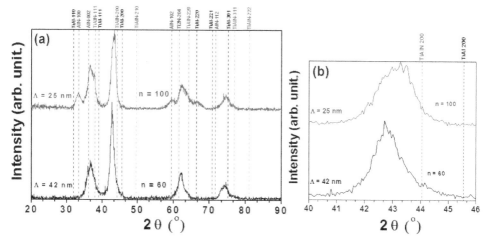

Fig. 3. XRD patterns of the [Ti-Al/Ti-Al-N]$_n$ multilayer coatings deposited on Si (100) substrates with a period of $\Lambda$ = 42 nm (n = 60) and $\Lambda$ = 25 nm (n = 100). Dash lines indicate the position of the peaks obtained from JCPDF 00-037-1140 (TiAlN), JCPDF 00-005-0678 (TiAl) and JCPDF 00-003-1144 (AlN) files from ICCD cards and (b) maximum peak shifts toward higher angles in relationship to increasing bilayer number (n).

| Phase | Ti-Al-N<br>*Cubic* | AlN<br>*Hexagonal* | | | TiAl<br>*Tetragonal* | | Ti2N<br>*Hexagonal* |
|---|---|---|---|---|---|---|---|
| Plane (*hkl*) | (200)<br>(111) | (100) | (002) | (102) | (200)<br>(220) | (301) | (204) |
| Diffraction angle (°) | 43.72<br>36.61 | 33.00 | 36.28 | 59.56 | 45.55<br>66.44 | 74.38 | 83.24 |

Table 2. Peaks present in diffraction patterns.

These preferential orientations are in agreement with JCPDF 00-037-1140 (TiAlN) JCPDF 00-005-0678 (TiAl) and JCPDF 00-003-1144 (AlN) from ICCD cards. The shift of the diffraction patterns toward high angles is in relation to the compressive residual stress characteristics for those multilayer systems deposited by magnetron sputtering technique. Therefore, it was observed that the Ti-Al-N (200) peak position suffers a great deviation from the bulk value indicating a possible stress evolution of Ti-Al/Ti-Al-N layers with the bilayer period (see Fig. 3b). The quasi-relaxed position observed for thinner bilayer periods was progressively shifted to higher compressive stress values when the bilayer period is reduced until the $\Lambda$ = 25 nm, therefore, this compressive effect is reached due to the multilayer mechanism is actuated on Ti-Al/Ti-Al-N coatings. For thinner period multilayers (n = 100), an abrupt change in Ti Al N (220) peak position was observed, presenting a stress relieving due to the movement of this peak toward higher angles compared to other multilayer, but close to bulk value (44.04°). The stress evolution in Ti Al-N (200) peak position comes together with a progressive and intense symmetric broadening and a increasing in its intensity. Moreover, in these patterns is shown clearly not presences of satellite peaks with the increasing in the bilayers number and the thickness reduction of individual layers, which indicate low structural coherency in relation to the lowest bilayer period ($\Lambda$ = 25 nm) (Tien & Duh., 2006).

### 2.2.2 [Cr-Al/Cr-Al-N]$_n$ multilayer

In Cr-Al/Cr-Al-N multilayer, the individual thickness was varied in function of bilayer number from n = 1 to n = 100, producing bilayer period ($\Lambda$) from 2000 nm to 20 nm. So, the Fig. 4a shows the XRD pattern of Cr-Al/Cr-Al-N coatings deposited with n = 60, $\Lambda$ = 33 nm; and n = 100, $\Lambda$ = 20 nm onto Si (100) substrate; the XRD pattern presents a cubic structure were the strongest peak corresponds to CrN (200) plane, indicating a light textured growth along this orientation.

The weak peaks correspond to diffractions of CrN (111), CrN (220), CrN (103), CrAl (514), CrAl (550), AlN (110) and (222) planes of the FCC structure. Moreover, the AlN (422), peak for the AlN within Cr-Al-N material are exhibited. On the other hand, it was observed that the CrN (200) peak position suffers a great deviation from bulk value indicating a possible stress evolution of Cr-Al/Cr-Al-N layers with the bilayer period (Fig. 4b). The quasi-relaxed position observed for thinner bilayer periods was progressively shifted to higher compressive stress values as the bilayer period is reduced until the $\Lambda$ = 20 nm, therefore, this compressive effect is reached due to multilayer mechanism actuating on Cr-Al/Cr-Al-N coatings. For thinner period multilayers (n =100), an abrupt change in CrN (200) peak position was observed, presenting a stress relieving due to the movement of this peak towards higher angles compared to other multilayer but close to bulk value (43.71°). The stress evolution in CrN (200) peak position comes together with a progressive and intense symmetric broadening and a increasing in its intensity. Moreover in these patterns is shown clearly not presences of satellite peaks with the increasing in the bilayers number and the thickness reduction of individual layers, which indicate low structural coherency in relation to the lowest bilayer period ($\Lambda$ = 20 nm) (Yashar & Sproul, 1999).

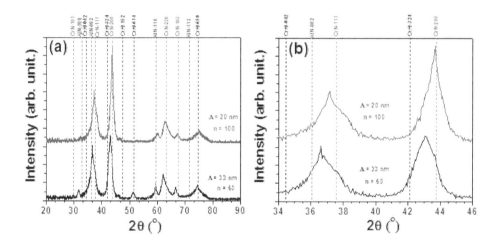

Fig. 4. XRD patterns of the [Cr-Al/Cr-Al-N]$_n$ multilayer coatings deposited on Si (100) substrates with a period of $\Lambda$ = 33 nm (n = 60) and $\Lambda$ = 20 nm (n = 100). (a) Dash lines indicate the position of the peaks obtained from JCPDF 00-001-1232 (CrN), JCPDF 00-002-1192 (CrAl) and JCPDF 00-003-1144 (AlN) files from ICCD cards and (b) maximum peak shifts toward higher angles in relationship to increasing bilayer number (n).

### 2.3 Compositional nature analyzed by X-ray photoelectron spectroscopy (XPS)

An exhaustive X-ray photoelectron spectroscopy (XPS) study was carried out for Ti-Al/Ti-Al-N and Cr-Al/Cr-Al-N multilayer films. XPS was used on [Ti-Al/Ti-Al-N]$_n$ and [Cr-Al/Cr-Al-N]$_n$ samples to determine the chemical composition and the bonding of titanium, aluminium, chromium and nitrogen atoms using ESCAPHI 5500 monochromatic Al K$\alpha$ radiation and a passing energy of 0.1 eV. The surface sensitivity of this technique is so high that any contamination can produce deviations from the real chemical composition and, therefore, the XPS analysis is typically performed under ultra high vacuum conditions with a sputter cleaning source to remove any undesired contaminants.

The XPS survey spectrum for TiAl/TiAlN films is shown in Fig. 5. The peaks at 532.8 eV, 457.6 eV, 463.5eV, 397 eV and 73.9 eV correspond to O1s, Ti2p, Ti2p1/2, N1s and Al2p3/2 binding energies, respectively. Calculation of the peak areas without Ti, Al and N contribution gives an atomic ratio of Ti:Al:N = 2.077:0.256:0.499.

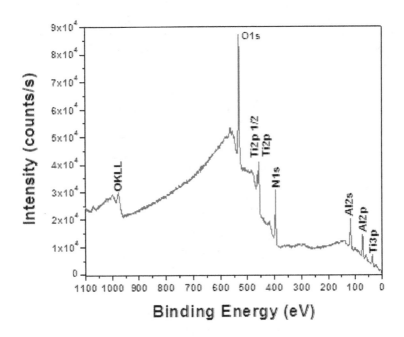

Fig. 5. XPS survey spectrum of Ti-Al-N coatings deposited on Si (100).

According to the XPS literature around Ti–Al–N materials (Fox-Rabinovich et al., 2008; Shum et al., 2003), when the peaks are fitted from experimental results, it is necessary to adjust first the N energy band because it is the element that provides greater reliability for XPS, then, take this first adjust as base, the other peaks related to the remaining elements are adjusted. The latter is indispensable due to the characteristic that is present in this kind of insulating materials (coatings) with respect to the incident signal, avoiding in this way the uncertainties caused for charging and shifts of the Fermi energy. Thus, the concentration measurements and the identification of the specific bonding configurations for the Ti-Al-N layer are more reliable. So, the core electronic spectra carry the information of the chemical composition and bonding characters of the Ti–Al–N films. The integral of O1s, Ti2p, Al2p3/2, and N1s spectra corrected by relevant sensitive factors can evaluate the concentrations of Ti, Al, and N elements in the Ti–Al–N films. The corresponding integral of the deconvoluted peaks can also be used to estimate the bond contents, which are described by the following equation (Moreno et al., 2011):

$$C_i = \sum (A_i / S_i) / (A_j / S_j) \qquad (1)$$

Where S is the sensitivity factor, A is the integral of deconvoluted peaks, and C is the atomic content. The numerator is the sum of the integral of one sort of bond; the denominator is the sum of the integral of all types of bonds decomposed from the whole peak of O1s, Ti2p, Al2p3/2, and N1s spectra in the sample. The atomic concentrations of Ti, Al, N, and O elements from XPS analysis of the coating with n = 1 and $\Lambda$ = 2.5 μm are listed in Table 3, where is observed that Ti–Al–N materials is an over- stoichiometric compound. For the all obtained Ti–Al–N films, the O concentration is less than that 3 at. %. Moreover when the substrate is heated to 250 °C during the deposition process, the atomic compositions in the films almost remain unchanged.

| Atomic composition (at.%) | | | (Ti+Al/N) |
|---|---|---|---|
| Ti | Al | N | 1.45 |
| 19.83 | 39.34 | 40.83 | |

Table 3. Atomic composition of [Ti-Al/Ti-Al-N]$_n$ coating with n = 1 and $\Lambda$ = 2.5 μm

The high-resolution spectra of Ti2p, N1s and Al2p3/2 were recorded from TiAlN coatings, as shown in Fig. 6. From Fig. 6(a), the Ti2p peak is composed of spin doublets, each separated by 6.1 eV. The XPS spectrum of Ti2p can be fitted well by two Gaussian functions. The values of binding energies obtained for the Ti2p1/2 peak were 463.7 eV and the lower value for Ti2p was 457.6 eV.

According to the literature (Tam et al., 2008; Kim et al., 2002) for the Ti2p1/2 peak, the first one (463.5 eV) and the second one (458.6 eV) can be assigned to Ti–O2 and Ti–N bonds. The appearance of the peak at 457.6 eV clearly shows that Ti has reacted with N; therefore, it can be assigned to Ti–N (Badrinarayanan et al., 1989; Moreno et al. 2011).

Fig. 6(b) depicts the N1s spectrum. The N1s peak can be deconvoluted in four peaks at 396.9 eV, 398.0 eV, and 400 eV characteristic for Ti–N$_X$, Ti–N and N-O, respectively (Shum et al., 2003; Shulga et al., 1976; Hironobu et al., 2003). Fig. 3(c) shows the Al2p3/2 spectrum. Four fitting peaks of the Al peak can be deconvoluted at 72.8 eV, 73.9 eV, 75.5 eV and 77.3 eV, moreover these peaks are associated to the Al–Al bond, Al–N, Al$_x$O and Al$_2$–O$_3$ with a single bond, respectively (Shum et al., 2003; Taylor & Rabalais, 1981; Lindsay et al., 1973). The XPS results demonstrate that Ti and the Al atoms have been bonded with N in the forms of nitride the elemental concentration of the Ti–Al–N film was controlled by adjusting the applied power of the Ti-Al binary target, and the Ar/N$_2$ gas mixture. Generally, formative N (Ti, Al) phase with good crystallization indicates that the aluminium and titanium activity and activation energy provided by the present deposition conditions are enough for the formation of a Ti–Al–N film. Although the surface temperature of the substrate during deposition of a Ti–Al–N film is around 250 °C, the substrate lies in a high-density plasma region and a high ion-to-atom ratio of titanium, aluminium and nitrogen can be propitious to the formation of Ti–Al–N phase proven by XPS results.

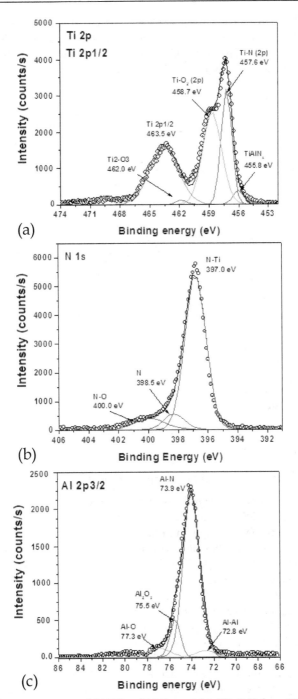

Fig. 6. High-resolution spectrum of (a) Ti2p, (b) N1s and (c) Al2p3/2 from Ti-Al/Ti–Al–N film.

On the other hand, the XPS survey spectrum for Cr-Al/Cr-Al-N films is shown in Fig. 7. The peaks at 532.8 eV, 397 eV, 73.9 eV, and 575.6 eV correspond to O1s, N1s, Al2p3/2 and Cr2p3/2 binding energies, respectively. Calculation of the peak areas without Cr, Al and N contribution gives an atomic ratio of Cr:Al:N=0.341:1.978:0.398.

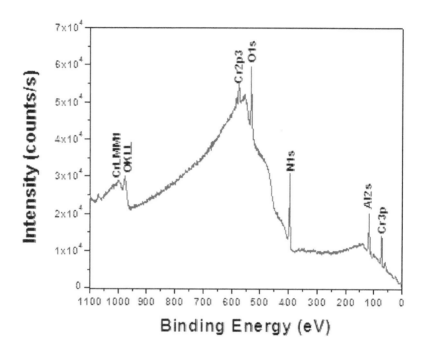

Fig. 7. XPS survey spectrum of Cr-Al-N coatings deposited on Si (100).

According to the XPS literature around Cr–Al–N materials (Barshilia et al., 2007; Del Re et al., 2003; Sanjinés et al., 2002), when the peaks are fitted from experimental results, it is necessary to adjust first the N energy band because it is the element that provides greater reliability for XPS, then, take this first adjust as base, the other peaks related to the remaining elements are adjusted. The latter is indispensable due to the characteristic that is present in this kind of insulating materials (films) with respect to the incident signal, avoiding in this way the uncertainties caused for charging and shifts of the Fermi energy. Thus, the concentration measurements and the identification of the specific bonding configurations for the Cr-Al-N layer are more reliable. So, the core electronic spectra carry the information of the chemical composition and bonding characters of the Cr-Al-N films. The integral of O1s, N1s, Al2p3/2, and Cr2p3/2 spectra corrected by relevant sensitive factors can evaluate the concentrations of Cr, Al, and N elements in the Cr–Al–N films. The corresponding integral of the deconvoluted peaks can also be used to estimate the bond contents, which are described by the following equation (Moreno et al., 2011):

$$C_i = \sum (A_i / S_i) / \sum (A_j / S_j) \qquad (2)$$

where S is the sensitivity factor, A is the integral of deconvoluted peaks, and C is the atomic content. The numerator is the sum of the integral of one sort of bond; the denominator is the sum of the integral of all types of bonds decomposed from the whole peak of O1s, N1s, Al2p3/2 and Cr2p3/2 spectra in the sample. The atomic concentrations of Cr, Al, N, and O elements from XPS analysis of the coating with n = 1 and Λ=2 μm are listed in Table 4. For all the Cr–Al–N films obtained, the O concentration is less than 3 at.%. Moreover when the substrate is heated to 250 °C, the atomic compositions in the films almost remain unchanged.

| Atomic composition (at.%) | | | (Cr+Al/N) |
|---|---|---|---|
| Cr | Al | N | |
| 38.27 | 29.71 | 32.46 | 2.09 |

Table 4. Atomic composition of [Cr-Al/Cr-Al-N]$_n$ coating with n = 1 and Λ = 2.0 μm

The high-resolution spectra of Cr2p, N1s and Al2p were recorded from CrAlN coatings, as shown in Fig. 8. From Fig. 8a, the Cr2p peak is composed of spin doublets, each separated by 9.9 eV. The XPS spectrum of Cr2p can be fitted well by two Gaussian functions. The values of binding energies obtained for the Cr2p3/2 peak were 575.6 eV, 578.3 eV associated at Cr-N bonds and the lower value for Cr2p1/2 was 585.5 eV.

According to the literature (Barshilia et al., 2007; Allen & Tucker, 1976; Geoffrey et al., 1973) for the Cr2p peak, the first composed of two peaks (575.6 eV and 578.3 eV) associated to Cr-N and Cr-O3 bonds and the second one (585.5 eV) can be assigned to Cr-N bond. The appearance of the peaks at 575.6 eV and 585.5 eV clearly shows that Cr has reacted with N.

Fig. 8b depicts the N1s spectrum. The N1s peak can be deconvoluted in two peaks at 396.4 eV and 397.2 eV characteristic for Cr-N and Cr$_2$-N, respectively (Del Re et al., 2003; Nishimura et al., 1989).

Fig. 5(c) shows the Al2p3/2 spectrum. Four fitting peaks of the Al peak can be deconvoluted in four peaks at 72.8 eV, 73.9 eV, 75.5 eV and 77.3 eV, moreover these peaks are associated to the Al-Al bond, Al-N, Al$_x$-O and Al$_2$O$_3$ with a single bond, respectively (Lindsay et al., 1973; Sanjinés et al., 2002; Briggs & Seah, 1994; Taylor, 1982; Kuo & Tsai, 2000; Barr, 1983; Endrino et al., 2007). The XPS results demonstrate that Cr and the Al atoms have been bonded with N in the forms of nitride. The elemental concentration of the Cr-Al-N film was controlled by adjusting the applied power of the Cr-Al binary target, and the Ar/N$_2$ gas mixture. Generally, formative N (Cr, Al) phase with good crystallization indicates that the aluminium and titanium activity and activation energy provided by the present deposition conditions are enough for the formation of a Cr-Al-N coating. Although the surface temperature of the substrate during deposition of a Cr-Al-N film is around 250 °C, the substrate lies in a high-density plasma region and a high ion-to-atom ratio of chromium, aluminium and nitrogen can be propitious to the formation of Cr-Al-N phase proven by XPS results.

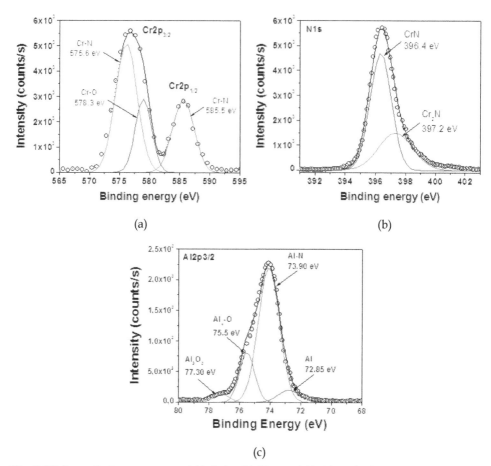

Fig. 8. High-resolution spectrum of (a) Cr2p, (b) N1s and (c) Al2p3/2 from Cr-Al/Cr–Al–N film

## 2.4 Raman spectroscopy analysis of Ti-Al-N and Cr-Al-N films

Raman spectroscopy has been widely used to analyze the present phases in thin films, in this work we used this technique to study ternary nitrides based in transition metals (Cr, Al and Ti). From structural nature of Ti-Al-N and Cr-Al-N, it is known that addition of Al to TiN films or Al to CrN films changes the structure of the dominating phases from the face-centered cubic (*fcc*) to the hexagonal wurtzite. In a perfect crystal with fcc structure, every ion is at a site of inversion symmetry, In consequence, vibration-induced changes in the electronic polarizability, necessary for Raman-scattering, are zero no first-order Raman effect can be observed. When the inversion symmetry is destroyed at the neighboring sites, certain atomic displacements have non-zero first-order polarizability derivatives and Raman scattering is induced. In the case of metal-nitride coatings, lattice defects, interstitial phases and "impurities" of the diatomic cubic coating structure render certain atomic displacements Raman active. Simple factor group analysis shows that both optic and acoustic branches give rise to Raman bands according to $\Gamma = A_{1g} + E_g + T_{2g}$ in the $O_h$ impurity-site symmetry (Montgomery et al., 1972; Constable et al., 1999).

### 2.4.1 Ti-Al-N films

In the Figure 9. shows Raman spectra of Ti-Al-N films in the range of 0 to 1500 cm$^{-1}$. Raman spectra can be fitted well with five Gaussian-Lorentzian peaks at approximately 210, 320, 476 and 620 cm$^{-1}$. The spectrum of Ti-Al-N showed two broad bands centered at 210 and 620 cm$^{-1}$. These bands originate due to the first-order transverse acoustic in the 150-350cm$^{-1}$ region (LA and TA) and the optic modes in the 400-650cm$^{-1}$ region (LO and TO). Additionally was observed at 476 cm$^{-1}$ the second-order acoustic (2A), at 735 cm$^{-1}$ the transverse optical (TO) mode (Constable et al., 1999; Shum et al., 2004).

Higher frequency spectral density then arises via second-order transitions (A+O) in the range of 700 to 900 cm$^{-1}$.

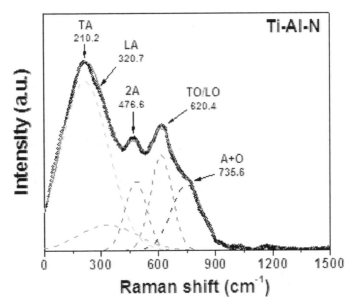

Fig. 9. Raman spectra of Ti-Al-N film.

### 2.4.2 Cr-Al-N films

Raman spectra of Cr-Al-N films can be fitted well with five Gaussian-Lorentzian peaks at approximately 236, 475, 650, 712 and 1410 cm$^{-1}$. The Raman shifts of the main bands and assignments are given in figure 10 spectra of binary aluminum nitride films have been described elsewhere (Shum et al., 2004; Barata et al., 2001) and are not analyzed here.

The spectrum of Cr-Al-N films shows a intense band at 236 cm$^{-1}$, this band originate due to acoustic transitions in the 150-300 cm$^{-1}$ region (LA and TA) because of vibrations of Cr ions. Additionally was observed two weak bands at 475 cm$^{-1}$ and 650 cm$^{-1}$, this bands were associated to longitudinal and transversal optic vibrations (LO and TO) in the 400 - 650 cm$^{-1}$ region. Bands at higher wave numbers arise via second-order transitions (A+O, 2O). With high probability, first-order optic and second-order acoustic modes (2A) strongly overlap in the wave number range between 400 - 1000 cm⁻¹.

Typical features of Raman spectra of Al-Cr-N coatings are three broad bands separated by low intensity "gaps" (Fig. 10). In concordance by literature (Kaindl et al., 2006) the spectrum of the Al-Cr-N coating, the low wave number band is assigned to TA and LA modes, the mid-wave number band to 2A, TO/LO and A+O modes and the high-wave number band to 2O modes.

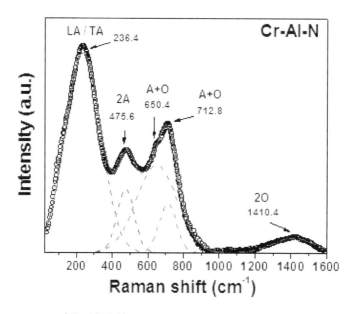

Fig. 10. Raman spectra of Cr-Al-N film.

## 3. Conclusion

The novel [Ti-Al/Ti-Al-N]ₙ and [Cr-Al/Cr-Al-N]ₙ nanometric multilayer structure was identified to be non-isostructural multilayer of Ti-Al, Cr-Al and Ti-Al-N, Cr-Al-N phases.

An exhaustive structural and compositional study was realized by X-ray photoelectron spectroscopy (XPS), Scanning electron microscopy (SEM), X-Ray Diffraction (XRD) and Raman spectroscopy. The preferential orientation for cubic TiAlN (200), AlN (002) and (301) tetragonal structure of TiAl single layer was shown via XRD. In [Cr-Al/Cr-Al-N]ₙ XRD pattern presents a cubic structure were the strongest peak corresponds to CrN (200) plane indicating a light textured growth along this orientation. From XPS results it was possible to identify the chemical composition in both single layer coatings. The SEM analysis confirmed well-defined multilayer structures and showed little variations in layer thickness ratio.

Raman spectra of Ti-Al-N and Cr-Al-N films show groups of bands in the acoustic and optical range, determined by Raman shifts and associated with typical phases present in these materials.

X-Ray spectroscopy techniques, such as, X-ray photoelectron spectroscopy (XPS), Scanning electron microscopy (SEM), X-Ray Diffraction (XRD) and Raman spectroscopy, are clearly powerful tools for the characterization of thin films and coatings with much scope for enhanced understanding of the structural and properties of future generations of PVD materials.

## 4. Acknowledgment

This research was supported by "El patrimonio Autónomo Fondo Nacional de Financiamiento para la Ciencia, la Tecnología y la Innovación Francisco José de Caldas" under contract RC-No. 275-2011 with Center of Excellence for Novel Materials (CENM). Moreover, the authors acknowledge the Serveis Científico-Técnics of the Centro de Investigación y de Estudios Avanzados del IPN, Unidad Querétaro, México and the Universitat de Barcelona for XPS and SEM analysis. L. Ipaz thanks Colciencias for the doctoral fellowship.

## 5. References

Allen G.C. & Tucker P.M. (1976). Multiplet splitting of X-ray photoelectron lines of chromium complexes. The effect of covalency on the 2p core level spin-orbit separation, *Inorganica Chimica Acta*, Vol. 16, pp. 41-45, ISSN 0020-1693.

Badrinarayanan, S.; Sinha, S. & Mandale, A.B. (1989). XPS studies of nitrogen ion implanted zirconium and titanium. *Journal of Electron Spectroscopy and Related Phenomena*, Vol. 49, No. 3, pp. 303-309, ISSN 0368-2048.

Barata, A.; Cunha, L. & Moura, C. (2001). Characterisation of chromium nitride films produced by PVD techniques, *Thin Solid Films*, Vol. 398-399, pp. 501-506, ISSN 0040-6090.

Barr, T.L. (1983). An XPS study of Si as it occurs in adsorbents, catalysts, and thin films, *Applications of Surface Science*, Vol. 15, No. 1-4, pp. 1-35, ISSN 0378-5963.

Barshilia, Harish C.; Deepthi, B.; Rajam, K.S. (2007). Deposition and characterization of CrN/Si$_3$N$_4$ and CrAlN/Si$_3$N$_4$ nanocomposite coatings prepared using reactive D.C. unbalanced magnetron sputtering, *Surface & Coatings Technology*, Vol. 201, No. 24, pp. 9468–9475, ISSN 0257-8972.

Briggs, D. & Seah, M. P. (1994). *Practical Surface Analysis: Auger and X-ray photoelectron spectroscopy*, 2 ed., Vol.1, pp. (641-644), John-Wiley & Sons, ISBN 0-471-920, New York, USA.

Chen, Y.I. & Duh, J.G. (1991). TiN coatings on mild steel substrates with electroless nickel as an interlayer. *Surface & Coatings Technology*, Vol. 48, No. 2, pp. 163-168, ISSN 0257-8972.

Cheng, Y. H.; Tay, B. K.; Lau S. P. & Shi X. (2001). Influence of substrate bias on the structure and properties of (Ti,Al)N films deposited by filtered cathodic vacuum arc. *Journal of Vacuum Science & Technology A*, Vol. 19, No. 3, pp. 736-743, ISSN 0734-2101.

Constable, C.P.; Yarwood, J. & Münz, W.-D. (1999). Raman microscopic studies of PVD hard coatings, *Surface & Coatings Technology*, Vol. 116-119, pp. 155-159, ISSN 0257-8972.

Del Re, M.; Gouttebaron, R.; Dauchot, J.-P.; Leclère, P.; Terwagne, G. & Hecq, M. (2003). Study of ZrN layers deposited by reactive magnetron sputtering, *Surface & Coatings Technology*, Vol. 174-175, pp. 240–245, ISSN 0257-8972.

Endrino, J.L.; Fox-Rabinovich, G.S.; Reiter, A.; Veldhuis, S.V.; Escobar Galindo, R.; Albella, J.M. & J.F. Marco. (2007). Oxidation tuning in AlCrN coatings, *Surface & Coatings Technology*, Vol. 201, No. 8, pp. 4505-4511, ISSN 0257-8972.

Fox-Rabinovich, G.S.; Yamamoto, K.; Kovalev, A.I.; Veldhuis, S.C.; Ning, L.; Shuster, L.S. & Elfizy, A. (2008). Wear behavior of adaptive nano-multilayered TiAlCrN/NbN coatings under dry high performance machining conditions. *Surface & Coatings Technology*, Vol. 202, No. 10, pp. 2015-2022, ISSN 0257-8972.

Geoffrey C. Allen, Michael T. Curtis, Alan J. Hooper & Philip M. Tucker. (1973). X-Ray photoelectron spectroscopy of chromium–oxygen systems, *Journal of the Chemical Society, Dalton Transactions*, No. 16, pp. 1675-1683, ISSN 1477-9226.

Hironobu Miya, Manabu Izumi, Shinobu Konagata & Takayuki Takahagi. (2003). Analysis of chemical structures of ultrathin oxynitride films by X-Ray Photoelectron Spectroscopy and Secondary Ion Mass Spectrometry, *Japanese Journal of Applied Physics*, Vol. 42, pp. 1119-1122, ISSN 0021-4922.

Ipaz, L.; Caicedo, J. C.; Alba de Sánchez, N.; Zambrano & G.; Gómez, M. E. (2010). Tribological characterization of Cr/CrN films deposited onto steel substrates by d.c. magnetrón co-sputtering method. *LatinAmerican Journal of Metallurgy and Materials*, Vol. 30, No. 1, pp. 82-88. ISSN 0255-6952.

Kaindl, R.; Franz, R.; Soldan, J.; Reiter, A.; Polcik, P.; Mitterer, C.; Sartory, B.; Tessadri, R. & O'Sullivan, M. (2006). Structural investigations of aluminum-chromium-nitride hard coatings by Raman micro-spectroscopy, *Thin Solid Films*, Vol. 515, No. 4, pp. 2197-2202, ISSN 0040-6090.

Kim, Soo Young; Jang, Ho Won; Kim, Jong Kyu; Jeon, Chang Min; Park, Won Il; Yi, Gyu-Chul & Jong-Lam LeeSoo. (2002). Low-resistance Ti/Al ohmic contact on undoped ZnO. *Journal of Electronic Materials*, Vol. 31, No. 8, pp. 868-871, ISSN 0361-5235.

Kuo, Hong-Shi & Tsai, Wen-Ta. (2000). Electrochemical Behavior of Aluminum during Chemical Mechanical Polishing in Phosphoric Acid Base Slurry, *Journal of The Electrochemical Society*, Vol. 147, No. 1, pp. 149-154, ISSN 0013-4651.

Lindsay, J.R.; Rose, H.J.; Swartz, W.E.; Watts, P.H. & Rayburn, K.A. (1973). X-ray Photoelectron Spectra of Aluminum Oxides: Structural Effects on the "Chemical Shift", *Applied Spectroscopy*, Vol. 27, No. 1, pp. 1-5, ISSN 0003-7028.

Montgomery Jr., G.P.; Klein, M.V.; Ganguly, B.N. & Wood, R.F. (1972). Raman scattering and far-infrared absorption induced by silver ions in sodium chloride, *Physical Review B*, Vol. 6, No. 10, pp. 4047-4060, ISSN 0163-1829.

Moreno, H.; Caicedo, J.C.; Amaya, C.; Cabrera, G.; Yate, L.; Aperador, W. & P. Prieto. (2011). Improvement of the electrochemical behavior of steel surfaces using a TiN[BCN/BN]$_n$/c-BN multilayer system. *Diamond and Related Materials*, Vol. 20, No. 4, pp. 588–595, ISSN 0925-9635.

Münz, W.D. (1986). Titanium aluminum nitride films: A new alternative to TiN coatings. *Journal of Vacuum Science & Technology A*, Vol. 4, No. 6, pp. 2717- 2726, ISSN 0734-2101.

Musil, J. & Hruby, H. (2000). Superhard nanocomposite $Ti_{1-x}Al_xN$ films prepared by magnetron sputtering. *Thin Solid Films*, Vol. 365, No. 1, pp. 104-109, ISSN 0040-6090.

Nishimura O., Yabe K. & Iwaki M. (1989). X-ray photoelectron spectroscopy studies of high-dose nitrogen ion implanted-chromium: a possibility of a standard material for chemical state analysis. *Journal of Electron Spectroscopy and Related Phenomena*, Vol. 49, No. 3, pp. 335-342, ISSN 0368-2048.

PalDey, S. & Deevi, S.C. (2003). Single layer and multilayer wear resistant coatings of (Ti,Al)N: a review. *Materials Science and Engineering: A*, Vol. 342, No. 1, pp. 58-79, ISSN 0921-5093.

Qu, X. X.; Zhang, Q. X.; Zou, Q. B.; Balasubramanian, N.; Yang, P. & Zeng, K. Y. (2002). Characterization of TiAl alloy films for potential application in MEMS bimorph actuators. *Materials Science in Semiconductor Processing*, Vol. 5, No. 1, pp. 35-38, ISSN 1369-8001.

Sanchéz, J.E.; Sanchez, O.M.; Ipaz, L.; Aperador, W.; Caicedo, J.C.; Amaya, C.; Hernandez-Landaverde, M.A.; Espinoza-Beltrán, F.; Muñoz-Saldaña, J. & Zambrano, G. (2010). Mechanical, tribological, and electrochemical behavior of $Cr_{1-x}Al_xN$ coatings deposited by r.f. reactive magnetron co-sputtering method. *Applied Surface Science*, Vol. 256, No. 8, pp. 2380-2387, ISSN 0169-4332.

Sanjinés, R.; Banakh, O.; Rojas, C.; Schmid, P.E. & Lévy, F. (2002). Electronic properties of $Cr_{1-x}Al_xN$ thin films deposited by reactive magnetron sputtering, *Thin Solid Films*, Vol. 420-421, pp. 312-317, ISSN 0040-6090.

Schönjahn, C.; Bamford, M.; Donohue, L.A.; Lewis, D.B.; Forder, S. & Münz, W.-D. (2000). The interface between TiAlN hard coatings and steel substrates generated by high energetic $Cr^+$ bombardment. *Surface & Coatings Technology*, Vol. 125, No. 1, pp. 66-70, ISSN 0257-8972.

Setsuhara, Y.; Suzuki, T.; Makino, Y.; Miyake, S.; Sakata, T. & Mori, H. (1997). Phase variation and properties of (Ti, Al)N films prepared by ion beam assisted deposition . *Surface & Coatings Technology*, Vol. 97, No. 1, pp. 254-258, ISSN 0257-8972.

Shieh, J. & Hon, M.H. (2001). Nanostructure and hardness of titanium aluminum nitride prepared by plasma enhanced chemical vapor deposition. *Thin Solid Films*, Vol. 391, No. 1, pp. 101-108, ISSN 0040-6090.

Shulga, Y.M.; Troitskii, V.N.; Aivazov, M.I. & Borodko. Y.G. (1976). X-ray Photoelectronspectra of scandium, titanium, vanadium and chromium mononitrides, *Zhurnal Neorganicheskoi khimii*, Vol. 21, No. 10, pp. 2621-2624, ISSN 0036-0236.

Shum, P.W.; Li, K.Y.; Zhou, Z.F. & Shen, Y.G. (2004). Structural and mechanical properties of titanium–aluminium–nitride films deposited by reactive close-field unbalanced magnetron sputtering, *Surface & Coatings Technology*, Vol. 185, No. 2-4, pp. 245–253, ISSN 0257-8972.

Shum, P.W.; Zhou, Z.F.; Li, K.Y. & Shen, Y.G. (2003). XPS, AFM and nanoindentation studies of $Ti_{1-x}Al_xN$ films synthesized by reactive unbalanced magnetron sputtering. *Materials Science and Engineering: B*, Vol. 100, No. 2, pp. 204-213, ISSN 0921-5107.

Tam, P.L.; Zhou, Z.F.; Shum, P.W. & Li, K.Y. (2008). Structural, mechanical, and tribological studies of Cr–Ti–Al–N coating with different chemical compositions. *Thin Solid Films*, Vol. 516, No. 16, pp. 5725-5731, ISSN 0040-6090.

Taylor, J.A. (1982). An XPS study of the oxidation of Al-As thin films grown by MBE. *Journal of Vacuum Science & Technology*, Vol. 20, No. 3, pp. 751-757, ISSN 0022-5355.

Taylor J.A. & Rabalais J.W. (1981). Reaction of $N_2^+$ beams with aluminum surfaces, *Journal of Chemical Physics*, Vol. 75, No. 4, pp. 1735-1746, ISSN 0021-9606.

Tien, S.-K. & Duh, J.-G. (2006). Comparison of microstructure and phase transformation for nanolayered CrN/AlN and TiN/AlN coatings at elevated temperatures in air environment. *Thin Solid Films*, Vol. 515, No. 3, pp. 1097-1101, ISSN 0040-6090.

Yashar, Philip C. & Sproul, William D. (1999). Nanometer scale multilayered hard coatings. *Vacuum*, Vol. 55, No. 3, pp. 179-190, ISSN 0042-207X.

# 3

# Analysis of the *K* Satellite
# Lines in X-Ray Emission Spectra

M. Torres Deluigi and J. Díaz-Luque
*Universidad Nacional de San Luis*
*Argentina*

## 1. Introduction

When electromagnetic radiation or particles interact with matter, and they are energetic enough, they can produce the ejection of electrons from the atoms, so the latter become ionized. The ionization can be achieved either by irradiation with a conventional x-ray tube (which emits the characteristic x-radiation of the anode material and a copious amount of white radiation over a wide wavelength range), or by impact with electrons (or heavier particles) accelerated in a suitable gun.

The emission of x-ray from an excited ion arises from a single electron transition between the states with final and initial vacancies. This is primarily because of the strong electric dipole selection rule, which remains dominant for all save the shortest wavelength x-rays (i.e. $\lambda \leq 100$ pm). This selection rule requires that the quantum number for the orbital angular momentum shall change by only one unit during the transition, i.e.:

$$\Delta l = \pm 1 \tag{1}$$

The probability that an incident photon (or particle) causes the ejection of a particular electron is proportional to the square of the integral:

$$\int \Psi_i P \Psi_f \tag{2}$$

where $P$ is the transition operator, and $\Psi_i$, $\Psi_f$ are the wave functions for the initial and final states of the system, respectively (Urch, 1985). An approach to this integral can be obtained by replacing the wave functions $\Psi_i$ and $\Psi_f$ by those of the orbitals directly involved in the transition, and considering that all the other orbitals are not affected. So, $\Psi_i$ should be replaced by the orbital $\varphi_i$ from which the electron was ejected, and $\Psi_f$ should be replaced by the orbital $\varphi_f$ from which comes the electron that will fill the initial vacancy. Furthermore, if we consider the electric dipole (*er*) approximation for $P$, the integral (2) can be replaced by the following:

$$\int \varphi_i (er) \varphi_f \tag{3}$$

If the electron is removed from an inner orbital (or core orbital), then the resulting ion may relax in two ways:

a.   either by emitting an x-ray that results from an electronic transfer from an external orbital ionization energy $E_j$ to an inner orbital ionization energy $E_i$. The energy of the emitted photon, which characterizes the levels involved, will be:

$$h\nu = E_i - E_j \tag{4}$$

b.   or ejecting an Auger electron, which will leave a doubly ionized atom with vacancies in orbitals $\varphi_j$ and $\varphi_k$. The energy of the Auger electron will be approximately:

$$E = E_i - E_j - E_{k'} \tag{5}$$

where $E_{k'}$ is the ionization energy for an electron from an orbital k in an atom with an atomic number which is one greater than the atom under consideration.

## 2. Characteristic x-rays

The x-ray spectrum emitted by the sample is produced by simple electronic transitions, which allow more straightforward interpretations than those of the Auger spectra. When only considering the electric dipole vector of radiation, x-ray emission from an excited ion is governed by simple rules of selection ($\Delta l = \pm 1$ and $\Delta j = 0, \pm 1$). The probability of relaxation via the magnetic vector or the electric quadrupole is less, by at least two orders of magnitude, than that for relaxation involving dipolar emission (Urch, 1979). The selection rules give rise to a series of x-rays which are called 'diagram lines' because the basic electronic transitions can be easily represented on a single line diagram as shown in Fig. 1.

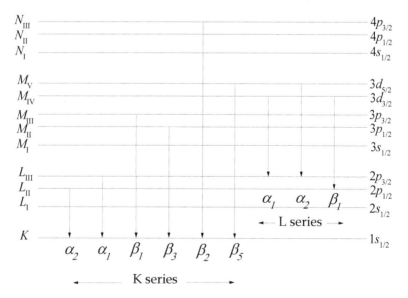

Fig. 1. Emission diagram of the x-ray lines of an atom. Lines $K$ and $L$.

When a $K$-shell electron is removed producing a $1s^{-1}$ vacancy, it can be replaced by a $p$ electron from layers $L$, $M$ or $N$ (see Fig. 1). If the transition occurs from a state $2p_{3/2}$, it emits a characteristic x-ray $K\alpha_1$; however, if the replacement is from $2p_{1/2}$, it emits a photon $K\alpha_2$. On the other hand, if the transition occurs from layers $M$ or $N$, i.e., $3p{\rightarrow}1s^{-1}$, $3d{\rightarrow}1s^{-1}$ or $4p{\rightarrow}1s^{-1}$, it produces a characteristic photon emission of the $K\beta$ spectrum.

The transition rules may also be extended to the lines originating from molecular orbitals of the valence band. The emitted spectrum will then have information on the contribution of atomic orbitals to molecular orbitals.

The intensity $I$ of x-rays emitted is given by the Einstein equation, which has the following form (Eyring, 1944):

$$I = cte \cdot \nu^3 \cdot \left[ \int \Psi_i \, P \, \Psi_f \right]^2 \qquad (6)$$

where $\Psi_i$ and $\Psi_f$ are complete wavefunctions for the initial and final states, before and after the x-ray emission respectively, and $P$ is the transition operator. In the one electron approximation, it is assumed that only the orbitals directly involved in the transition need to be considered. The effects of relaxation and electronic reorganization in other molecular orbitals attendant upon the creation of the initial vacancy, and also in the final state after the x-ray has been emitted, are ignored. Since other orbitals are assumed unaffected, this approach is also called the 'frozen orbital' approximation.

Thus, as in the case of equation (2), $\Psi_i$ will be replaced by the atomic orbital for the initial vacancy $\varphi_i$, and $\Psi_f$ by the (atomic or molecular) orbital with the final vacancy. When $\Psi_f$ is to be replaced by a molecular orbital ($\psi_f$), the simplest form will be the Linear Combination of Atomic Orbitals (LCAO):

$$\psi_f = \sum a_{f,\lambda} \varphi_\lambda \qquad (7)$$

where $a_{f,\lambda}$ is a coefficient describing the contribution of atomic orbitals $\varphi_\lambda$ of the atom $\lambda$ to the molecular orbital $\psi_f$.

These molecular orbitals, which are solutions of the Schrödinger equation, can be calculated in various degrees of approximation using tools of quantum mechanics. The energies of the molecular orbitals (occupied and unoccupied) and its composition in terms of electronic populations of the constituent atomic orbitals can be obtained by theoretical models.

There are several computational methods that calculate the molecular orbitals; some of them are, in order of increasing sophistication: the extended Huckel method, the method of Fenske-Hall and the methods $X\alpha$ (Cotton and Wilkinson, 2008). Among them, the one that shows greater agreement with the experimental spectra is the Discrete Variational $X\alpha$ (DV-$X\alpha$), that was shown for sulfur compounds by Mogi (Mogi et al., 1993), Uda (Uda et al., 1993) and Kawai (Kawai, 1993), and for manganese and chromium by Mukoyama (Mukoyama et al., 1986). This method first calculates the atomic orbitals $\varphi_\lambda$ using the approximation of Hartree-Fock-Slater (HFS) for each constituent atom. The atomic wave functions obtained numerically are used as base functions to construct the molecular wave functions. The elements of the secular matrix are calculated using the DV method and this matrix is then diagonalized to obtain eigenvalues and eigenfunctions of the molecular orbital (Adachi et al., 1978).

The detail with which an x-ray peak will reflect the molecular orbital structure of a molecule or solid will be, apart from experimental considerations, a function of the line widths of the constituent peaks. In addition, the line widths will be determined by the lifetimes of the initial and final states of the ion between which the transition occurs.

## 3. Satellite lines

An x-ray emission line (or diagram line) resulting from a transition between two levels in the energy-level diagram is frequently accompanied by satellites lines (or non-diagram lines), i.e., x-ray lines whose energies do not correspond to the difference of two energy levels of the same atom. The term 'satellite' means weak lines close to the strong parent (or diagram) lines. Particle induced x-ray emission (PIXE) spectra, electron probe microanalysis (EPMA) x-ray spectra, or x-ray fluorescence (XRF) spectra of materials exhibit intensity modifications of satellite lines from one compound to another. The satellite lines are classified into three groups by its origin: 1) multivacancy satellites; 2) charge-transfer satellites for late transition-metal compounds; 3) molecular-orbital splitting satellites (Kawai, 1993).

### 3.1 Multivacancy satellites

The $K\alpha$ lines usually show high-energy satellite lines, corresponding to the existence of a specific number (i) of 'spectator' vacancies in the layer $L$, which affect the subsequent transitions. These peaks can be identified generally as $K\alpha L^i$, and they appear on the side of higher energies of the main line $K\alpha_{1,2}$. This kind of satellite lines are called multivacancy satellites lines, and among them we find the double ionization caused by $K\alpha L^1$ (or simply $K\alpha L$): when $1s$ and $2p$ vacancies are created simultaneously, the $2p$ vacancy has a relatively long life-time compared to that of the $1s$ vacancy. Thus, the inner vacancy de-excites in presence of a 'spectator hole' which produces a change in the electrostatic potential, leading to shifts in the energy levels, affecting as a result the energy of the photon emitted (Fig. 2). In this chapter we will study the satellite lines produced by double and triple vacancies in the $K\alpha$ spectrum of aluminium.

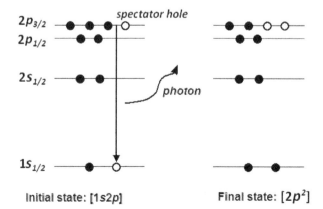

Fig. 2. Emission of a satellite line by double ionization (orbitals with vacancies are enclosed in brackets).

The high energy satellite lines have been intensively studied since the 1930s to the 1940s, beginning with the detailed works of Parrat (Parratt, 1936; 1959; Randall and Parratt, 1940).

In the case of the $K\alpha$ emission line, the high energy satellites are usually labeled as $K\alpha_{3,4}$. The high energy satellite resulting from the de-excitation in presence of two outer vacancies is referred as $K\alpha_{5,6}$ and exhibits a very weak amplitude. The energy separation distance between the satellite band and the $K\alpha$ line ranges from about 10 eV up to about 40 eV for atomic numbers $12 < Z < 30$. Aberg (Aberg, 1927) presents an extensive set of values for the relative intensity of the satellite, which decreases from about 30 % for $Z = 10$ to 0.5 % at $Z = 30$.

The $KL^n$ satellite intensity has a close relation to the electronegativity of the neighboring atoms of the x-ray emitting atom, as shown by Watson et al. (Watson et al., 1977), Uda et al. and Endo et al. (Uda et al., 1979; Endo et al., 1980). This is because the $KL^n$ satellite is stronger for ionic compounds than it is for covalent compounds: The valence electrons are delocalized for covalent compounds, thus the perturbation due to the creation of the core-hole is small for said compounds. Consequently, the multielectron ionization probability is small for covalent compounds. On the other hand, for ionic compounds, the valence electrons are localized, thus the perturbation due to the core-hole creation is large. Therefore, the satellites are strong for ionic compounds. Hence, the satellite intensity is a measure of valence electron delocalization, so we can determine the covalence/ionicity of compounds or delocalization/localization of valence electrons by measuring the multivacancy satellites of x-ray emission spectra (Kawai, 1993).

The de-excitation of an $L$ or $M$ level in presence of outer holes may also lead to the presence of high energy satellites associated with $L$ or $M$ x-ray peaks. The additional outer vacancies may result from Coster-Kronig transitions or shake-off mechanisms. The Coster-Kronig transitions result from an Auger process between sub-shells of the same shell. For instance, the hole created on the $L_1$ sub-shell may be filled by an electron originating from the $L_2$ or $L_3$ subshell. According to the selection rules, these transitions are not radiative and the excess of energy $L_1$-$L_2$ or $L_1$-$L_3$ is dissipated by the emission of an Auger electron from the $M$ or $N$ levels. The transition rate of non-radiative Coster-Kronig transitions $f_{ij}$, where i and j are two subshells within the same energy level, is not permitted for all elements (Rémond, 2002).

The double excitation threshold is the value of energy needed to cause double vacancies in the atoms, so the probability to observe satellite lines $K\alpha$ is zero for incident energies below this value. Therefore, the formation of double vacancy is only possible for values equal or superior to the threshold, and the satellite spectra are variable according to the energy that excites the sample. The effects of variation of the $K\alpha$ satellite spectrum depending on the excitation energy can be seen, for example, in the work of Oura et al. (Oura et al., 2003). These authors bombarded NaF samples and measured the intensity of the lines $KL^1$ in the F, from the double excitation threshold to saturation. Fig. 3 shows the different spectra obtained with the different energies for the incident photons (indicated on the right vertical axis). The double excitation threshold is 706.7 eV for F, energy value around which their measures began, and for which the satellite lines are not even seen.

The multivacancy satellites in x-ray emission spectra are lines of importance in PIXE spectra: When heavy ions (such as $Ar^{5+}$, $N^+$ or $O^{2+}$, with an energy of several tens of MeV) bombard the sample material to ionize the core electron, these satellites turn out to be stronger than the diagram lines.

In Fig. 4 it is shown the proton-excited spectrum, which consists of the normal $K\alpha_{1,2}$ peak and $K\alpha_{3,4}$ satellite group. The $K\alpha_{5,6}$ group and other satellites have lower intensity and, in consequence, were not measured with proton excitation. In the case of nitrogen-ion excitation, the spectrum is radically different: A relatively weak line appears at the position of the normal $K\alpha_{1,2}$. The five peaks on the high-energy side of the $K\alpha_{1,2}$ line are $K\alpha$ satellites lines from atoms with one through five vacancies in the $L$ shell (Knudson, 1971). We can also see the increasing intensity of the lines of the spectrum $K\beta$, and the emergence of new satellite lines in this region.

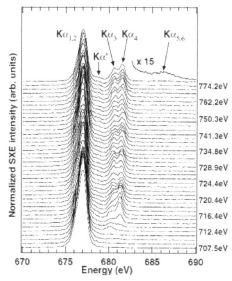

Fig. 3. Spectral variation of the F $K\alpha$ emission for NaF. Typical energies of the exciting photons are indicated beside each spectrum (Oura et al., 2003).

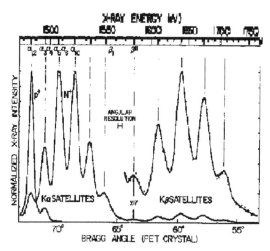

Fig. 4. Al $K\alpha$ x-ray spectra excited by impact of 5 MeV protons and nitrogen ions on Al metal, (Knudson, 1971).

### 3.1.1 Double and triple vacancy satellite lines in Al $K\alpha$

At higher energies than $K\alpha_{1,2}$ we find the satellite spectrum $KL^1$ consisting of lines (in increasing order of energy value) $K\alpha'$, $K\alpha_3$ and $K\alpha_4$. At even higher energies we can also observe satellite lines by triple ionization ($KL^2$), called $K\alpha_5$ and $K\alpha_6$. In this section, we present the results of our measurements for these lines of multiple vacancies for pure aluminum and alumina ($Al_2O_3$), with XRF and EPMA. The $KL^2$ satellites are the highest order of ionization obtained by bombarding with photons and electrons, and they have even lower intensities than $KL^1$. All these lines can be seen in Fig. 5, which presents the experimental intensities of the Al $K\alpha$ spectrum on a logarithmic scale to facilitate its appreciation.

The spectra obtained with XRF and EPMA were fitted with the software Peakfit. This software employs the method of least squares to approximate peaks of the Voigt Gaussian / Lorentzian type to the lines, by setting residuals. Comparing Fig. 6 and Fig. 7 we see that there is an increase in the intensities of the satellite lines by double and triple vacancy with respect to the main lines, when measured with EPMA.

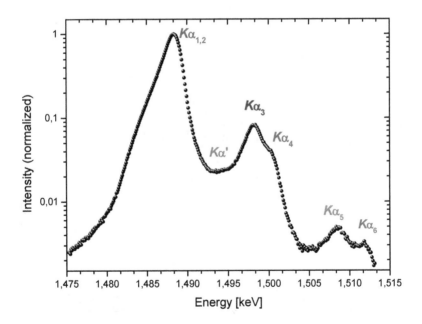

Fig. 5. Al $K\alpha$ spectrum measured through XRF.

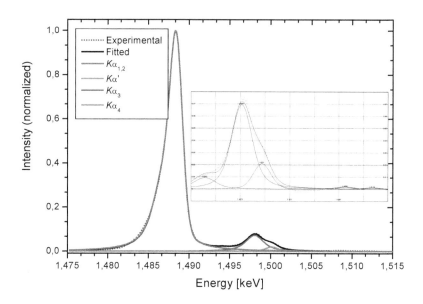

Fig. 6. Al Kα spectrum measured through XRF. Expanded Kα satellites in the top right.

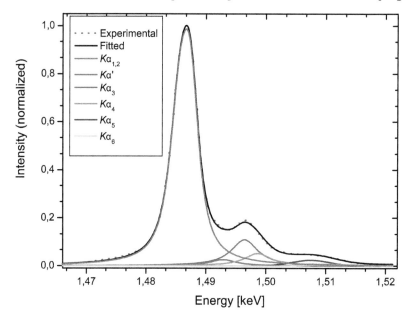

Fig. 7. Al Kα spectrum measured through EPMA. Expanded Kα satellites in the top right.

| Lines | XRF (This work) | | XRF (Other Authors) | |
|-------|------|-------|--------|-----------|
| | Al | Al$_2$O$_3$ | Al [*] | Al$_2$O$_3$ [**] |
| $K\alpha_{1,2}$ | 1486,7 | 1487,0 | - | - |
| $K\alpha'$ | 1492,6 | 1494,0 | 1492,6 | 1492,94 |
| $K\alpha_3$ | 1496,5 | 1497,2 | 1496,4 | 1496,85 |
| $K\alpha_4$ | 1498,5 | 1499,2 | 1498,4 | 1498,70 |
| $K\alpha_5$ | 1507,2 | 1507,8 | - | 1507,4 |
| $K\alpha_6$ | 1510,0 | 1510,4 | - | 1510,9 |

Table 1. Energy of Al $K\alpha$ lines in Al and Al$_2$O$_3$ (in eV) for single and double ionization, obtained with XRF in this work and by different authors: [*]: (Mauron and Dousse, 2002); [**]: (Wollman et al., 2000).

Table 1 shows that the concordance of our measurements for the energies of the lines Al $KL^1$ with Mauron and Dousse (Mauron and Dousse, 2002) is remarkable (are equal considering the experimental error), especially if one considers that the latter is a recent work that uses a Von Hamos spectrometer, which was specially designed for this purpose. The same applies to the work of Wollman et al. (Wollman et al., 2000) on the Al$_2$O$_3$, whose energy values differ at most in 1 eV for $K\alpha'$ (which is the line with the greatest uncertainty due to its low intensity), and in general the energies of the other lines differ by less than 0.5 eV of ours. In Table 1 we can also see that the energies obtained for the triple vacancy $KL^2$ lines are very similar to those reported by other authors.

Also in Table 1 we show that the Al $K\alpha$ satellite spectrum changes significantly in the aluminum oxide with regard to the metal: there are shifts in the energy positions of the lines and changes in their relative intensities (see Fig. 8). The shift towards higher energies in the oxide (Al$_2$O$_3$) is in agreement with the results found by Liu et al. (Liu, 2004).

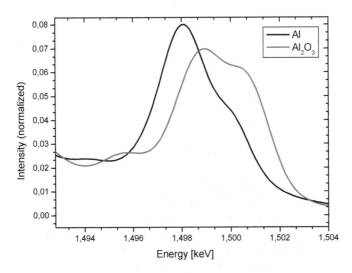

Fig. 8. Satellite spectra of Al and Al$_2$O$_3$ measured through XRF.

In Table 2 there are the energies of the lines $KL^0$ and $KL^1$ in Al measured by XRF and EPMA in this work, and also those reported in other studies which use different experimental techniques. The major significance of the correlation between the energies measured by us and the ones measured by other authors -whose results are shown comparatively in Table 2-, is the fact that these references relate to works performed with different excitation techniques. Thus, the fact that our results on Al are equal (considering the uncertainties of the measurements) to those of Burkhalter et al. (Burkhalter et al., 1972), obtained with PIXE, reveals the intrinsic nature of the energies of the double-ionization lines, and its independence on the technique used to measure them. Identical energy positions are also observed in our own measurements with XRF and EPMA, confirming this independence on the excitation method.

| Line | EPMA+ | XRF+ | EPMA* | EPMA** | XRF* | PIXE* |
|------|-------|------|-------|--------|------|-------|
| $K\alpha_{1,2}$ | 1486,7 | 1486,7 | 1486,7 | - | - | 1486,6 |
| $K\alpha'$ | 1492,9 | 1492,6 | 1492,3 | 1492,8 | 1492,6 | 1492,8 |
| $K\alpha_3$ | 1496,5 | 1496,5 | 1496,4 | 1496,5 | 1496,4 | 1496,4 |
| $K\alpha_4$ | 1498,7 | 1498,5 | 1498,4 | 1498,5 | 1498,4 | 1498,6 |

Table 2. Energy in eV of the Al $K\alpha$ lines for single and double ionization, obtained with different techniques and different authors: EPMA+ and XRF+: this work; EPMA*: (Fischer and Baun, 1965); EPMA** and XRF*: (Mauron and Dousse, 2002) ;PIXE*: (Burkhalter et al., 1972).

Fig. 9 and Fig.10 show the double ionization satellite lines measured in Al and in $Al_2O_3$ with XRF. In these figures, we can see the changes that occur in every satellite line: In the case of $K\alpha'$, it is shifted to higher energy and increases its intensity in the oxide, while the intensities of the lines $K\alpha_4$ and $K\alpha_3$ change from the pure element to an oxide. In particular, it is noted that the intensity ratio between these two lines is $IK\alpha_4 / I\ K\alpha_3 \cong 0.5$ in the metal, while in the oxide it is $IK\alpha_4 / I\ K\alpha_3 \cong 1$.

The intensity distribution of the satellite spectrum analyzed, in all spectra measured with both techniques, behaves as follows: $K\alpha'$ is the less intense (10% of $K\alpha_3$), $K\alpha_3$ is the most intense (approximately 10% of $K\alpha_{1,2}$), followed by $K\alpha_4$ (5% of $K\alpha_{1,2}$), except in the aluminum oxide, when the latter two are almost of the same intensity. The ratio of intensities between them ($IK\alpha_4/ IK\alpha_3$) varies in our measurements with EPMA from about 0.5 (in the metal) to 1 in the oxide.

The total intensity of the satellite spectrum with respect to the main one ($I_{satélites}/IK\alpha_{1,2}$, where $I_{satélites} = IK\alpha' + IK\alpha_3 + IK\alpha_4$), is of 10% for XRF and of the order of 20% for EPMA. In Al $K\alpha$ spectrum of oxide, there is a decrease of this intensity with EPMA, and there are no significant changes with XRF. The double vacancy lines are more intense with EPMA, compared to XRF (approximately the double). This confirms the higher probability of double ionization when the sample is excited by charged particles (electrons in our case) than when excited by photons.

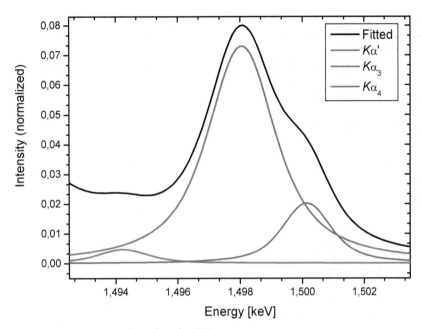

Fig. 9. Satellite lines measured in Al with XRF.

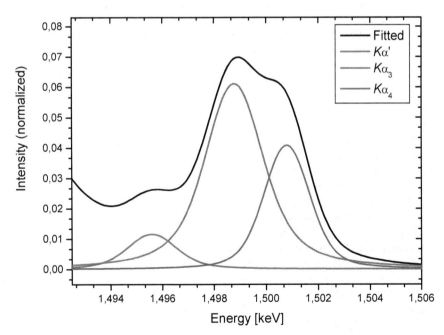

Fig. 10. Satellite lines measured in $Al_2O_3$ with XRF.

## 3.2 Satellite lines produced by charge transfer

The effect of charge transfer is important in the compounds formed by the last transition metals. Usually, in these compounds, the $3d$ orbital is the outermost. Then there is the ligand energy level (e.g. in $NiF_2$, the level of F $2p$) which is a few eV below of $3d$ of the main atom. So if there was a vacancy in the $1s$ level of the transition element, the $3d$ level is far deeper than the valence bond. This situation is quite unstable, so some electrons from the decay of the F $2p$ level to the level of Ni $3d$ (Kawai, 1993).

## 3.3 Satellite lines caused by splitting of molecular orbitals

X-rays from transitions from the valence layer to internal orbitals (as the lines $K$ and $L$) of elements that are linked to other lighter ones have satellite lines of lower energy than the main peak, with a relative intensity of 5-30% (Urch, 1970). In consequence, these low-energy satellite lines are closely associated with bond formation. The energy separation between these two peaks is a function of the ligand atom (about 20 eV for F, 15 eV for O, 11 eV for N and 8 eV for C) (Esmail et al., 1973). The origin of these lines is just the splitting of molecular orbitals, and can be understood by the simple model of the Molecular Orbital (MO) that was described in the introduction.

Let $p$ and $d$ orbitals of the valence shell of a central atom (A) in a compound, or a crystalline medium, which is surrounded by x binder atoms (L). The transition $K$ ($p \rightarrow 1s$) now displays a structure that reflects a number of molecular orbitals of the form ALx. Light elements, such as C, N, O and F, use their $2s$ and $2p$ orbitals in bond formation. There are therefore two types of molecular orbitals involving electrons A$p$: (A$p$, L$2p$) and (A$p$, L$2s$). Similar arguments can be made for the $L$ lines. The energy difference between these two orbitals can be directly measured in appropriate experiments. The presence of these light elements cannot be easily detected using conventional x-ray detectors, which justifies the additional interest in studying this kind of satellite lines.

The occurrence of low energy satellite peaks in the x-ray emission spectrum of the element A, which identify the ligand L, not only shows that L is present in the sample, but also that L is bound to A. The word 'bound' is used in a broad sense, however covalent bonds give more intense satellite peaks. But it is true that ionic bonds are never completely ionic, and even a small percentage of covalent bond is sufficient to generate significant emission peaks. Among this class of satellite lines, the main one is the $K\beta'$, which is located on the side of lower energies than the $K\beta_{1,3}$, and it originates in the manner described above for those elements of the third period with valence electrons in the $3p$ subshell (Al, Si, P, S and Cl). It is interesting to analyze the $K\beta$ spectrum of the mineral called Topaz, which is an aluminosilicate formed by chains of octahedra of $AlO_4F_2$. In Topaz, aluminum presents two $K\beta'$ satellite lines, one for each covalent union: Al–O and Al–F (see Fig. 11) (Torres Deluigi et al., 2006). The identification of these satellite lines was done taking into account the energetic separation $\Delta E$ between the ionization energies of the $L2s$ and $L2p$ orbitals: $\Delta E \cong 20$, 14, 9 and 5 eV, for F, O, N and C, respectively (Eyring et al., 1944).

In the case of transition metals, the physical origin of the $K\beta'$ line is different. It has been assigned to a strong exchange coupling between unpaired electrons of the valence subshell $3d$ and the unpaired electron of state $3p^5$. The vacancy $3p^{-1}$ is generated by transitions $3p \rightarrow 1s$ that give rise to the line $K\beta_{1,3}$. In contrast, the satellite line $K\beta''$ of the transition metals have a similar origin to the $K\beta'$ in elements of the third period.

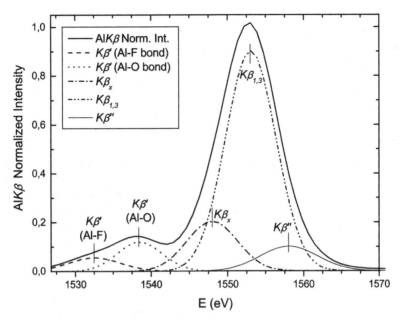

Fig. 11. Al $K\beta$ spectrum of Topaz. The two $K\beta'$ lines (due to the Al–F and Al–O covalent bonds) and the Al $K\beta_x$, Al $K\beta_{1,3}$ and Al $K\beta''$ lines have been deconvoluted with Gaussian functions and its energies are pointed out. The spectrum is normalized to the height at the maximum intensity point. (Torres Deluigi et al., 2006).

Several other theories are available to describe the $K\beta'$ low energy feature associated with the $K\beta_{1,3}$ emission resulting from transitions involving the partially filled $3d$ shells of transition elements and their oxides (Rémond, 2002). The Radiative Auger Effect (RAE) produces a broad structure at a lower energy than the characteristic diagram line. The RAE process results from a de-excitation of a $K$ vacancy, similar to an Auger process with simultaneous emission of a bound electron and an x-ray photon. For transition elements, the low energy structures associated with the $K\beta_{1,3}$ diagram line can be interpreted in terms of $KMM$ Radiative Auger Emission (Keski-Rakhonen and Ahopelto, 1980).

According to Salem and Scott (Salem and Scott, 1974), the interaction between the electrons in the incomplete $3d$ shell and the hole in the incomplete $3p$ shell splits both $3p$ and $3d$ levels causing a demultiplication of transitions.

The $K\beta'$ satellite has also been explained in terms of the plasmon oscillation theory (Tsutsumi et al., 1976). During the x-ray emission process, the transition valence electron excites a plasmon in the valence band. The transition energy of the $K\beta_{1,3}$ line will thus be shared between the plasmon and the emitting photon, which will be deprived of an energy equal to the plasmon energy.

## 4. Conclusion

High-resolution measurements of the $K$ satellite lines in the x-ray emission spectra obtained by XRF and EPMA techniques, provide detailed information on the electronic structure of the analyzed samples. The profile of the peaks, their energy shifts, the rise of new satellite peaks and the changes in their intensities are all related to chemical and structural properties, such as valence electronic states, the type of bond and the nature of the ligand atom. In particular, the intensity of satellite lines originated by multiple vacancies characterizes the excitation technique used, allowing at the same time to infer information about the internal electronic mechanisms that explain the production of multiple vacancies.

## 5. Acknowledgments

The authors gratefully acknowledge the involvement of the *Laboratorio de Microscopía Electrónica y Microanálisis* and the *Laboratorio de FRX de Química Analítica* of the *Universidad Nacional de San Luis* (UNSL), Argentina, where measurements were carried out. They also acknowledge the financial support received from the *Secretaría de Ciencia y Tecnología* and the *Proyecto PROICO 2-7502* of the UNSL of the Argentine Republic, which made this publication possible.

## 6. References

Aberg, T. (1927). Theory of multiple ionization processes, *Proc. Int. Conf. Inner Shell Ionization Phenomena*, USAEC Conf. 720404 USA EC, Atlanta, GA, 1927, pp. 1509-1542

Adachi, H., Tsukada, M. & Satoko, C. (1978). Discrete Variational Xα Cluster Calculations. I. Application to Metal Clusters. *J. Phys. Soc. Jpn.*, Vol. 45, N°3, (September 1978), pp. (875-883), ISSN 0031-9015

Burkhalter, P. G., Knudson, A. R., Nagel, D. J. & Dunning, K. L. (1972). Chemical Effects on Ion-Excited Aluminum K X-Ray Spectra. *Phys. Rev. A*, Vol. 6, (December 1972), pp. (2093-2101), ISSN 1050-2947

Cotton, F. A. & Wilkinson, G. (2008). *Química Inorgánica Aplicada* = *Advanced inorganic chemistry* (4th Ed.), Limusa, ISBN: 978-968-18-1795-4 , Mexico

Endo, H., Uda, M. & Maeda, K. (1980). Influence of the chemical bond on the intensities of F Kα x-ray satellites produced by electron and photon impacts. *Phys. Rev. A*, Vol. 22, (October 1980), pp. (1436-1440), ISSN 1050-2947

Esmail, E. I., Nicholls, C. J. & Urch, D. S. (1973). The detection of light elements by X-ray emission spectroscopy with use of low-energy satellite peaks. *Analyst*, Vol. 98, (October 1973), pp. (725-731), ISSN 0003-2654

Eyring, H., Walter, J. & Kimball, G. E. (1944). *Quantum Chemistry*, John Wiley & Sons Inc., New York

Fischer, D. W. & Baun, W. L. (1965). Diagram and Nondiagram Lines in K Spectra of Aluminum and Oxygen from Metallic and Anodized Aluminum. *J. Appl. Phys.*, Vol. 36, (February 1965), pp. (534-537), ISSN 0021-8979

Kawai, J. (1993). Chemical effects in the satellites of X-ray emission spectra. *Nucl. Instr. and Meth. in Phys. Res. B*, Vol. 75, (April 1993), pp. (3- 8), ISSN 0168-583X

Keski-Rakhonen, O. & Ahopelto, J. (1980). The K-M2 radiative Auger effect in transition metals. *J. Phys. C: Solid State Phys.*, Vol. 13, N° 4, (February 1980), pp. (471-482), ISSN 0022-3719.

Knudson, A. R., Nagel, D. J., Burkhalter, P. G. & Dunning, K. L. (1971). Aluminum X-Ray Satellite Enhancement by Ion-Impact Excitation. *Phys. Rev. Lett.*, Vol. 26, (May 1971), pp. (1149-1152), ISSN 0031-9007

Lui, Z., Sugata, S., Yuge, K., Nagasono, M., Tanaka, K. & Kawai, J. (2004). Correlation between chemical shift of Si Kα lines and the effective charge on the Si atom and its application in the Fe-Si binary system. *Phys. Rev. B*, Vol. 69, (January 2004), pp. (035106-035110), ISSN 1098-0121

Mauron, O. & Dousse, J.-Cl. (2002). Double KL ionization in Al, Ca, and Co targets bombarded by low-energy electrons. *Phys. Rev. A*, Vol. 66, (October 2002), pp. (042713-042726), ISSN 1050-2947

Mogi, M., Ota, A., Ebihara, S., Tachibana, M. & Uda, M. (1993). Intensity analysis of S Kß emission spectra of Na2SO3 by the use of DV-Xα MO method. *Nucl. Instr. and Meth. in Phys. Res. B*, Vol. 75, (April 1993), pp. (20-23), ISSN 0168-583X

Mukoyama, T., Taniguchi, K. & Adachi, H. (1986). Chemical effect on Kβ:Kα x-ray intensity ratios. *Phys. Rev. B*, Vol. 34, N° 6, (September 1986), pp. (3710-3716), ISSN 1098-0121

Oura, M., Mukoyama, T., Taguchi, M., Takeuchi, T., Haruna, T. & Shin, S. (2003). Resonant Double Excitation Observed in the Near-Threshold Evolution of the Photoexcited F Kα Satellite Intensity in NaF. *Phys. Rev. Lett.*, Vol. 90, (May 2003), pp. (173002-173005), ISSN 0031-9007

Parratt, L. G. (1936). K satellites. *Phys. Rev.*, Vol. 50, (July 1936), pp. (1-15)

Parratt, L. G. (1959). Electronic band structure of solids by x-ray spectroscopy. *Rev. Mod. Phys.*, Vol. 31, N°. 3, (July 1959), pp. (616-645), ISSN 0034-6861

Randall, C. A. & Parratt, L. G. (1940). Lα satellite lines for elements Mo(42) to Ba(56). *Phys. Rev. A*, Vol. 57, (May 1940), pp. (786-791), ISSN 1050-2947

Rémond, G., Myklebust, R., Fialin, M., Nockolds, C., Phillips, M. & Roques-Carmes, C., J. (2002). Decomposition of wavelength dispersive X-ray spectra. *J. Res. Natl. Inst. Stand. Technol.*, Vol. 107, N° 6, (November–December 2002), pp. (509–529), ISSN 0160-1741

Salem, S. I. & Scott, B. L. (1974). Splitting of the 4d 3/2 and 4d 5/2 levels in rare-earth elements and their oxides. *Phys. Rev. A*, Vol. 9, (February 1974), pp. (690-696), ISSN 1050-2947

Torres Deluigi, M., Strasser, E. N., Vasconcellos, M. A. Z. & Riveros, J. A. (2006). Study of the structural characteristics of a group of natural silicates by means of their Kβ emission spectra. *Chem. Phys.*, Vol. 323, (April 2006), pp. (173-178), ISSN 0301-0104

Tsutsumi, K., Nakamori, H. & Ichikawa, K. (1976). X-ray MnKβ emission spectra of manganese oxides and manganates. *Phys. Rev. B*, Vol. 13, (January 1976), pp. (929-933), ISSN 1098-0.121

Uda, E., Kawai, J. & Uda, M. (1993). Calculation of sulfur Kβ X-ray spectra. *Nucl. Instr. and Meth. in Phys. Res. B*, Vol. 75, (April 1993), pp. (24-27), ISSN 0168-583X

Uda, M., Endo, H., Maeda, K., Awaya, Y., Kobayashi, M., Sasa, Y., Kumagai, H., Tonuma, T. (1979). Bonding Effect on F Kα Satellite Structure Produced by 84-MeV N4+. *Phys. Rev. Lett.*, Vol. 42, (May 1979), pp. (1257-1260), ISSN 0031-9007

Urch, D. S. (1970). The origin and intensities of low energy satellite lines in X-ray emission spectra: a molecular orbital interpretation. *J. Phys. C: Solid State Phys.*, Vol. 3, (June 1970), pp. (1275-1291), ISSN 1361-6455

Urch, D. S. (1979). X-ray Emission Spectroscopy, In: *Electron Spectroscopy: Theory, Techniques and Applications, Vol. 3,* C. R. Brundle, A. D. Baker, pp. (1-39), Academic Press, Londres

Urch, D. S. (1985). X-ray Spectroscopy and Chemical Bonding in Minerals, In: *Chemical Bonding and Spectroscopy in Mineral Chemistry,* F. J. Berry, D. J. Vaughan, pp. (31-61), Chapman & Hall, Londres

Wollman, D. A., Nam, S. W., Newbury, D. E., Hilton, G. C., Irwin, K. D., Bergren, N. F., Deiker, S., Rudman, D. A. & Martinis, J. M. (2000). Superconducting transition-edge-microcalorimeter X-ray spectrometer with 2 eV energy resolution at 1.5 keV. *Nucl. Instr. and Meth. in Phys. Res. A,* Vol. 444, (April 2000), pp. (145-150), ISSN 0168-9002

# 4

# High Resolution X-Ray Spectroscopy with Compound Semiconductor Detectors and Digital Pulse Processing Systems

Leonardo Abbene and Gaetano Gerardi
*Dipartimento di Fisica, Università di Palermo*
*Italy*

## 1. Introduction

The advent of semiconductor detectors has revolutionized the broad field of X-ray spectroscopy. Semiconductor detectors, originally developed for particle physics, are now widely used for X-ray spectroscopy in a large variety of fields, as X-ray fluorescence analysis, X-ray astronomy and diagnostic medicine. The success of semiconductor detectors is due to several unique properties that are not available with other types of detectors: the excellent energy resolution, the high detection efficiency and the possibility of development of compact detection systems. Among the semiconductors, silicon (Si) detectors are the key detectors in the soft X-ray band (< 15 keV). Si-PIN diode detectors and silicon drift detectors (SDDs), with moderate cooling by means of small Peltier cells, show excellent spectroscopic performance and good detection efficiency below 15 keV. Germanium (Ge) detectors are unsurpassed for high resolution spectroscopy in the hard X-ray energy band (>15 keV) and will continue to be the choice for laboratory-based high performance spectrometers. However, there has been a continuing desire for ambient temperature and compact detectors with the portability and convenience of a scintillator but with a significant improvement in resolution. To this end, numerous high-Z and wide band gap compound semiconductors have been exploited. Among the compound semiconductors, cadmium telluride (CdTe) and cadmium zinc telluride (CdZnTe) are very appealing for hard X-ray detectors and are widely used for the development of spectrometer prototypes for medical and astrophysical applications.

Beside the detector, the readout electronics also plays a key role in the development of high resolution spectrometers. Recently, many research groups have been involved in the design and development of high resolution spectrometers based on semiconductor detectors and on digital pulse processing (DPP) techniques. Due to their lower dead time, higher stability and flexibility, digital systems, based on directly digitizing and processing of detector signals (preamplifier output signals), have recently been favored over analog electronics ensuring high performance in both low and high counting rate environments.

In this chapter, we review the research activities of our group in the development of high throughput and high resolution X-ray spectrometers based on compound semiconductor detectors and DPP systems. First, we briefly describe the physical properties and the signal formation in semiconductor detectors for X-ray spectroscopy. Second, we introduce the main properties and critical issues of a X-ray detection system, highlighting the characteristics of both analog and digital approaches. Finally, we report on the spectroscopic performance of a high resolution spectrometer based on a CdTe detector and a custom DPP system. As an application, direct measurements of mammographic X-ray spectra by using the digital CdTe detection system are also presented.

## 2. Compound semiconductor detectors

Silicon (Si) and germanium (Ge) are traditional semiconductors used for radiation detectors in a wide range of applications (Knoll, 2000). The growing field of applications stimulated the development of detectors based on compound semiconductors (Knoll, 2000; McGregor & Hermon, 1997; Owens & Peacock, 2004). Compound semiconductors were first investigated as radiation detectors in 1945 by Van Heerden, who used AgCl crystals for detection of alpha particles and gamma rays. The great advantage of compound semiconductors is the possibility to grow materials with a wide range of physical properties (band gap, atomic number, density) making them suitable to almost any application. Interests in radiation detectors operating at room temperature gave rise to development of compound semiconductors with wide band gaps, in comparison to Si and Ge. Moreover, for X-ray and gamma ray detection, compound semiconductors with high atomic number were preferred in order to emphasize photoelectric interaction. It is well known that, among the various interaction mechanisms of X rays with matter, only the photoelectric effect results in the total absorption of the incident energy giving useful information about the photon energy.

Compound semiconductors are generally derived from elements of groups III and V (e.g. GaAs) and groups II and VI (e.g. CdTe) of the periodic table. Besides binary compounds, ternary materials have been also produced, e.g. CdZnTe and CdMnTe. Table 1 reports the physical properties of common compound semiconductors typically used for radiation detection.

Among the compound semiconductors, CdTe and CdZnTe attracted growing interests in the development of X-ray detectors (Del Sordo et al., 2009; Takahashi & Watanabe, 2001). Due to the high atomic number, the high density and the wide band gap (Table 1), CdTe and CdZnTe detectors ensure high detection efficiency, good room temperature performance and are very attractive for X-ray and gamma ray applications. Figure 1(a) shows the linear attenuation coefficients, calculated by using tabulated interaction cross section values (Boone & Chavez, 1996), for photoelectric absorption and Compton scattering of Si (green line), Ge (blu line) and CdTe (red line); as shown in Figure 1(a), photoelectric absorption is the main process up to about 200 keV for CdTe. Figure 1(b) shows the total and photoelectric efficiency for 1 mm thick CdTe detectors, compared with those of traditional semiconductors (Si and Ge with same thickness).

Difficulties in producing detector-grade materials and in growing chemically pure and structurally perfect crystals are the critical issues of CdTe and CdZnTe detectors. In fact, the great potentialities of these compounds has not been exploited for many decades due mainly to the limited commercial availability of high-quality crystals. This situation has changed dramatically during the mid nineties with the emergence of few companies committed to the advancement and commercialization of these materials.

| Material | Si | Ge | GaAs | CdTe | $Cd_{0.9}Zn_{0.1}Te$ | $HgI_2$ | TlBr |
|---|---|---|---|---|---|---|---|
| Crystal structure | Cubic | Cubic | Cubic (ZB) | Cubic (ZB) | Cubic (ZB) | Tetragonal | Cubic (CsCl) |
| Growth method* | C | C | CVD | THM | HPB | VAM | BM |
| Atomic number | 14 | 32 | 31, 33 | 48, 52 | 48, 30, 52 | 80, 53 | 81, 35 |
| Density $(g/cm^3)$ | 2.33 | 5.33 | 5.32 | 6.20 | 5.78 | 6.4 | 7.56 |
| Band gap (eV) | 1.12 | 0.67 | 1.43 | 1.44 | 1.57 | 2.13 | 2.68 |
| Pair creation energy (eV) | 3.62 | 2.96 | 4.2 | 4.4 | 4.6 | 4.2 | 6.5 |
| Resistivity $(\Omega\ cm)$ | $10^4$ | 50 | $10^7$ | $10^9$ | $10^{10}$ | $10^{13}$ | $10^{12}$ |
| $\mu_e\tau_e$ $(cm^2/V)$ | >1 | >1 | $10^{-4}$ | $10^{-3}$ | $10^{-3}$-$10^{-2}$ | $10^{-4}$ | $10^{-5}$ |
| $\mu_h\tau_h$ $(cm^2/V)$ | ~1 | >1 | $10^{-6}$ | $10^{-4}$ | $10^{-5}$ | $10^{-5}$ | $10^{-6}$ |

Table 1. The physical properties of Si, Ge and principal compound semiconductors. The abbreviations are related to the most common growth methods: C = Czochralski, CVD = chemical vapor deposition, THM = traveler heater method, BM = Bridgman method, HPB = high-pressure Bridgman and VAM = vertical ampoule method.

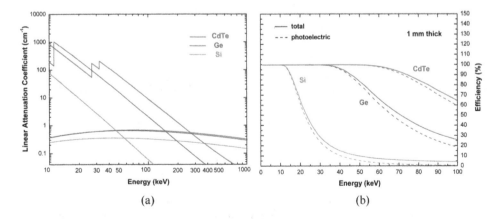

(a)     (b)

Fig. 1. (a) Linear attenuation coefficients for photoelectric absorption and Compton scattering of CdTe, Si, and Ge. (b) Total and photoelectric efficiency for 1 mm thick CdTe detectors compared with Si and Ge.

## 3. Principles of operation of semiconductor detectors for X-ray spectroscopy

Semiconductor detectors for X-ray spectroscopy behaves as solid-state ionization chambers operated in pulse mode (Knoll, 2000). The simplest configuration is a planar detector i.e. a slab of a semiconductor material with metal electrodes on the opposite faces of the semiconductor (Figure 2). Photon interactions produce electron-hole pairs in the semiconductor volume through the above discussed interactions. The interaction is a two-step process where the electrons created in the photoelectric or Compton process loose their energy through electron-hole ionization. The most important feature of the photoelectric absorption is that the number of electron-hole pairs is proportional to the photon energy. If $E_0$ is the incident photon energy, the number of electron-hole pairs $N$ is equal to $E_0/w$, where $w$ is the average pair creation energy. The generated charge cloud is $Q_0 = e\ E_0/w$. The electrons and holes move toward the opposite electrodes, anode and cathode for electrons and holes, respectively (Figure 2). The movement of the electrons and holes, causes variation $\Delta Q$ of induced charge on the electrodes. It is possible to calculate the induced charge $\Delta Q$ by the Shockley–Ramo theorem (Cavalleri et al., 1971; Ramo, 1939; Shockley, 1938) which makes use of the concept of a weighting potential $\varphi$. The weighting potential is defined as the potential that would exist in the detector with the collecting electrode held at unit potential, while holding all other electrodes at zero potential. According to the Shockley–Ramo theorem, the induced charge by a carrier $q$, moving from $x_i$ to $x_f$, is given by:

$$\Delta Q = -q\left[\varphi\ (x_f) - \varphi(x_i)\right] \tag{1}$$

where $\varphi\ (x)$ is weighting potential at position $x$. It is possible to calculate the weighting potential by analytically solving the Laplace equation inside a detector . In a semiconductor, the total induced charge is given by the sum of the induced charges due both to the electrons and holes. For a planar detector, the weighting potential $\varphi$ of the anode is a linear function of distance $x$ from the cathode:

$$\varphi(x) = \frac{x}{L} \qquad\qquad 0 \le \frac{x}{L} \le 1 \tag{2}$$

where $L$ is the detector thickness. Neglecting charge loss during the transit time of the carriers, the charge induced on the anode electrode by $N$ electron-hole pairs is given by:

$$\Delta Q\ = \Delta Q_h + \Delta Q_e = -\frac{(Ne)}{L}(0 - x) + \frac{(Ne)}{L}(L - x) = Ne = Q_0$$

$$t > t_h = \frac{x}{\mu_h E} \qquad\qquad t > t_e = \frac{L - x}{\mu_e E} \tag{3}$$

where $t_h$ and $t_e$ are the transit times of holes and electrons, respectively.

Charge trapping and recombination are typical effects in compound semiconductors and may prevent full charge collection. For a planar detector, having a uniform electric field, neglecting charge de-trapping, the charge collection efficiency (CCE), i.e. the induced charge normalized to the generated charge, is given by the Hecht equation (Hecht, 1932):

$$CCE = \frac{Q}{Q_0} = \left[\frac{\lambda_h}{L}\left(1-e^{-\frac{x}{\lambda_h}}\right) + \frac{\lambda_e}{L}\left(1-e^{-\frac{L-x}{\lambda_e}}\right)\right] \qquad (4)$$

where $\lambda_h = \mu_h \tau_h E$ and $\lambda_e = \mu_e \tau_e E$ are the mean drift lengths of holes and electrons, respectively. The CCE depends not only on $\lambda_h$ and $\lambda_e$, but also on the incoming photon interaction position. Small $\lambda/L$ ratios reduce the charge collection and increase the dependence by the photon interaction point. So, the random distribution of the interaction point increases the fluctuations on the induced charge and thus produces peak broadening in the energy spectra. The charge transport properties of a semiconductor, expressed by the hole and electron mobility lifetime products ($\mu_h \tau_h$ and $\mu_e \tau_e$) are key parameters in the development of radiation detectors. Poor mobility lifetime products result in short $\lambda$ and therefore small $\lambda/L$ ratios, which limit the maximum thickness and energy range of the detectors. Compound semiconductors, generally, are characterized by poor charge transport properties, especially for holes, due to charge trapping. Trapping centers are mainly caused by structural defects (e.g. vacancies), impurities and irregularities (e.g. dislocations, inclusions). In compound semiconductors, the $\mu_e \tau_e$ is typically of the order of $10^{-5}$-$10^{-2}$ cm²/V while $\mu_h \tau_h$ is usually much worse with values around $10^{-6}$-$10^{-4}$ cm²/V, as reported in Table 1. Therefore, the corresponding mean drift lengths of electrons and holes are 0.2-200 mm and 0.02-2 mm, respectively, for typical applied electric fields of 2000 V/cm.

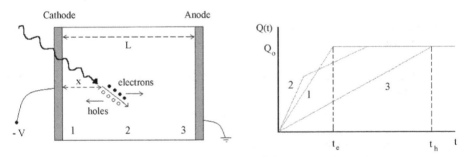

Fig. 2. Planar configuration of a semiconductor detector (left). Electron–hole pairs, generated by radiation, are swept towards the appropriate electrode by the electric field. (right) The time dependence of the induced charge for three different interaction sites in the detector (positions 1, 2 and 3). The fast rising part is due to the electron component, while the slower component is due to the holes.

As pointed out in the foregoing discussions, poor carrier transport properties of CdTe and CdZnTe materials are a critical issue in the development of X-ray detectors. Moreover, the significant difference between the transport properties of the holes and the electrons produces well known effects as spectral distortions in the measured spectra, i.e. peak asymmetries and long tails. To overcome the effects of the poor transport properties of the holes, several methods have been employed (Del Sordo et al., 2009). Some techniques concern the particular irradiation configuration of the detectors. *Planar parallel field* (PPF) is the classical configuration used in overall planar detectors, in which the detectors are irradiated through the cathode electrode, thus minimizing the hole trapping probability. In an alternative configuration, denoted as *planar transverse field* (PTF), the irradiation direction is orthogonal (transverse) to the electric field. In such configuration different detector thicknesses can be chosen, in order to fit the detection efficiency required, without modifying the inter-electrode distance and then the charge collection properties of the detectors. Several techniques are used in the development of detectors based on the collection of the electrons (single charge carrier sensitive), which have better transport properties than that of the holes. Single charge carrier sensing techniques are widely employed in compound semiconductor detectors by developing careful electrode designs (Frisch-grid, pixels, coplanar grids, strips and multiple electrodes) and by using electronic methods (pulse shape analysis).

As it is well clear from the above discussions, the charge collection efficiency is a crucial property of a radiation detector that affects the spectroscopic performance and in particular the energy resolution. High charge collection efficiency ensures good energy resolution which also depends by the statistics of the charge generation and by the noise of the readout electronics. Generally, the energy resolution of a radiation detector, estimated through the full-width at half maximum (FWHM) of the full-energy peaks, is mainly influenced by three contributes:

$$\Delta E = \sqrt{(2.355)^2 (F \cdot E_0 \cdot w) + \Delta E_{el}^2 + \Delta E_{coll}^2} \tag{5}$$

The first contribute is the Fano noise due to the statistics of the charge carrier generation. In compound semiconductors, the Fano factor $F$ is much smaller than unity (0.06-0.14). The second contribute is the electronic noise which mainly depends on the readout electronics and the leakage current of the detector, while the third is the contribute of the charge collection process.

## 4. Electronics for high resolution spectroscopy: The digital pulse processing (DPP) approach

Nowadays, the dramatic performance improvement of the analog-to-digital converters (ADC) stimulated an intensive research and development on digital pulse processing (DPP) systems for high resolution X-ray spectroscopy. The availability of very fast and high precision digitizers has driven physicists and engineers to realize electronics in which the analog-to-digital conversion is performed as close as possible to the detector. This approach is reversed with respect to more traditional electronics which were made out of mainly analog circuits with the A/D conversion at the end of the chain. Figure 3 shows the simplified block diagrams of analog and DPP electronics for X-ray detectors. In a typical analog electronics, the detector signals are amplified by a charge sensitive preamplifier (CSP), shaped and filtered by an analog shaping amplifier and finally processed by a multichannel analyzer (MCA) to generate the energy spectrum. In a DPP system, the preamplifier output signals are directly digitized by an ADC and so processed by using digital algorithms. A DPP system leads to better results than the analog one, mainly due to (i) stability, (ii) flexibility and (iii) higher throughput (i.e. the rate of the useful counts in the energy spectrum). With regard to the improved stability of a DPP system, the direct digitizing of the detector signals minimizes the drift and instability normally associated with analog signal processing. In terms of flexibility, it is possible to implement complex algorithms, that are not easily implementable through a traditional analog approach, for adaptive processing and optimum filtering. Moreover, a DPP analysis require considerably less overall processing time than the analog one ensuring lower dead time and higher throughput, very important under high rate conditions. The dead time of an analog system is mainly due to the pulse processing time of the shaper and to the conversion time of MCA. The pulse processing time is generally related to the temporal width of the pulse which depends on the shaping time constant of the shaper and can be described through the well known paralyzable dead time model (Knoll, 2000). The MCA dead time, generally described by nonparalyzable dead time model (Knoll, 2000), is often the dominant contributor to overall dead time. In a DPP system there is no additional dead time associated with digitizing the pulses and so the equivalent to MCA dead time is zero. Therefore the overall dead time of a digital system is generally lower than that of the analog one.

An another positive aspect of the DPP systems regards the possibility to perform off-line analysis of the detector signals: *since that signals are captured, more complex analyses can be postponed until the source event has been deemed interesting.*

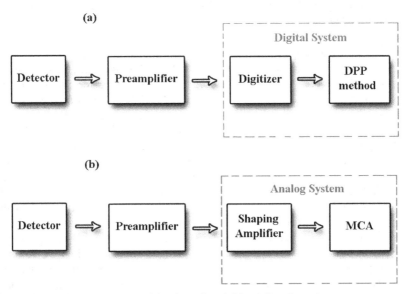

Fig. 3. Simplified block diagrams of (a) digital and (b) analog detection systems.

## 5. A digital CdTe X-ray spectrometer for both low and high counting rate environments

In this section, we report on the spectroscopic performance of a CdTe detector coupled to a custom DPP system for X-ray spectroscopy. We first describe the main characteristics of the detector and the DPP system and then we present the results of the characterization of the overall detection system at both low (200 cps) and high photon counting rates (up to 800 kcps) by using monoenergetic X-ray sources ($^{109}$Cd, $^{152}$Eu, $^{241}$Am, $^{57}$Co) and a nonclinical X-ray tube with different anode materials (Ag, Mo). This work was carried out in the sequence of previously developed DPP systems (Abbene et al., 2007, 2010a, 2010b, 2011; Gerardi et al., 2007) with the goal of developing a digital spectrometer based on DPP techniques and characterized by high performance both at low and high photon counting rate environments.

## 5.1 CdTe detector

The detector is based on a thin CdTe crystal (2 x 2 x 1 mm³), wherein both the anode (indium) and the cathode (platinum) are planar electrodes covering the entire detector surface. The Schottky barrier at the In/CdTe interface ensures low leakage current even at high bias voltage operation (400 V), thus improving the charge collection efficiency. The thickness of the detector guarantees a very good photon detection efficiency (~99%) up to 40 keV. A Peltier cell cools both the CdTe crystal and the input FET of the charge sensitive preamplifier (A250, Amptek, U.S.A.) at a temperature of -20 °C. Cooling the detector reduces the leakage current, allowing the application of higher bias voltages to the electrodes; moreover, cooling the FET increases its transconductance and reduces the electronic noise. The detector, the FET and the Peltier cooler are mounted in a hermetic package equipped with a light-vacuum tight beryllium window (modified version of Amptek XR100T-CdTe, S/N 6012). To increase the maximum counting rate of the preamplifier, a feedback resistor of 1 GΩ and a feedback capacitor of 0.1 pF were used. The detector is equipped with a test input to evaluate the electronic noise.

## 5.2 Digital pulse processing system

The DPP system consists of a digitizer and a PC wherein the digital analysis of the detector pulses (preamplifier output pulses) was implemented. The detector signals are directly digitized by using a 14-bit, 100 MHz digitizer (NI5122, National Instruments). The digital data are acquired and recorded by a Labview program on the PC platform and then processed off-line by a custom digital pulse processing method (C++ coded software) developed by our group. The analysis time is about 3 times the acquisition time. The DPP method, implemented on the PC platform, performs a height and shape analysis of the detector pulses. Combining fast and slow shaping, automatic pole-zero adjustment, baseline restoration and pile-up rejection, the digital method allows precise pulse height measurements both at low and high counting rate environments. Pulse shape analysis techniques (pulse shape discrimination, linear and non linear pulse shape corrections) to compensate for incomplete charge collection were also implemented. The digitized pulses are shaped by using the classical single delay line (SDL) shaping technique (Knoll, 2000). Each shaped pulse is achieved by subtracting from the original pulse its delayed and attenuated fraction. The attenuation of the delayed pulse allows to eliminate the undesirable undershoot following the shaped pulse, i.e. acting as pole-zero cancellation. For a single digitized pulse, consisting of a defined number of samples, this operation can be represented by the following equation:

$$V^{shaped}(nT_s) = V^{preamp}(nT_s) - \left[V^{preamp}(nT_s - n_d T_s)\right] \cdot A \tag{6}$$

where $T_s$ is the ADC sample period, $n_d T_s = T_d$ is the delay time, $A$ is the attenuation coefficient, $V^{shaped}(nT_s)$ is the shaped sample at the discrete time instant $nT_s$ and $V^{preamp}(nT_s)$ is the preamplifier output sample at the discrete time instant $nT_s$. The width of each shaped pulse is equal to $T_d + T_p$, wherein $T_p$ is the peaking time of the related preamplifier output pulse.

Our DPP method is characterized by two shaping modes: a "*fast*" SDL shaping mode and a "*slow*" SDL shaping mode, operating at different delay times. The "*fast*" shaping operation, characterized by a short delay time $T_{d, fast}$, is optimized to detect the pulses and to provide a pile-up inspection. If the width of the shaped pulses exceeds a maximum width threshold then the pulse is classified as representative of pile-up events; whenever it is possible, each overlapped event is recognized through a peak detection analysis. Obviously, these events are not analyzed by the *"slow"* shaping procedure. The delay time of the *"fast"* shaping operation is a dead time for the DPP system (paralyzable dead time) and it must be as small as possible, depending of detector and ADC characteristics. With regard to the paralyzable model, the true photon counting rate $n$ is related to the measured photon counting rate $m$ through the following equation (Knoll, 2000):

$$m = n \cdot \exp\left[-nT_{d,fast}\right] \tag{7}$$

It is possible to evaluate the true rate $n$ from the measured rate $m$ by solving the equation (7) iteratively. The DPP system, through the *"fast"* shaping operation, gives the estimation of the true rate $n$ through the equation (7) and the measured rate $m$. We used $T_{d,fast} = 50$ ns.

The *"slow"* shaping operation, which has a longer delay time $T_{d,slow}$ than the *"fast"* one, is optimized to shorten the pulse width and minimize the ballistic deficit. To obtain a precise pulse height measurement, a convolution of the shaped pulses with a Gaussian function was performed. The slow delay time $T_{d,slow}$ acts as the shaping time constant of an analog shaping amplifier: the proper choice depends on the peaking time of the preamplifier pulses, the noise and the incoming photon counting rate.

To ensure good energy resolution also at high photon counting rates, a standard detection system is typically equipped with a baseline restorer which minimize the fluctuations of the baseline. The digital method performs a baseline recovery by evaluating the mean value of the samples, within a time window equal to $T_{d,slow}/2$, before and after each shaped pulse. This operation sets a minimum time spacing between the pulses equal to $T_a = 2 \cdot T_{d,slow} + T_p$ for which no mutual interference must exist in the baseline measurement. The minimum time spacing $T_a$ is used to decide whether the events must be discarded, in particular if the time spacing does not exceed $T_a$ the two events are rejected. It is clear that a $T_{d,slow}$ value too long reduces the number of the counts in the measured spectrum (analyzed events) and its optimum value is the best compromise between the required energy resolution and throughput. The time $T_a$ is paralyzable dead time for the *slow* shaping operation. Both *fast* and *slow* procedures gave consistent values of the incoming photon counting rate.

### 5.2.1 Pulse shape discrimination

A pulse shape discrimination (PSD) technique was implemented in our DPP system by analyzing the peaking time distribution of the pulses and their correlations with the energy spectra. Pulse shape discrimination, first introduced by Jones in 1975 (Jones & Woollam, 1975), is a common technique to enhance spectral performance of compound semiconductor detectors. Generally, this technique is based on the selection of a range of peaking time values of the pulses that are less influenced of incomplete charge collection. As previously discussed, incomplete charge collection, mainly due to the poor transport properties of the holes, is a typical drawback of compound semiconductor detectors, producing long tailing and asymmetry in the measured spectra. As well known, the pulses mostly influenced by the hole contribution are generally characterized by longer peaking times. These effects are more prominent increasing the energy of radiation (i.e. the depth of interaction of radiation); the events, with a greater depth of interaction, take place close to the anode electrode producing pulses mostly due to the hole transit.

To perform the pulse shape analysis, we carried out the measurement of the peaking time of the analyzed pulses. We first evaluate the rise time of the pulses, i.e. the interval between the times at which the shaped pulse reaches 10% and 90% of its height (after baseline restoration). The times, corresponding to the exact fractions (10% and 90%) of the pulse height, are obtained through a linear interpolation. We estimate the peaking time equal to 2.27 times the rise time (i.e. about five times the time constant). Due to the precise measurement of the pulse height-baseline and interpolation, the method allows fine peaking time estimations (with a precision of 2 ns) both at low and high photon counting rates.

Figure 4(a) shows the pulse peaking time distribution of $^{57}$Co events measured with the CdTe detector. The distribution has an asymmetric shape and suffers from a tail, which is attributed to the slow peaking time events. The correlation between the peaking time and the height of the pulses is pointed out by the bi-parametric distribution ($^{57}$Co source), shown in Figure 4(b). It is clearly visible that for longer peaking times, the photopeak shifts to lower energies, as expected. This distribution is very helpful to better understand the tailing in the measured spectra and implement correction methods.

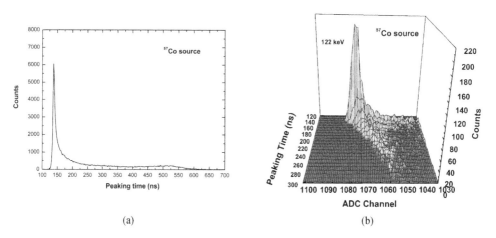

(a)                                                                                (b)

Fig. 4. (a) Pulse peaking time distribution of the CdTe detector ($^{57}$Co source). The peaking time is equal to 2.27 times the rise time of the pulses. (b) 3D plot of $^{57}$Co spectra measured for different peaking time values (bi-parametric distribution).

As proposed by Sjöland in 1994 (Sjöland & Kristiansson, 1994) pulse shape discrimination can also be used to minimize peak pile-up events, i.e. overlapped preamplified pulses within the peaking time that are not detectable through the "*fast*" shaping operation. Because the shape (peaking time) of a peak pile-up pulse differs from that of a pulse not affected by pile-up, analyzing the measured spectra at different peaking time regions (PTRs) in the peaking time distribution is helpful to reduce peak pile-up. Figure 5(a) shows some selected PTRs in the peaking time distribution of the pulses from the $^{109}$Cd source (at 820 kcps), while in Figure 5(b) are shown the $^{109}$Cd spectra for each PTR. These results point out the characteristics of the peak pile-up events, which have a longer peaking time than the correct events, and then the potentialities of the PSD technique to minimize these spectral distortions.

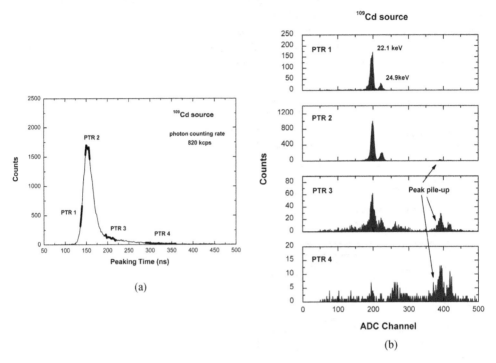

Fig. 5. (a) Pulse peaking time distribution of the CdTe detector ($^{109}$Cd source) at a photon counting rate of 820 kcps; the selected peaking time regions (PTRs) are also visible. (b) $^{109}$Cd spectra for the selected PTRs (820 kcps). It is evident that the peak-pile events are characterized by longer peaking times than the correct events.

## 5.2.2 Linear and non linear pulse shape corrections

Despite the potentiality of the PSD technique, the choice of the optimum peaking time region is a trade-off between the energy resolution and the counts in the measured spectrum. The strong correlation between the peaking time and the height of the pulses, as shown in Figure 4(b), opens up the possibility of charge loss correction.

Parallel to the PSD technique, we implemented linear and non linear pulse shape correction (PSC) methods in our DPP system, based on the measurement of both the peaking time and the height of the pulses. As introduced by Keele et al. (Keele et al., 1996), the method corrects all pulses to a hypothetical zero peaking time. The method requires a preliminary calibration procedure, strictly depending on the characteristics of the detector, based on the analysis of the behaviour of the centroid of photopeaks versus the peaking time. Figure 6(a) shows the photopeak centroid vs. the peaking time values for some photopeaks of the measured spectra ($^{109}$Cd, $^{152}$Eu, $^{241}$Am and $^{57}$Co). We analyzed the photopeak centroid shift by using the following linear function:

$$E_{\text{det}} = m_E \cdot T_p + E_{cor} \tag{8}$$

where $E_{det}$ is the photopeak centroid, $m_E$ is the slope of the linear function, $T_p$ is the peaking time and $E_{cor}$ is the corrected centroid at zero peaking time. The corrected centroid, $E_{cor}$, is the result of correcting $E_{det}$ to an ideal point of zero peaking time and it is the desired height for a pulse. It is interesting to note that both the slope $m_E$ and $E_{cor}$ are linear functions of the photon energy $E$, as shown in Figures 6(b) and 6(c).

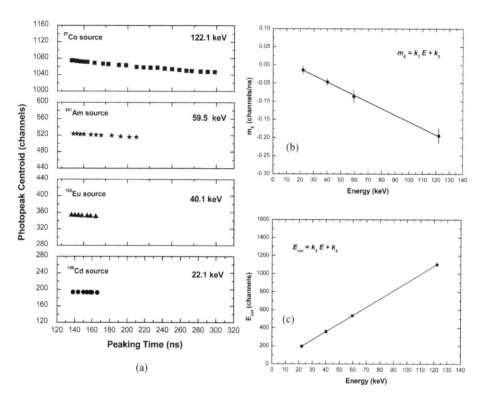

Fig. 6. (a) Photopeak centroid vs. the peaking time values for some photopeaks of the measured spectra ($^{109}$Cd, $^{152}$Eu, $^{241}$Am and $^{57}$Co). (b) Slope $m_E$ and (c) corrected centroid $E_{cor}$ vs. the radiation energy.

The fitting equations are:

$$m_E = k_1 \cdot E + k_2 \tag{9}$$

$$E_{cor} = k_3 \cdot E + k_4 \tag{10}$$

where $k_1$, $k_3$ and $k_2$, $k_4$ are the slopes and the y-intercepts of the linear functions, respectively. Combining the equations (9) and (10), yields:

$$m_E = \frac{k_1}{k_3} \cdot (E_{cor} - k_4) + k_2 \tag{11}$$

i.e.:

$$m_E = A \cdot E_{cor} + B \tag{12}$$

with:

$$A = \frac{k_1}{k_3} \qquad\qquad B = \frac{k_2 \cdot k_3 - k_1 \cdot k_4}{k_3}. \tag{13}$$

Combining the equations (8) and (12), yields:

$$E_{cor} = \frac{(E_{det} - B \cdot T_p)}{1 + A \cdot T_p} \tag{14}$$

By using equation (14) is possible to adjust the pulse height $E_{det}$ of a pulse through the knowledge of the bi-parametric distribution and of the constants $A$ and $B$ obtained by the calibration procedure.

We also implemented a non linear pulse shape correction method, by using the following function:

$$E_{det} = n_E \cdot T_p^2 + m_E \cdot T_p + E_{cor} \tag{15}$$

where the coefficients $n_E$, $m_E$ and $E_{cor}$ are yet linear functions of the photon energy $E$:

$$n_E = k_5 \cdot E + k_6 \tag{16}$$

$$m_E = k_7 \cdot E + k_8 \tag{17}$$

$$E_{cor} = k_9 \cdot E + k_{10} \tag{18}$$

So, combining the equations (16), (17), (18) with (15) yields:

$$E_{cor} = -\frac{-E_{det} \cdot k_9 + (k_9 \cdot k_8 - k_{10} \cdot k_7) \cdot T_p + (k_9 \cdot k_6 - k_{10} \cdot k_5) \cdot T_p^2}{k_9 + k_7 \cdot T_p + k_5 \cdot T_p^2} \tag{19}$$

The bi-parametric correction method presents an important limitation: it is only applicable to pure photoelectric interactions, i.e. when the energy of each incident photon is deposited at a single point in the detector. If the photon Compton scatters at one depth in the detector and then undergoes photoelectric absorption at a second depth, the height-peaking time relationship can vary from that due to a single interaction. For high atomic number compound semiconductors, such as CdTe and CdZnTe, photoelectric absorption is the main process up to about 200 keV (Del Sordo et al., 2009).

## 5.3 Spectroscopic characterization

To investigate on the spectroscopic performance of the system, we used X-ray and gamma ray calibration sources ([109]Cd: 22.1, 24.9 and 88.1 keV; [241]Am: 59.5, 26.3 keV and the Np L X-ray lines between 13 and 21 keV; [152]Eu: 121.8 keV and the Sm K lines between 39 and 46 keV;

[57]Co: 122.1, 136.5 keV and the W fluorescent lines, $K_{\alpha 1}$=59.3 keV, $K_{\alpha 2}$=58.0 keV, $K_{\beta 1}$=67.1 keV, $K_{\beta 3}$=66.9 keV, produced in the source backing). The 14 keV gamma line ([57]Co) is shielded by the source holder itself. For high rate measurements, we also used another [241]Am source, with the Np L X-ray lines shielded by the source holder. To obtain different rates (up to 820 kcps) of the photons incident on the detector (through the cathode surface), we changed the irradiated area of the detector by using collimators (Pb and W) with different geometries.

To better investigate on the high-rate performance of the system, we also performed the measurement of Mo-target X-ray spectra at the "Livio Scarsi" Laboratory (LAX) located at DIFI (Palermo).

(a)                                                                                     (b)

Fig. 7. (a) An overview of the LAX facility. (b) The detector chamber located at 10.5 m from the X-ray tube; the detection system was mounted on a XYZ microtranslator system.

The facility is able to produce X-ray beams with an operational energy range of 0.1–60 keV (tube anodes: Ag, Co, Cr, Cu, Fe, Mo, W), collimated on a length of 10.5 m with a diameter at full aperture of 200 mm (Figure 7). In this work, we used Mo and Ag targets. No collimators were used for these measurements. To characterize the spectroscopic performance of the system, we evaluated, from the measured spectra, the *energy resolution* (*FWHM*) and the *FW.25M to FWHM ratio*, defined in agreement with the IEEE standard (IEEE Standard, 2003). We also evaluated the area of the energy peaks (photopeak area) through the *high side area* (*HSA*) (IEEE Standard, 2003), i.e. the area between the peak center line and the peak's high-energy toe; the photopeak area was calculated as twice the *HSA*. The measured spectra were analyzed by using a custom function model, which takes into account both the symmetric and the asymmetric peak distortion effects (Del Sordo et al., 2004). Statistical errors on the spectroscopic parameters with a confidence level of 68% were associated.

### 5.3.1 Low count rate performance

In this paragraph we present the performance of the system at low photon counting rate (200 cps), by using PSD, linear and non linear PSC techniques. We first used the PSD technique looking for the best performance despite the high reduction of the photopeak area (about 90%). With regard to the PSD technique, we obtained the following results: energy resolution (*FWHM*) of 2.05 %, 0.98 % and 0.68 % at 22.1, 59.5 and 122.1 keV, respectively. Figure 8 shows the measured [241]Am and [57]Co spectra using PSD and no PSD techniques. To better point out the spectral improvements of the PSD technique, we report in Figures 8(b) and 8(d) a zoom of the 59.5 and 122.1 keV photopeaks, normalized to the photopeak centroid counts. As widely shown in several works, PSD produced a strong reduction of peak asymmetry and tailing in the measured spectra: the 122.1 keV photopeak of [57]Co spectrum, after PSD, is characterized by an energy resolution improvement of 57 % and low tailing; the *FW.25M to FWHM ratio* is reduced up to 1.46, quite close to the ideal Gaussian ratio (*FW.25M/FWHM*$_{Gaussian}$=1.41).

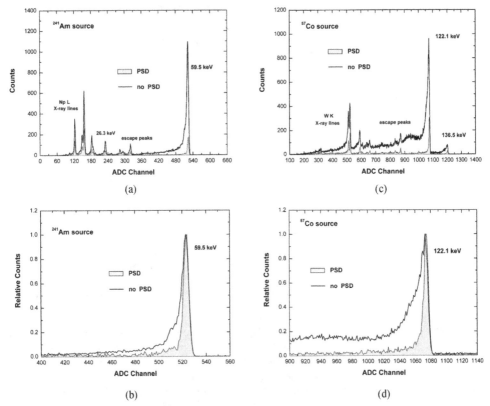

Fig. 8. Measured (a) [241]Am and (c) [57]Co spectra using PSD and no PSD techniques. After PSD, we obtained an energy resolution of 0.98 % and 0.68 % *FWHM* at 59.5 keV and 122.1 keV, respectively. Zoom of the (b) 59.5 and (d) 122.1 keV photopeaks, normalized to the photopeak centroid counts.

The reduction of trapping effects allows to use some semi-empirical models of the energy resolution function that can be used to estimate some characteristic parameters of compound semiconductors, such as the Fano factor $F$ and the average pair creation energy

$w$. For low trapping, the energy resolution, as proposed by Owens (Owens & Peacock, 2004), can be described by the following equation:

$$\Delta E = \sqrt{(2.355)^2 (F \cdot E \cdot w) + \Delta E_{el}^{2} + a \cdot E^b}$$                                 (20)

where $a$ and $b$ are semi-empirical constants that can be obtained by a best-fit procedure. The equation (20) could be used to estimate the Fano factor $F$ and the average pair creation energy $w$. In our case, we used a tabulated value of $w$ (4.43 eV) and obtained $F$, $a$ and $b$ by a best-fit procedure; we also measured the energy resolution of the 17.77 keV Np L X-ray line to obtain almost one degree of freedom (*dof*).

Best fitting equation (20) to the measured energy resolution points (with no PSD) resulted in a bad fit, due to the high trapping contribute. We obtained a good fit with the measured data points after PSD [$\chi^2/dof$ = 1.21; *dof* = 1)], as shown in Figure 9.

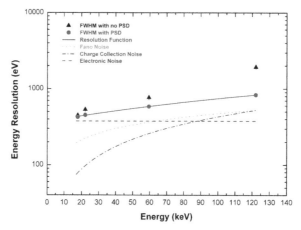

Fig. 9. Energy resolution (*FWHM*) using no PSD and PSD techniques. The solid line is the best-fit resolution function (equation 20). The components of the energy resolution are also shown: the noise due to carrier generation or Fano noise, the electronic noise and charge collection or trapping noise.

The fitted value for the Fano factor (0.09±0.03) is in agreement with the literature values (Del Sordo et al., 2009). Figure 9, also, shows the individual components of the energy resolution (after PSD). It is clearly visible that electronic noise (dashed line) dominates the resolution function below 60 keV, whereas Fano noise (dotted line) dominates the charge collection noise (dot-dashes line) within the overall energy range (up to 122 keV). After PSD, the charge collection noise is mainly due to electron trapping and diffusion. We stress that this analysis was performed in order to better point out the performance enhancements of the PSD technique and not for precise measurements of Fano factor. Nevertheless, the potentialities of the technique for precise Fano factor estimations are well evident: It would be sufficient to measure a greater number of monoenergetic X-ray lines to ensure more precise Fano factor estimations.

Table 2 shows the performance of the detector with no correction, after PSD, linear and non linear PSC.

| Spectroscopic parameter | Energy (keV) | | | | |
| --- | --- | --- | --- | --- | --- |
| | 59.5 | | | | |
| | No correction | PSD 100 ≤PTR≤ 170 ns | Linear PSC and PSD 100 ≤PTR≤ 178 ns | Linear PSC | Non linear PSC |
| Energy resolution FWHM (%) | 1.29 ± 0.05 | 1.19 ± 0.04 | 1.12 ± 0.04 | 1.19 ± 0.04 | 1.17 ± 0.04 |
| FW.25M to FWHM ratio (Gaussian ratio 1.41) | 2.00 ± 0.05 | 1.53 ± 0.05 | 1.51 ± 0.05 | 1.70 ± 0.04 | 1.64 ± 0.05 |
| Percentage deviation of photopeak area (%) | 0 | 0 | 0 | +5 | +10 |
| Percentage deviation of total counts (%) | 0 | -23 | -20 | 0 | 0 |
| | 122.1 | | | | |
| | No correction | PSD 100 ≤PTR≤ 198 ns | Linear PSC and PSD 100 ≤PTR≤ 180 ns | Linear PSC | Non linear PSC |
| Energy resolution FWHM (%) | 1.58 ± 0.05 | 1.21 ± 0.07 | 0.73 ± 0.04 | 0.89 ± 0.04 | 0.87 ± 0.04 |
| FW.25M to FWHM ratio (Gaussian ratio 1.41) | 2.30 ± 0.05 | 1.69 ± 0.07 | 1.65 ± 0.07 | 1.93 ± 0.04 | 1.89 ± 0.06 |
| Percentage deviation of photopeak area (%) | 0 | 0 | 0 | +51 | +65 |
| Percentage deviation of total counts (%) | 0 | -44 | -50 | 0 | 0 |

Table 2. Spectroscopic results for the CdTe detector at low photon counting rate (200 cps) using pulse shape analysis techniques. Changes of the photopeak area and total counts are calculated respect to the spectra with no correction. The photopeak area was calculated as twice the HSA. We used $T_{d, slow}$ = 15 µs.

We first used (i) the PSD technique selecting the PTR which produced no reduction of the photopeak area, (ii) we applied both linear PSC and PSD to obtain no reduction of the photopeak area, and finally (iii) we applied both linear and non linear PSC techniques to all peaking time values obtaining no reduction of the total counts. Figure 10(a) shows the enhancements in [57]Co spectra after linear PSC and the PSD techniques, without any photopeak area reduction. Figure 10(b) shows the enhancements in [57]Co spectra after non linear PSC, applied to all peaking time values (with no reduction of the total counts). No spectral improvements are obtained in [109]Cd spectrum with the PSC methods.

Despite the similar results for linear and non linear corrections, the implementation of the non linear correction opens up to the possibility for charge collection compensation for thicker detectors wherein hole trapping effects are more severe. We stress as the flexibility of the digital pulse processing approach also allows the easy implementation of more complicated correction methods for the minimization of incomplete charge collection.

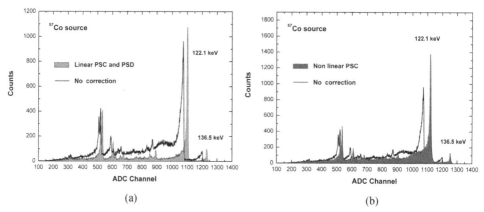

(a)                                                                    (b)

Fig. 10. (a) Measured [57]Co spectra with no correction and using both linear PSC and PSD. The linear PSC was applied to a selected PTR which ensured no photopeak area reduction. After linear PSC, we obtained an energy resolution of 0.73% *FWHM* at 122.1 keV. (b) Measured [57]Co spectra with no correction and using non linear PSC. The non linear PSC was applied to all peaking time values obtaining no reduction of the total counts. After non linear PSC, we obtained an energy resolution of 0.87% *FWHM* at 122.1 keV.

### 5.3.2 High count rate performance

Figure 11 shows the performance of the detection system (with $T_{d, slow}$= 3 μs), irradiated with the [109]Cd source, at different photon counting rates (up to 820 kcps). The throughput of the system (i.e. the rate of the events in the spectrum or the rate of the analyzed events by the *"slow"* pulse shaping), the 22.1 keV photopeak centroid and the energy resolution (FWHM) at 22.1 keV were measured. The photopeak centroid shift was less than 1 % and low energy resolution worsening characterized the system. Despite the low throughput of the system, it is possible to measure the input photon counting rate through the *"fast"* pulse shaping. The system, through both *"slow"* and *"fast"* pulse shaping is able to determine the input count rate and the energy spectrum with high accuracy and precision even at high photon counting rates.

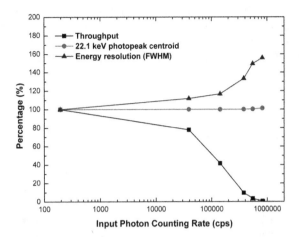

Fig. 11. Performance of the detection system, irradiated with a $^{109}$Cd source: throughput, 22.1 keV photopeak centroid and energy resolution (FWHM) at 22.1 keV at different input photon counting rates .

Figure 12 shows the measured $^{109}$Cd spectra at (a) 200 cps with no correction, (b) at 820 kcps with no correction and (c) at 820 kcps after PSD (100 ≤PTR≤ 148 ns). The results (summarized in Table 3) highlight the excellent high rate capability of our digital system. Moreover, PSD allowed a strong  reduction (96%) of the number of peak pile-up events in the measured spectrum, as shown in Figure 12(c).

| | Energy | | | |
|---|---|---|---|---|
| | 22.1 keV | | 59.5 keV | |
| | No correction | PSD 100≤PTR≤148 ns | No correction | Linear PSC and PSD 100≤PTR≤154 ns |
| Energy resolution FWHM (%) | 4.77 ± 0.04 | 4.52 ± 0.04 | 2.12 ± 0.04 | 1.87 ± 0.04 |
| Total rate (kcps) | 820 | 820 | 255 | 255 |
| Analyzed rate or Spectrum rate (kcps) | 4.4 | 1.1 | 53.5 | 28.4 |

Table 3. Spectroscopic results from high rate measurements. The energy resolution at low rate 200 cps is (with $T_{d, slow}$= 3 µs): 3.1% at 22.1 keV and 1.9% at 59.5 keV. Both PSD and linear PSC give a peak pile-up reduction of  96%.

High-rate [241]Am spectrum measurements (Figures 12(d), (e), (f)) also show as both PSD and linear PSC can be used for compensation of charge trapping and peak pile-up. With regard to [241]Am spectra, we first minimized peak pile-up (with a reduction of about 96%) by selecting a proper PTR ($100 \leq PTR \leq 154$ ns) and then we applied the linear PSC in the selected PTR.

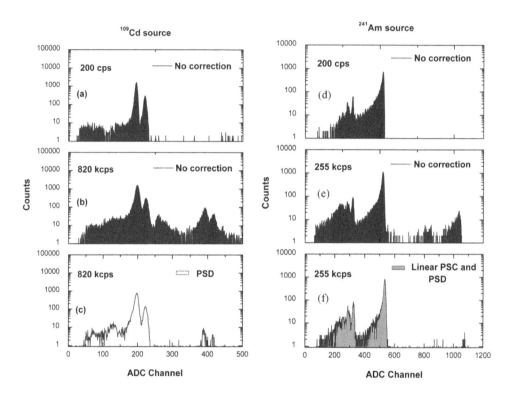

Fig. 12. Measured [109]Cd spectra at (a) 200 cps with no correction, (b) at 820 kcps with no correction and (c) at 820 kcps after PSD. Measured [241]Am spectra at (d) 200 cps with no correction, (e) at 255 kcps with no correction and (f) at 255 kcps after linear PSC.

We also measured X-ray spectra from a non clinical X-ray tube with different anode materials (Ag, Mo). In Figure 13 are shown the measured Ag-target X-ray spectra (32 kV) at 8.8 kcps with no correction, at 258 kcps with no correction and at 258 kcps after PSD. At high photon counting rate, the measured Ag spectrum, despite the good energy resolution of the peaks (22.1 and 24.9 keV), is characterized by a high background beyond the end point energy, due to the peak pile-up; while, after PSD, this background is quite similar to the spectrum at low photon counting rate. These results open up the possibility of precise estimations of the end point energy, i.e. the peak voltage of a X-ray tube, even at high photon counting rates. As well known, precise peak voltage measurements are essential for accurate quality controls on clinical X-ray tubes. Figures 13(d), (e) and (f) also show the measured Mo X-ray spectra.

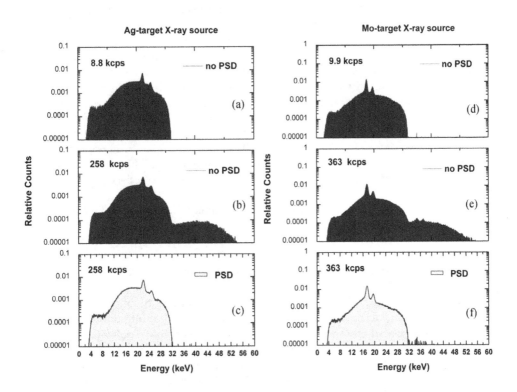

Fig. 13. Ag-target X-ray spectra (32 kV) at (a) 8.8 kcps with no correction, (b) at 258 kcps with no correction and (c) at 258 kcps after PSD. Mo-target X-ray spectra (32 kV) at (d) 9.9 kcps with no correction, (e) at 363 kcps with no correction and (f) at 363 kcps after PSD.

## 6. Direct measurement of mammographic X-ray spectra with the digital CdTe spectrometer: A medical application

Knowledge of energy spectra from X-ray tubes is a key tool for quality controls (QCs) in mammography. As well known, X-ray spectra directly influence the dose delivered to the patients as well as the image quality. X-ray spectra can be used for accurate estimations of the peak voltage (KVp) of the tubes, the energy fluence rate, the inherent filtration, the beam-hardening artifacts and for the correct implementation of the new dual-energy techniques.

By way of example, the peak voltage of a diagnostic X-ray tube should be routinely monitored, since small KVp changes can modify both absorbed dose and image contrast in mammography. With regard to dosimetric investigations, X-ray spectra can be also used to estimate the exposure, the air kerma and the absorbed energy distribution inside a breast tissue or a test phantom, overcoming the well known problem of the energy dependence of the response of the dosimeters (solid state detectors and ionization chambers) which are commonly used for the measurements of the absorbed energy distribution. Dosimeter calibrations, which usually involve complicated and time-consuming procedures, are a critical issue for routine investigations.

The spectrum emitted by a mammographic X-ray tube is, typically, obtained by analytical procedures based on semi-empirical models and Monte Carlo methods. In routine quality controls, insufficient information about some characteristic parameters of the X-ray tubes, such as the anode angle, the filters and the exact value of the applied tube voltage, could compromise the precision and the accuracy of the spectra. Of course, measurement of X-ray spectra would be the best procedure for accurate quality controls in mammography. Currently, routine measurement of mammographic X-ray spectra is quite uncommon due to the complexity of the measurement procedure. The measurement of mammographic X-ray spectra is a difficult task because of limitations on measurement with high energy resolution at high photon counting rates as well as geometrical restrictions, especially in a hospital environment.

In this section, we report on direct measurements of clinical molybdenum X-ray spectra by using the digital CdTe detection system.

## 6.1 Experimental set-up

Mo-target X-ray spectrum measurements were performed under clinical conditions (Istituto di Radiologia, Policlinico, Palermo). We used a Sylvia mammographic unit (Gilardoni) with a Mo anode tube (MCS, 50MOH), characterized by an additional filtration of 0.03 mm Mo and a 10° anode angle. The compression paddle was removed during the measurements. The detector was placed on the cassette holder with a 59.5 cm focal spot-detector distance. To reduce the photon counting rate on the detection system to an acceptable level, we used a pinhole collimator: a tungsten collimator disk, 1 mm thick with a 100 μm diameter circular hole, placed in front of the detector (over the beryllium window). Using this collimation setup, we measured X-ray spectra with a photon counting rate up to 453 kcps. It is well known that the choice of the proper collimation system is a critical issue for accurate measurements of X-ray spectra. An excessive reduction of the aperture and the thickness of the collimator can produce several distortions in the measured spectra (peaks and continuum events). These distortions are mainly due to (i) the penetration of photons through the collimator material, (ii) scattered photons from the collimator edges and (iii) characteristic X rays from the collimator material. The first effect can be reduced by choosing a proper collimator material and thickness. In the investigated energy range (1-30 keV), the penetration of photons through the 1 mm thick tungsten collimator is negligible, as demonstrated by the estimated values of the transmission equivalent aperture ($TEA$). By using the tabulated tungsten mass attenuation coefficient values (Boone & Chavez, 1996), we obtained a $TEA$ equals to the collimator aperture area, showing that no photon penetration occurs. The other collimation distortions mainly depend on the alignment of the collimator with the X-ray beam and on the energy of the X-ray beam. Misalignment between the X-ray beam and the collimator can produce scattered photons and characteristic X rays from the collimator edge. Obviously, accurate alignment becomes more difficult as the thickness of the collimator increases and its aperture diameter decreases. The optimum collimation set-up should be a trade-off between the reduction of collimation distortions and the photon counting rate. To optimize the beam-detector alignment, the detector was mounted on an aluminum plate equipped with three micrometric screws. A preliminary focal spot-detector alignment was carried out with a laser pointer taking into account the reference marks positioned on both sides of the tube head, while a more accurate alignment was obtained by changing the plate orientation looking for the maximum photon counting rate and the absence of distortions in the measured spectra.

We also performed the measurement of attenuation curves through the measured Mo spectra. This curve, which is usually measured to characterize the spectral properties of a beam, was compared with that measured by using a standard mammographic ionization chamber (Magna 1cc together with Solidose 400, RTI Electronics) and a solid state dosimeter (R100, RTI Electronics). Exposure values from spectral data were obtained through the estimation of the energy fluence and the air mass energy absorption coefficients, as described in our previous work (Abbene et al., 2007). To measure the attenuation curves, we used a standard aluminum filter set (type 1100, Al 99,0 % purity, RTI electronics). To minimize the effects of scattered radiation, we performed the measurements in a "good geometry" condition, as suggested by several authors (Johns & Cunninghan, 1983); in particular, the experimental set-up for these measurements was characterized by a filter-detector distance of 42 cm.

## 6.2 Clinical X-ray spectra measurements

Figure 14(a) shows the measured Mo-target X-ray spectra under clinical conditions. The tube settings were: tube voltages of 26, 28 and 30 kV and tube current-time product of 20 mAs. The photon counting rates are 278, 362 and 453 kcps at 26, 28 and 30 kV, respectively.

Figure 14(b) shows the attenuation curves (28 kV and 20 mAs) obtained from the spectra measured with the digital system, from simulated spectra (IPEM Report 78) and by using the exposure values directly measured with the ionization chamber (Magna 1cc together with Solidose 400, RTI Electronics) and with the solid state dosimeter (R100, RTI Electronics). It is evident the good agreement among the curves obtained from the detector, the simulation and the ionization chamber. The disagreement with the attenuation curve obtained from the solid state dosimeter, points out the energy dependence of the dosimeter response. Since aluminum filters harden the X-ray beam and alter the energy spectrum, if dosimeter does not has a *flat* response for different spectra, the attenuation curve will be in error. The correction of the energy dependence of the dosimeter response need accurate calibrations which involve complicated and time-consuming procedures, critical for routine investigations.

These comparisons highlight two main aspects: (i) the ability of the digital system to perform accurate mammographic X-ray spectra without excessive time-consuming procedures and (ii) the possible use of this system both as dosimeter and for calibrations of dosimeters.

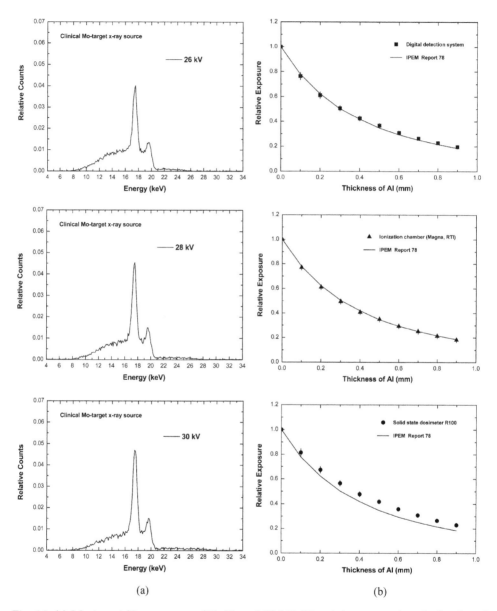

Fig. 14. (a) Mo-target X-ray spectra (26, 28 and 30 kV; 20 mAs) measured with the digital system under clinical conditions; the counts were normalized to the total number of detected events. (b) Attenuation curves obtained from measured and simulated spectra and from direct exposure measurements (ionization chamber and solid state detector). The tube settings were: 28 kV and 20 mAs.

## 7. Conclusion

High-Z and wide band gap compound semiconductors are very promising materials for the development of portable high resolution spectrometers in the hard X-ray energy band (>15 keV). In particular, CdTe and CdZnTe detectors, due to the high atomic number, the high density and the wide band gap, ensure high detection efficiency, good room temperature performance and are very attractive for several X-ray applications. CdTe/CdZnTe detectors coupled to digital readout electronics show excellent performance and are very appealing for high-rate X-ray spectroscopy. The digital pulse processing (DPP) approach is a powerful tool for compensation of the effects of incomplete charge collection (typical of CdTe and CdZnTe detectors) and the effects due to high-rate conditions (baseline fluctuations, pile-up).

The performance of the presented prototype, based on a CdTe detector and on a custom DPP system, highlight the high potentialities of these systems especially at critical conditions. The digital system, combining fast and slow shaping, automatic pole-zero adjustment, baseline restoration, pile-up rejection and some pulse shape analysis techniques (pulse shape discrimination, linear and non linear corrections), is able to perform a correct estimation of the true rate of the impinging photons, a precise pulse height measurement and the reduction of the spectral distortions due to pile-up and incomplete charge collection. High-rate measurements (up to 800 kcps) highlight the excellent performance of the digital system: (i) low photopeak centroid shift, (ii) low worsening of energy resolution and (iii) the minimization of peak pile-up effects.

Measurements of clinical X-ray spectra also show the high potentialities of these systems for both calibration of dosimeters and advanced quality controls in mammography.

The results open up the development of new detection systems for spectral X-ray imaging in mammography, based on CdTe/CdZnTe pixel detectors coupled with a DPP system. Recently, single photon counting detectors are very appealing for digital mammography allowing the implementation of dual-energy techniques and improvements on the quality of images. In this contest, CdTe/CdZnTe pixel detectors can ensure better performance (energy resolution <5% at 30 keV) than the current prototypes based on silicon detectors (energy resolution of about 15% at 30 keV).

Future works will regard the development of a *real time* system, based on the digital method, by using a field programmable gate array (FPGA)technology.

## 8. References

Abbene, L. et al. (2007). X-ray spectroscopy and dosimetry with a portable CdTe device. *Nucl. Instr. and Meth. A*, Vol. 571, No. 1, (February 2007), pp. 373-377, ISSN 0168-9002

Abbene, L. et al. (2010). Performance of a digital CdTe X-ray spectrometer in low and high counting rate environment. *Nucl. Instr. and Meth. A*, Vol. 621, No. 1, (September 2010), pp. 447-452, ISSN 0168-9002

Abbene, L. et al. (2010). High-rate x-ray spectroscopy in mammography with a CdTe detector: A digital pulse processing approach. *Med. Phys.*, Vol. 37, No. 12, (December 2010), pp. 6147-6156, ISSN 0094-2405

Abbene, L., Gerardi, G. (2011). Performance enhancements of compound semiconductor radiation detectors using digital pulse processing techniques. *Nucl. Instr. and Meth. A*, in press, ISSN 0168-9002

Boone, J. M, Chavez, A. E. (1996). Comparison of x-ray cross sections for diagnostic and therapeutic medical physics. *Med. Phys.*, Vol. 23, No. 12, (December 1996), pp. 1997-2005, ISSN 0094-2405

Cavalleri, G. et al. (1971). Extension of Ramo theorem as applied to induced charge in semiconductor detectors. *Nucl. Instr. and Meth.*, Vol. 92, pp. 137-140, ISSN 0168-9002

Del Sordo, S. et al. (2004). Spectroscopic performances of 16 x 16 pixel CZT imaging hard-X-ray detectors. *Nuovo Cimento B.*, Vol. 119, No. 3, (March 2004) pp. 257-270, ISSN 1594-9982

Del Sordo, S. et al. (2009). Progress in the Development of CdTe and CdZnTe Semiconductor Radiation Detectors for Astrophysical and Medical Applications. *Sensors*, Vol. 9, No. 5, (May 2009), pp. 3491-3526, ISSN 1424-8220

Gerardi, G. et al. (2007). Digital filtering and analysis for a semiconductor X-ray detector data acquisition. *Nucl. Instr. and Meth. A*, Vol. 571, No. 1, (February 2007), pp. 378-380, ISSN 0168-9002

Hecht, K. (1932). Zum Mechanismus des lichtelektrischen Primärstromes in isolierenden Kristallen. *Z. Phys.*, Vol. 77, pp. 235-245

IEEE Standard (2003), ANSI N42.31, pp. 1-33

Jones, L. T., Woollam, P. M. (1975). Resolution improvement in CdTe gamma detectors using pulse-shape discrimination. *Nucl. Instr. and Meth.*, Vol. 124, No. 2, (March 1975), pp. 591-595, ISSN 0168-9002

Johns, H. E., Cunninghan, J. R. (1983). *The Physics of Radiology* (4th Ed.), C. C. Thomas Publisher, ISBN 0-398-04669-7, Springfield

Keele, B. D. et al. (1996). A method to improve spectral resolution in planar semiconductor gamma-ray detectors. *IEEE Trans. Nucl. Sci.*, Vol. 43, No. 3, (June 1996), pp. 1365-1368, ISSN 0018-9499

Knoll, G. F. (2000). *Radiation Detection and Measurement* (3rd Ed.), Wiley, ISBN 978-047-1073-38-3, New York

McGregor, D. S., Hermon, H. (1997). Room-temperature compound semiconductor radiation detectors. *Nucl. Instr. and Meth. A*, Vol. 395, No. 1, (August 1997), pp. 101-124, ISSN 0168-9002

Owens, A., Peacock, A. (2004). Compound semiconductor radiation detectors. *Nucl. Instr. and Meth. A*, Vol. 531, No. 1, (September 2004), pp. 18-37, ISSN 0168-9002

Ramo, S. (1939). Currents induced by electron motion. *Proceedings of the I.R.E.*, pp. 584-585

Shockley, W. (1938). Currents to conductors induced by a moving point charge. *J. Appl. Phys*, Vol. 9, pp. 635-636, ISSN 0021-8979

Sjöland, K. A., Kristiansson, P. (1994). Pile-up and defective pulse rejection by pulse shape discrimination in surface barrier detectors. *Nucl. Instr. and Meth. B*, Vol. 94, No. 3, (November 1994), pp. 333-337, ISSN 0168-583X

Takahashi, T., Watanabe, S. (2001). Recent progress in CdTe and CdZnTe detectors. *IEEE Trans. Nucl. Sci.*, Vol. 48, No. 4, (August 2001), pp. 950-959, ISSN 0018-9499

# 5

# Application of Wavelength Dispersive X-Ray Spectroscopy in X-Ray Trace Element Analytical Techniques

Matjaž Kavčič

*J. Stefan Institute*

*Slovenia*

## 1. Introduction

The basic purpose of the x-ray trace element analytical techniques is to detect with high sensitivity elemental constituents of the target (including elements in trace amounts) and determine quantitatively the elemental composition of the investigated sample. These techniques are based on the detection of x-rays following the atomic inner shell ionization. An inner shell (core) electron can be removed from the atom in different ways, for the analytical purposes x-ray absorption and proton scattering are most commonly used. In the case of X-Ray Fluorescence (XRF) method the x-ray tube is usually applied to irradiate the sample while in the case of Proton Induced X-ray Emission (PIXE) the proton beam accelerated by the electrostatic accelerator to typical energy of MeV is used to ionize the target atoms. The atom with a hole in the inner shell (core hole) is extremely unstable with characteristic lifetime in the order of $10^{-15}$ s. Consequently, the inner shell ionization is followed via subsequent x-ray or Auger electron emission. Both, PIXE and XRF method exploit the radiative decay channel. The energy of the emitted x rays is given by the energy difference of the electron states involved in the transition which is characteristic of the target element atomic number. In order to identify the elemental composition of the target, energy analysis of the emitted radiation is required with energy resolution high enough to resolve characteristic spectral contributions from different elements in the sample. Such resolution is achieved by the energy dispersive solid state detectors in which electric signals are proportional to the incident x-ray energy and they are commonly used in x-ray trace element analytical techniques. Besides good enough energy resolution they also provide an excellent efficiency, which is crucial to collect weak signals from trace elements in the sample. It was in fact the development of the semiconductor detectors in the seventies that has triggered the development of x-ray analytical techniques. Today both PIXE and XRF techniques being a multi-element, sensitive, fast, non destructive and relatively inexpensive, have established as a routine analytical tool in a variety of fields such as material analysis, environmental and biomedical research, archeological and art studies,…

However, in some special cases the energy resolution of solid state energy dispersive detectors, nowadays reaching the order of 130 – 150 eV for the x-ray energies of few keV, is not enough and significantly higher energy resolution is required to enhance the analytical capabilities of the x-ray techniques. In order to increase further the energy resolution of x-ray detection a wavelength dispersive x-ray (WDX) spectroscopy is usually applied. In this case x-rays are analyzed according to their wavelength using the Bragg diffraction on the crystal plane. While WDX spectroscopy has been traditionally used in combination with the electron excitation for major and minor element analysis it has usually not been considered in PIXE and XRF trace element analysis mainly due to low detection efficiency. Compared to the simplest wavelength dispersive spectrometers employing flat crystals the efficiency was enhanced significantly in modern spectrometers employing cylindrically or even spherically curved crystals in combination with position sensitive x-ray detectors. The energy resolution of such a spectrometer may exceed the resolution of the energy dispersive detector by two orders of magnitude while keeping the efficiency at a high enough level to perform trace element analysis.

In this chapter we will address two particular issues where we can apply successfully the WDX spectroscopy in order to enhance the capabilities of the x-ray trace element analytical techniques. The first part of the chapter deals with the chemical speciation of some light elements (P, S) via high resolution x-ray emission measurements. In the second part we will discuss the improvement of the detection limits in the case of analysis of trace elements with atomic number neighboring the predominant matrix element. In both cases the limiting factor is actually the restricted energy resolution of the energy dispersive solid state detectors. Application of WDX spectroscopy with energy resolution reaching towards the natural linewidths of the measured spectral lines overcomes this limitation and spreads the analytical capabilities of the x-ray techniques even further.

## 2. Chemical speciation of light elements via high resolution x-ray emission measurements

While PIXE and XRF technique yield with high accuracy elemental concentrations they are not sensitive to the chemical environment of the x-ray emitting atom and consequently they can not provide the chemical speciation of the elements in the sample. The most commonly used x-ray spectroscopic technique to perform chemical speciation is probably the x-ray absorption spectroscopy (XAS), which probes directly the unoccupied states just above the Fermi level. Experimental requirement for XAS spectroscopy is of course an intense monochromatic tunable x-ray source, which is available by a synchrotron. On the other hand x-ray emission spectroscopy (XES) probes the occupied states below the Fermi level. If we can achieve the energy resolution in the emission channel comparable to the resolution of the synchrotron beamline monochromator, information complementary to absorption spectra is expected from the high resolution XES spectra for the case of core-valence electron transitions. This principle is nicely illustrated in Figure 1 representing the direct comparison of the S K edge XAS spectrum and the Kβ emission spectrum recorded with the high energy resolution. Similarly as for the absorption spectrum the chemical environment of an element affects and modifies also the various characteristics of its x-ray emission spectrum. Usually, the influence of the chemical environment results in energy shifts of the characteristic lines, formation of satellite lines and changes in the emission linewidths and relative intensities (Gohshi et al. 1973, Perino et al. 2002, Tamaki 1995, Yasuda et al. 1978). A very important qualitative difference with the XAS spectroscopy requiring synchrotron source is that the x-ray emission spectrum can be recorded also in a smaller laboratory employing an x-ray tube (XRF) or proton accelerator (PIXE) in combination with wavelength dispersive x-ray spectrometer. Provided that the energy resolution of the x-ray emission spectrometer is high enough a quantitative chemical state analysis is therefore feasible also in a small scale laboratories as it was demonstrated for the first time by Gohshi and co-workers by means of a double crystal spectrometer (Gohshi & Ohtsuka 1973). In order to perform chemical state analysis of trace elements in the sample a single crystal focusing type spectrometer is advantageous due to higher efficiency, which is crucial to collect weak signal from the sample.

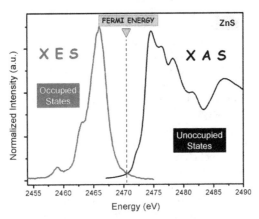

Fig. 1. S K-edge x-ray absorption near edge (XANES) spectrum and high resolution Kβ emission spectrum of ZnS (sphalerite) (Alonso Mori et al. 2010).

## 2.1 Experimental set-up

Here we present the wavelength dispersive x-ray spectrometer constructed and built at the J. Stefan Institute (Kavčič et al. 2004, Žitnik et al. 2009) that has been used extensively in the last couple of years to perform high resolution XES studies. The spectrometer is equipped with the cylindrically curved crystal in Johansson geometry, the Rowland circle radius is 50 cm.

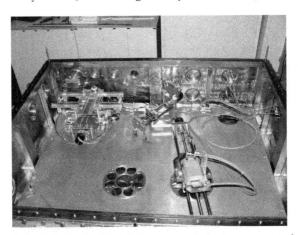

Fig. 2. Inside view of the J. Stefan Institute's WDX spectrometer vacuum chamber with the target holder on a goniometer, crystal holder, and the CCD camera. A mechanical theta/2theta stage serves to adjust the Bragg angle.

The target holder, which is positioned well inside the Rowland circle, is mounted on the special custom made goniometer enabling two separate perpendicular translations (within beam direction and perpendicular to it) as well as rotation around vertical axis. The target goniometer serves for precise alignment of the target within the Rowland circle. The position of the target inside the Rowland circle closer to the crystal analyzer has two main advantages. First it enables efficient collection of x-rays emitted from large (extended) x-ray source without any loss of resolution. The second important point is that in combination with position sensitive x-ray detector it enables collection of diffracted x-rays over certain bandwidth given by the detector size. In our case the diffracted x-rays are detected by a Peltier cooled charged coupled device (CCD) detector, with dimensions 17.3 x 25.9 mm². The detector consists of 770 x 1152 pixels each having a size of 22.5 x 22.5 μm². The typical working temperature is -40 °C. Generally, the diffracted x-rays detected by the CCD detector form a two dimensional image on the detector plane. This image is projected on the horizontal axis corresponding to the energy axis of the spectrum. The spectrometer is equipped with three crystals, Quartz (1010) crystal with 2d lattice spacing of 8.510 Å, a Si(111) crystal with 6.271 Å, and a Si(220) crystal with 3.840 Å respectively. The whole spectrometer is enclosed in a 1.6 x 1.3 x 0.4 m³ stainless-steel chamber evacuated by a turbomolecular pump down to 10⁻⁶ mbar. The proton incidence and x-ray emission angles are perpendicular to each other. With a given mechanics the Bragg angles of ~ 30⁰ – 65⁰ can be reached, this range being constrained by the size of the vacuum chamber. With the crystals mentioned above the energy range of 1.6 – 6.5 keV can be reached in first order of reflection.

The main characteristic of such wavelength dispersive setup is the energy resolution, which should, combined with good enough efficiency, enable the chemical speciation of elements in the sample. Figure 3 shows the $K\alpha$ emission spectrum of yellow sulfur recorded after excitation with the 2.52 keV monochromatic photon beam. The $K\alpha_{1,2}$ doublet due to spin orbit splitting of the 2p shell, which is slightly above 1 eV is clearly resolved in the measured spectrum. The energy resolution is estimated to ~ 0.3 - 0.4 eV, which is well below the natural linewidth of the measured line at 0.61 eV (Campbell & Papp 2001). At such high experimental resolution the spectrum exhibit almost pure Lorentzian lineshape.

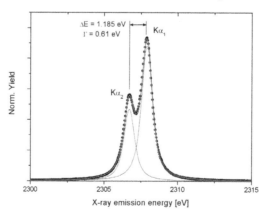

Fig. 3. High resolution $K\alpha$ x-ray emission spectrum of yellow S excited by a monochromatic focused (200 x 50μm²) photon beam with an 2.52 KeV energy. The spectrum was recorded at the ID26 beamline of the ESRF synchrotron using the wavelength dispersive spectrometer of the J. Stefan Institute.

The experimental setup enables detailed studies of fine structure within the XES spectra and accurate measurements of the energy shifts at the order of tenths of eV, which is relevant for the quantitative chemical speciation studies.

## 2.2 Chemical speciation of light elements via high resolution Kα XES measurements

As already explained in the introduction the core hole state induced by the proton scattering or photon absorption is unstable and decay via radiative or Auger transition. In case of 1s ionized atoms the fluorescence exhibit characteristic K x-ray lines. The most intense K spectral line named Kα corresponds to the 2p to 1s electron transition (core-core transition). Since only core electrons are involved in the transition it is generally expected that Kα spectral lines should be mostly free from chemical bond effects except from small energy shifts. These shifts are related to the valence electron population only indirectly through slight changes in the screening of the effective nuclear potential. We have performed extensive study of the energy shifts of the Kα lines for the various S compounds using excitation with monochromatic photon beam (Alonso Mori et al. 2009) and also MeV proton excitation (Kavčič et al. 2004, 2005). While the shape of the spectrum is not affected by the chemical environment, clear energy shifts of the measured Kα spectral lines are observed and they can be correlated directly with the oxidation state of sulfur in the sample (Figure 4). The energy position of the Kα line enables therefore a simple and reliable determination of the sulfur oxidation state.

In the next step the high resolution measurement of the Kα line energy is applied for the chemical state analysis of sulfur in a typical sample used in the analytical work. High resolution PIXE analysis of sulfur in an aerosol sample was chosen since PIXE is well established and commonly used method to determine the elemental concentrations in aerosol samples collected under specific time and particle size constraints. Among different elements in the atmosphere sulfur and its compounds present one of the most important pollutants in the atmosphere with the anthropogenic emission to account for the 75% of the total. The chemical reactivity and consequently its impact on the environment depend on the chemical state of sulfur so the information of oxidation state of sulfur in aerosol samples

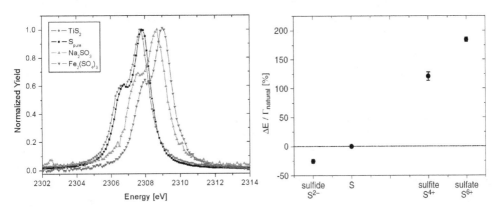

Fig. 4. (left) High resolution proton induced Kα x-ray emission spectra of several S compounds (Kavčič et al. 2004). (right) Measured energy shifts versus the oxidation state of S in the target. The energy shifts are given in units of the natural linewidth of the principal Kα line. A linear dependence of the Kα energy shift on the S oxidation state can be observed (Kavčič et al. 2005).

is highly relevant. Our measurements were performed on the aerosol sample collected on a Pallflex (2500Q AT-UP) tissue quartz filter using a 'Ghent type' sampling station. The total aerosol loading mass was ~ 0.4 mg/cm². The total S concentration of this specific sample was determined by the standard PIXE technique as 28.1 µg/cm². The measurements were performed at the Microanalytical Centre of the J. Stefan Institute in Ljubljana using the 2 MV tandem proton accelerator. The sample was exposed to 2 MeV proton beam with the 50 nA proton current. The measured $K\alpha$ signal of sulfur in the sample is presented in Figure 5. The energy shift of the $K\alpha_{1,2}$ doublet was determined as $1.19 \pm 0.06$ eV. According to the $K\alpha$ diagram energy shifts measured previously for various S compounds, the chemical state of sulfur in the aerosol sample was identified as a sulfate ($[SO_4]^{2-}$).

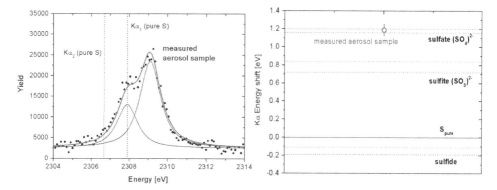

Fig. 5. (left) High-resolution proton-induced S $K\alpha$ x-ray (HR-PIXE) spectrum from an aerosol sample (yield represents raw, digitized data from the detector's A/D converter; absorption of one S $K\alpha$ photon corresponds to ~ 70 counts) and (right) graphical presentation of the chemical state speciation of S in aerosols employing HR-PIXE measurements of the S $K\alpha$ diagram line. For each measured chemical state of sulfur, the horizontal lines define the discriminating ability of the HR-PIXE setup (Kavčič et al. 2005).

The same method can be applied also to other light elements. Recently we have performed a high resolution PIXE study on several P compounds in order to establish and apply the PIXE-based wavelength dispersive spectroscopy to determine the chemical state of various phosphorus compounds. The phosphorus is the key element in soil and bio-solid science. The speciation of P-mixtures is extensively studied in many aspects in order to understand the relationship between its chemical state and rate-reactivity, long-term solubility and accumulation, mobility and transfer to ground water. A similar procedure as in the case of sulfur was followed also for the case of P compounds. As the energy of the emitted fluorescence is slightly lower in this case the Quartz(1010) crystal was employed in the spectrometer to record the P $K\alpha$ lines. Also in this case the target fluorescence was induced with the 2 MeV proton beam. The recorded $K\alpha$ spectral lines for different P compounds with oxidation states ranging from -3 to +5 are presented in Figure 6. Also in this case clear energy shifts are observed, which are again correlated directly to the oxidation state of P in the sample. The extracted dependence is well reproduced by a linear fit as also shown in Figure 6.

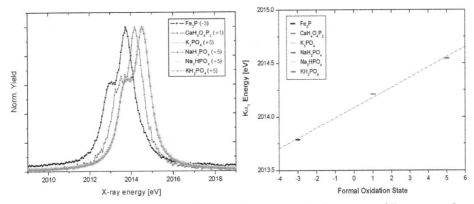

Fig. 6. (left) High-resolution proton-induced P Kα x-ray spectra from several P compounds. (right) Similar as for the S compounds a linear dependence of the Kα energy shift on the S oxidation state can be observed (Szlachetko & Kavčič 2011)

Based on this we can conclude that high resolution Kα x-ray emission measurements serve as a kind of chemical "ruler" for low Z elements oxidation. The spectral shape is practically independent on the oxidation state making the analysis simple and robust. A simple linear combination deconvolution procedure can be applied to mixed-valence system to perform also a quantitative study of the compound mixtures (Kavčič et al. 2004). The Kα line represent the strongest K x-ray spectral component which can be detected also for elements in low concentrations making it therefore feasible to perform chemical speciation of minor and trace elements in the sample.

## 2.3 Chemical and structural analysis of light elements via high resolution Kβ XES measurements

Compared to the Kα diagram line, which corresponds to the core–core (1s - 2p) transition the Kβ emission lines of light elements originate from valence-core transition. The high resolution Kβ emission measurements probes directly the p-density of occupied valence states and should therefore reflect with higher sensitivity the chemical environment of element in the sample. In fact most of the experimental as well as theoretical work on the chemical speciation of light elements, such as sulfur, has been focused on Kβ emission measurements (Kavčič et al. 2007, Maeda et al. 1998, Perino et al. 2002, Sugiura et al. 1974, Uda et al. 1999). All these works show pronounced effects in the emitted Kβ x-ray spectra, however, in contrast with the Kα spectra exhibiting independent characteristic lineshape the measured Kβ spectra are much more complex and consequently it is very difficult to give a simple parameter as it was the energy shift in the case of Kα spectra, which would correlate directly with the oxidation state and enable simple, reliable identification of the latter. Additional drawback is the fact that the Kβ line is much weaker than the Kα one making it more difficult to apply for chemical analysis of minor and trace elements. On the other hand the p-density of occupied valence states which is reflected in the emitted spectrum depend on the local coordination around central atom so using the theoretical modeling based on the quantum chemical calculations much more detailed analysis is possible yielding not only the oxidation state but also local coordination of ligand atoms.

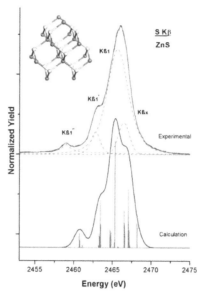

Fig. 7. S Kβ X-ray experimental spectrum of ZnS along with a 4.5Å cluster theoretical spectrum calculated within the density functional theory (DFT) (Alonso Mori et al. 2010)

We can therefore conclude that the high resolution Kβ x-ray emission spectra provide quantitative information about ligand environment. Comparison with the theoretically calculated spectra yields an opportunity to extract information about the electronic structure of element in the studied sample complementing this way the x-ray absorption near edge spectroscopy (XANES) used commonly at the synchrotron sources for this purpose.

## 3. Detection limits for trace elements with atomic number close to the target matrix element

One of the most important quantities for any x-ray trace element analytical technique is the minimum detection limit (MDL) yielding the minimum concentration of the trace element in the sample that we can detect. The principal factor governing the detection limits is the level of continuous background in the measured x-ray spectrum. Different physical processes contribute to this background. For charged particle excitation (electrons, protons, α particles) the bremsstrahlung from the projectile or secondary electrons and atomic bremsstrahlung (Ishii & Morita, 1984) are the most important sources of the continuous x-ray background. For heavier projectiles and higher energies the Compton tails of the γ-rays from nuclear reaction in the target specimen start to contribute significantly. When primary x-rays are used to excite the target (XRF method) the background in the measured spectra is produced mainly by elastic and Compton scattering of the primary photons. In this section we will discuss the analysis of trace elements with atomic numbers very close to the atomic number of the predominant matrix element. In this specific case the limitation comes from the strong x-ray fluorescence of the major matrix element which prevails the continuous x-ray background and represents a major limitation to detect a signal from neighboring trace elements. In this case the MDLs are limited by the energy resolution rather than by the efficiency of the detection system. In order to drastically reduce the contribution of the tails of the dominant diagram lines from the target matrix element and consequently improve the detection limits the WDX spectroscopy can be successfully applied.

A typical example of such analytical problem comes from the field of archeology and cultural heritage. In the archeological studies of gold objects Pt and Pb, both with atomic numbers very close to Au, are among the most informative trace elements. Particularly Pt is extremely important since it provides information on the origin and provenance of ancient gold. Using standard proton induced x-ray emission (PIXE method), which is one of the most commonly applied techniques in this field, the detection limit of 2000 ppm was found for Pt. That was lowered to 1000 ppm by using selective Zn filter to absorb the Au L-lines but this value is still too high for most of the archeological needs. In order to further improve the detection limit, the energy selective photoexcitation (PIXE-XRF) was employed and the method was finally reported to achieve the 80 ppm detection limit (Guerra 2005). As expected, this value was constrained by the resonant Raman scattering on Au.

In our work the particular case of Pd and Cd in silver matrix was chosen to demonstrate the basic principle how the application of WDX spectroscopy helps to improve the detection limits for trace elements neighboring the target matrix element. Similar as gold also silver archeological objects are frequently studied by x-ray analytical methods (PIXE, XRF). Especially coins are objects of great historical value and x-ray based techniques can provide a proper determination of their manufacturing technology, age, authenticity and give useful information about the monetary history of a certain period (Rodrigues 2011). Pd and Cd are both archeologically relevant elements, Cd being particularly important for authenticity verification of silver antiques since it is being used as an additive in silver solders and alloys only recently and its presence clearly indicates a fairly modern origin of a silver alloy sample (Devos 1999).

## 3.1 Improved PIXE detection limits of Pd and Cd in silver matrix employing high resolution measurements of Lα spectral lines

Generally the absolute minimum detectable yield of the spectral contribution corresponding to particular target element is usually defined as

$$Y_{MDL} = 3\sqrt{bckgr} \tag{1}$$

where *bckgr* represents the background yield taken within the full width at half maximum (FWHM) of the measured line. In order to determine the minimum detectable yields for Pd and Cd we need to measure the widths of the corresponding Lα lines as well as the background level in the position of these lines. In order to determine quantitatively the MDLs that can be achieved for PIXE method employing WDX spectroscopy we have recorded the high resolution Lα x-ray emission spectra of pure Ag, Pd, and Cd target using our wavelength dispersive spectrometer (presented in section 2.1) yielding sub eV resolution. For this experiment the spectrometer was equipped with the Si(111) crystal and 3 MeV protons were used for target excitation. We have first recorded the Lα diagram line of a pure Ag target. Keeping the same target the spectrometer was tuned to the energy of the Pd and later also to the Cd Lα diagram line and the corresponding part of the

Ag x-ray spectrum was recorded (Fig. 8). The acquisition time for each measurement was set to 1000 seconds. After that the proton induced Lα diagram lines of Pd and Cd target were also recorded in order to determine the FWHM values (the acquisition time was reduced to 100 seconds) and compared directly with the corresponding x-ray spectral region collected on pure Ag target.

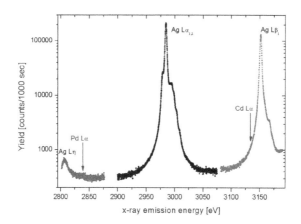

Fig. 8. High-resolution Ag Lα (L$_3$ – M$_{4,5}$) X-ray emission spectra induced with 3 MeV protons and the neighboring spectral regions centered around the energies of the Pd and Ag L lines. The Lβ$_1$ (L$_2$ – M$_4$) diagram line is situated in the high energy part of the measured spectrum and a weak Lη (L$_2$ – M$_1$) line is observed at the low energy edge. Arrows indicate the expected positions of the Pd and Cd Lα lines (Kavčič 2010).

The FWHM values of the Pd and Cd Lα line were determined by fitting the measured spectra with Voigt lines representing the Lorentzian natural lineshape convoluted with the Gaussian-like response of the spectrometer. Two Voigt profiles were used to model the Lα$_{1,2}$ lines. The proton induced spectra exhibit additional LM shell double ionization satellite structure on the high energy side of the diagram line which was fitted with several arbitrary Voigt profiles to account for the complex multiplet structure. The final result of the fitting procedure yielding the FWHM of the Lα$_1$ diagram line is presented in Figure 9. for Pd. In the next step the background level at the position of the expected Pd (or Cd) Lα signal was determined. The corresponding part of the measured proton induced Ag spectrum corresponding to the position of expected Pd signal was modeled by a single Voigt line representing the Ag Lη line (L$_2$-M$_1$ transition) and a constant background. The result of the fitting procedure which yielded the background level at position of the Pd line is also presented in Figure 9. A similar procedure that enables the determination of $Y_{MDL}$ (Eq. 1) was followed also for Cd.

In order to reach final MDL values the absolute minimum detectable yields should be normalized relative to the absolute Lα yield from a pure target. In our analysis we have used the measured Ag Lα yield for normalization and than employ theory to scale appropriately this yield for each particular element (Pd or Cd). Theoretically the thick target Lα yield is given by

$$Y_{TT}^{proton} \propto \omega_{L_3} \frac{\Gamma_{L\alpha}}{\Gamma_{L_3}^{Rad}} \int_0^R \left[ \sigma_{L_3}^I(E(x)) + f_{23}\sigma_{L_2}^I(E(x)) + (f_{12}f_{23} + f_{13} + f_{13}')\sigma_{L_1}^I(E(x)) \right] e^{-\mu_{L\alpha}x} dx \qquad (2)$$

where $E(x)$ is the proton energy as a function of the penetration depth, $R$ the range of protons in the target, $\sigma^I$ the $L$- subshell ionization cross sections, $\Gamma_{L\alpha}/\Gamma_L^{Rad}$ the relative emission rates for the $L\alpha$ radiative transition, $f_{ij}$ Coster-Kronig (CK) yields and $\omega_3$ the fluorescence yield for the $L_3$ subshell. The integration over $x$ is usually converted into the energy integrals using the well known transformation $dx = dE / S(E)$, with $S$ being the proton stopping power. We can calculate the $Y_{TT}$ yield for any particular element using the corresponding ionization cross sections, emission rates and CK yields, while for the stopping power and attenuation we always consider the values for the Ag matrix. The scaling factor $F^{th}$ is than given as a thick target yield $Y_{TT}$ calculated for particular element (Pd, Cd) incorporated homogeneously in the Ag matrix relative to the value calculated for pure Ag target, which was used for normalization

$$F^{th} = \frac{Y_{TT}(Pd, Cd \text{ within } Ag \text{ matrix})}{Y_{TT}(\text{pure } Ag \text{ target})} \qquad (3)$$

and the final MDL values are than obtained as

$$MDL = \frac{Y_{MDL}}{Y_{Ag}F^{th}} \qquad (4)$$

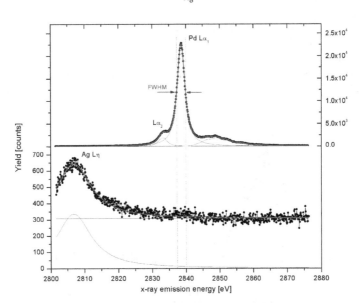

Fig. 9. (top) High-resolution Pd L$\alpha$ X-ray emission spectrum induced with 3 MeV protons fitted with the model explained in the text in order to determine FWHM of the L$\alpha_1$ line. (bottom) Part of the high-resolution Ag L$\alpha$ X-ray emission spectra corresponding to the energy region of the Pd L$\alpha$ line. The Ag signal level defines the minimum detectable yield of Pd. (Kavčič 2010).

The final detection limits obtained from our high resolution proton induced Lα x-ray emission measurements are summarized in Table 1, together with the $Y_{MDL}$ and FWHM values extracted from the measured Lα spectra, as well as calculated $F^{th}$ scaling factors. For Pd very low MDL value reaching only 40 ppm was obtained, which is a direct consequence of excellent experimental energy resolution. In case of Cd slightly larger value of MDL was obtained (~60 ppm) mainly due to the proximity of Ag L$β_1$ line whose tail extends to the region of Cd Lα line.

These detection limits can be compared directly with the corresponding values obtained in the standard PIXE analysis based on energy dispersive detectors. The latter were obtained using the model Pd/Ag PIXE spectra collected with Si(Li) detector within our standard PIXE set-up by adding a short accumulation of Pd signal to the long accumulation of Ag. These model spectra were evaluated with the GUPIX program package (Maxwell et al. 1989) yielding also limits of detection for each particular element. Consistent detection limits were obtained four all four model PIXE spectra combining Ag (1000 sec) and Pd spectra with different exposure times (10, 20, 30, 50 sec) yielding final detection limit for Pd in Ag of 1140ppm ± 10ppm. Despite the fact that the K x-ray lines were used to obtain this result, as it is usually the case in standard PIXE analysis of mid-Z elements, this value is still almost 30 times larger than the corresponding detection limit obtained from Lα x-ray spectra collected with wavelength dispersive x-ray spectrometer.

Finally the same procedure was used to produce a model Pd/Ag spectrum this time recorded by a WDX spectrometer. One single CCD exposure (1 sec) collected from Pd target was added to the total Ag spectrum corresponding to total exposure time of 1000 seconds producing a model spectrum which can serve as a close approximation to the spectrum recorded on a standard sample with 1000 ppm of Pd within Ag matrix. While Pd signal is below the detection limit of standard PIXE analysis we have determined previously, the high resolution model spectrum presented in Figure 10 exhibit strong Pd signal demonstrating directly the capability to detect concentrations well below this level by applying WDX spectroscopy.

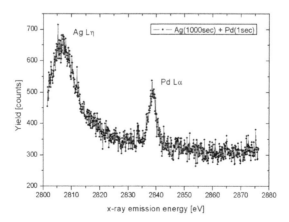

Fig. 10. A model high resolution PIXE spectrum measured with the WDX spectrometer. The spectrum was obtained by adding spectra from Ag and Pd target with exposition times of 1000 and 1 second, respectively (Kavčič 2010).

|  | Pd | Cd |
|---|---|---|
| $Y_{MDL}$ [counts] | 292 ± 2 | 477 ± 8 |
| FWHM [eV] | 2.78 ± 0.06 | 3.04 ± 0.10 |
| $F^{th}$ | 0.974 | 1.040 |
| **MDL [ppm]** | **39.8 ± 0.4** | **61.0 ± 1.1** |

Table 1. The PIXE minimum detection limits (MDL) for Pd and Cd within the Ag matrix achieved by applying WDX spectroscopy. Given errors are calculated considering the background, $Y_{Ag}$ and the Lα FWHM fitting errors. Tabulated are also minimum Pd and Cd detectable yields, FWHMs of the measured Lα spectra, and also calculated values of the scaling factor $F^{th}$ used to obtain final detection limits.

## 3.2 Further improvement of detection limits achieved by combining a WDX spectroscopy with the tunable monochromatic synchrotron radiation

For trace elements with atomic number just below the matrix element, very good detection limits can be achieved using the tunable synchrotron radiation with excitation energy tuned below the absorption edge of the matrix element. In this case the majority of the unwanted fluorescence signal is eliminated, but there remains a part of it due to the resonant Raman scattering (RRS) on the target matrix element (Jaklevic et al. 1988). Also in this case the background due to the RRS signal can be removed effectively by applying the WDX spectroscopy. We have therefore performed additional high resolution measurements on Ag and Pd employing the tunable monochromatic synchrotron radiation. Two different photon beam energies were employed for excitation, namely 3400 eV, being well above the $L_3$ absorption edge of both Ag and Pd, and 3172.8 eV corresponding to the top of the strong photoexcited resonance at the $L_3$ absorption edge of Pd. To locate the resonance, $L_3$ absorption spectrum of Pd was previously recorded in the fluorescence mode. The Pd Lα spectra were then recorded at both selected energies, and finally also the part of the spectrum from the Ag target at excitation energy 3172.8 eV, corresponding to the maximum of the Pd $L_3$ cross section (Figure 11). The overall acquisition time for this measurement was 4000 s.

As a direct consequence of high energy resolution of the measured x-ray emission and the proper choice of photon probe energy the resonant Raman scattering (RRS) signal from Ag is effectively separated from the region of the expected Pd signal (Fig. 11). The remaining almost flat background originates mainly from the scattering of the strong primary photon beam. Unfortunately, the Ag target foil used in the photon probe measurements was slightly contaminated with Cl, which is manifested by the relatively strong Cl Kβ peak around 2816 eV. In order to determine the background level at the position of the expected Pd Lα signal, which is needed to extract the MDL, we have fitted the part of the spectrum above the Ag RRS signal with the combination of flat background, the Pd Lα$_{1,2}$ profile as obtained from the spectrum of Pd target and an additional peak to account for the Cl impurity Kβ line. The result of this fit yielding the background level needed to extract the minimum detectable yield (Equation 1) is presented in Figure 11. Besides the background level also the FWHM of the Pd Lα$_1$ line is needed and was determined by fitting the measured Pd Lα spectrum with Voigt lines similar as in the case of proton excitation. Compared to proton induced spectra a significantly smaller FWHM was obtained. This is partly due to slightly better experimental resolution originating from the very small focused photon beam, but most of the broadening of the proton induced Lα lines is due to the strong LN shell multiple ionization induced by proton beam resulting in the line broadening. The final MDL was deduced using the same procedure as for the PIXE measurements (Equations 1, 3, 4). The only difference is the theoretical thick target yield which is in case of excitation with the monochromatic photon beam given by

$$Y_{TT}^{photo} \propto \omega_{L_3} \frac{\Gamma_{L\alpha}}{\Gamma_{L_3}^{Rad}} \left( \sigma_{L_3}^{ph}(E_{ph}) + f_{23}\sigma_{L_2}^{ph}(E_{ph}) + (f_{12}f_{23} + f_{13} + f_{13}')\sigma_{L_1}^{ph}(E_{ph}) \right) \frac{1}{\mu_{matrix}^{E_{ph}} + \mu_{matrix}^{L\alpha}} \quad (5)$$

In this equation $E_{ph}$ is the energy of the photon beam, $\sigma^{ph}$ corresponds to L-subshell photoionization cross sections and $\mu_{matrix}$ denotes the attenuation coefficient of the target matrix for the incoming ($\mu_{matrix}^{E_{ph}}$) and outgoing ($\mu_{matrix}^{L\alpha}$) photon energies. The finally extracted MDL value is given in Table 2 together with the minimum detectable yield and the FWHM of the photoinduced Pd L$\alpha_1$ line. Tabulated is also the scaling factor calculated as a theoretical thick target yield for Pd in Ag matrix induced with 3172.8 eV photon beam relative to the yield of pure Pd excited with 3400 eV beam used for normalization.

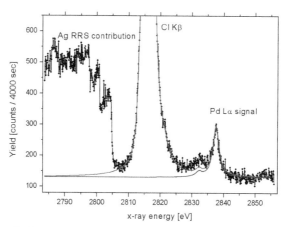

Fig. 11. The high resolution x-ray emission spectrum of Ag target measured after excitation with the 3172.8 eV photon beam tuned to the maximum of the Pd L$_3$ absorption edge. The Ag resonant Raman scattering contribution exhibiting characteristic high energy cutoff is completelly separated out of the region of the Pd Lα line. A strong signal at around 2816 eV corresponds to the Cl Kβ line and is due to the contamination of the Ag foil. An extremely high sensitivity of the set up is demonstrated by a clear signal corresponding to Pd, which is expected to be present in the Ag foil only in very low trace amount (Kavčič et al. 2011).

From the final MDL value given in Table 1 we can conclude that in the case of $Z_{matrix}$ - 1 trace element (the case of Pd in Ag) the detection limits are improved even further compared to the previously determined PIXE detection limits. Finally the ppm level was reached by combining the WDX spectroscopy with the energy selective target photoexcitation. A small part of this improvement is achieved also on account of the extremely high photon flux but most importantly, the WDX spectroscopy enables to separate completely the resonant Raman scattering signal from the Ag target at proper setting of the photon probe energy. The position of the characteristic high energy cutoff of the Ag RRS signal is clearly displayed in Figure 11 and its position depends linearly on the energy of the photon beam used for excitation. The energy displacement of this cutoff edge with respect to the position of Ag Lα fluorescence line corresponds to the actual energy detuning (the difference between the position of Ag $L_3$ absorption edge and the actual excitation energy). Since the lowest photon energy that can be used for such analysis is given by the Pd $L_3$ absorption edge, the energy difference between the Ag and Pd $L_3$ absorption edges (~ 178 eV (Deslattes et al. 2003)) sets the upper limit for this energy displacement. If we compare this value with the difference between Ag and Pd Lα fluorescence lines being ~ 146 eV (Deslattes et al. 2003) it is clear that a high energy resolution is mandatory to separate completely the two signals. It is very important to point out that this is a general problem in the analysis of $(Z-1)_{trace} / Z_{matrix}$ element combinations. It can be solved completely only by a combination of energy selective photoionization and WDX spectroscopy as demonstrated for the case of Pd/Ag combination reaching finally a minimum detection limits on the ppm level.

|  | Pd |
|---|---|
| $Y_{MDL}$ [counts] | 170 ± 1 |
| FWHM [eV] | 2.35 ± 0.01 |
| Fth | 2.55 |
| MDL [ppm] | 1.37 ± 0.01 |

Table 2. The minimum detection limit (MDL) for Pd within the Ag matrix obtained for excitation with monochromatic synchrotron radiation tuned just below the Ag $L_3$ edge. Tabulated are also minimum detectable yield, FWHM of the measured $Lα_1$ line, and also calculated value of the scaling factor $F^{th}$ used to obtain final MDLs. The final error is calculated considering the background, $Y_{Norm}$ and the Lα FWHM fitting errors.

## 4. Conclusion

The high resolution x-ray emission spectroscopy employing wavelength dispersive spectrometers with resolving power on the order of $E/\Delta E$ ~ 5000 can be used efficiently in the x-ray trace element analytical techniques. At such resolution a chemical speciation can be performed in addition to the standard XRF and PIXE elemental analysis. The energy shifts of the Kα lines of several light elements depend linearly on the oxidation state and the measurement of the absolute energy can be exploited to determine the oxidation state of the element in the sample. The Kβ line reflects directly the p-density of occupied states and through comparison with the theoretically calculated spectra provides quantitative information about ligand environment. Therefore high resolution XES measurements with a modern wavelength dispersive spectrometer opens possibility to perform reliable chemical speciation studies also in the small laboratory employing x-ray tube or electrostatic proton accelerator for target excitation.

Another aspect where we can improve the performance of the x-ray analysis by applying the WDX spectroscopy is the detection sensitivity for the trace elements neighboring the major target matrix element. A significant improvement of the detection limits for the case of Pd and Cd in silver matrix has been demonstrated. Because of extremely high experimental energy resolution reaching below the natural linewidths of the measured L x-ray lines, the detection limits of few tens of ppm were reached in case of proton excitation. For energy selective photoexcitation employing synchrotron radiation the detection limit for Pd was lowered even further. Due to high experimental energy resolution in both, the photon-in and photon-out channels, the Ag RRS contribution could be completely separated from the Pd signal yielding the final MDL value on the ppm level. This result demonstrates the general potential of WDX spectroscopy to obtain substantially lower detection limits for x-ray analysis of trace elements neighboring the target matrix element.

## 5. Acknowledgment

The author acknowledges the support of the Slovenian Ministry of Education, Science and Sport through the research program P1-0112. A part of the work on chemical speciation has been supported also by the European Community as an Integrating Activity 'Support of Public and Industrial Research Using Ion Beam Technology (SPIRIT)' under EC contract no. 227012.

## 6. References

Alonso Mori, R.; Paris, E.; Giuli, G.; Eeckhout, S. G.; Kavčič, M.; Žitnik, M.; Bučar, K.; Pettersson, L. G. M.; Glatzel, P. (2009) , Electronic Structure of Sulfur Studied by X-ray Absorption and Emission Spectroscopy, Analytical Chemistry Vol. 81, Issue 15, pp. 6516-6525

Alonso Mori, R.; Paris, E.; Giuli, G.; Eeckhout, S. G.; Kavčič, M.; Žitnik, M.; Bučar, K.; Pettersson, L. G. M.; Glatzel, P. (2010) , Sulfur-Metal Orbital Hybridization in Sulfur-Bearing Compounds Studied by X-ray Emission Spectroscopy, Inorganic Chemistry Vol. 49, No. 14, pp. 6468-6473

Campbell, J.L.; Papp, T. (2001), Widths of the atomic K-N7 levels, Atomic Data and Nuclear Data Tables, Vol. 77, Issue 1, pp 1-56

Deslattes, R.D.; Kessler E.G.; Indelicato, P.; De Billy, L.; Lindroth, E.; Anton, J. (2003), X-ray transition energies: new approach to a comprehensive evaluation, Reviews of Modern Physics, Vol. 75, Issue 1, pp. 35-99

Devos, W.; Moor, Ch.; Lienemann, P. (1999), Determination of impurities in antique silver objects for authentication by laser ablation inductively coupled plasma mass spectrometry (LA-ICP-MS), Journal of Analytical Atomic Spectroscopy, Vol. 14, Issue 4, pp. 621-626

Gohshi, Y.; Ohtsuka , A. (1973), The application of chemical effects in high resolution X-ray spectrometry, Spectrochimica Acta B, Vol. 28, Issue 5, pp. 179-188

Guerra, M.F.; Calligaro, T.; Radtke, M.; Reiche, I.; Riesemeier, H. (2005), Fingerprinting ancient gold by measuring Pt with spatially resolved high energy Sy-XRF, Nuclear Instruments and Methods in Physics Research B, Vol. 240, Issues 1-2, pp. 505-511

Ishii, K.; Morita, S. (1984), Continuum x rays produced by light-ion−atom collisions, Physical Review A, Vol. 30, Issue 5, pp. 2278-2286

Jaklevic, J.M.; Giaugue, R.D.; Thompson, A.C. (1988), Resonant Raman scattering as a source of increased background in synchrotron excited x-ray fluorescence, Analytical Chemistry Vol. 60, Issue 5,  pp. 482-484

Kavčič, M.; Karydas, A.; Zarkadas, Ch. (2004), Chemical state analysis of sulfur in samples of environmental interest using high resolution measurement of Kα diagram line, Nuclear Instruments and Methods in Physics Research. B, Vol. 222, Issues 3-4, pp. 601-608

Kavčič, M.; Karydas, A.; Zarkadas, Ch. (2005), Chemical state analysis employing sub-natural linewidth resolution PIXE measurements of Kα diagram lines, X-Ray Spectrometry, Vol. 34, Issue 4, pp. 310-314

Kavčič, M.; Dousse, J.-Cl.; Szlachetko, J., Cao, W (2007), Chemical effects in the Kβ X-ray emission spectra of sulfur, Nuclear Instruments and Methods in Physics Research B, Vol. 260, Issue 2, pp. 642-646

Kavčič, M. (2010), Improved detection limits in PIXE analysis employing wavelength dispersive X-ray spectroscopy, Nuclear Instruments and Methods in Physics Research B, Vol. 268, Issue 22, pp. 3438-3442

Kavčič, M.; Žitnik, M.; Bučar, K.; Mihelič, A.; Szlachetko, J. (2011), Application of wavelength dispersive X- ray spectroscopy to improve detection limits in X-ray analysis, X-Ray Spectrometry, Vol. 40, Issue 1, pp. 2-6

Maeda, K.; Hasegawa, K.; Hamanaka H., Maeda, M. (1998), Chemical state analysis in air by high-resolution PIXE, Nuclear Instruments and Methods in Physics Research B, Vol. 136-138, pp. 994-999

Maxwell, J.A.; Campbell, J.L.; Teasdale, W.J. (1989), The Guelph PIXE software package, Nuclear Instruments and Methods in Physics Research. B, Vol. 43, Issue 2, pp. 218-230

Perino, E.; Deluigi, MT.; Olsina, R.; Riveros, JA. (2002), Determination of oxidation states of aluminium, silicon and sulfur, X-Ray Spectrometry, Vol. 31, Issue 2, pp. 113-187

Rodrigues, M.; Schreiner, M.; Melcher, M.; Guerra, M.; Salomon , J.; Radtke, M.;Alram, M.; Schindel, N. (2011), Characterization of the silver coins of the Hoard of Beçin by X-ray based methods, Nuclear Instruments and Methods in Physics Research B, doi:10.1016/j.nimb.2011.04.068

Sugiura, C; Gohshi, Y.; Suzuki, I. (1974) Sulfur Kβ x-ray emission spectra and electronic structures of some metal sulfides, Physical Review B, Vol. 10, Issue 2, pp. 338-343

Szlachetko, J.; Kavčič, M. (2011), to be published

Tamaki, Y. (1995), Chemical effect on intensity ratios of K-series x-rays in vanadium, chromium and manganese compounds, X-Ray Spectrometry, Vol. 24, Issue 5, pp. 235-240

Uda, M.; Yamamoto, T.; Tatebayashi T. (1999), Theoretical prediction of S Kβ fine structures in PIXE-induced XRF spectra, Nuclear Instruments and Methods in Physics Research B, Vol. 150, Issues 1-4, pp. 55-59

Yasuda, S.; Kakiyama, H. (1978), X-ray K emission spectra of vanadium in various oxidation states, X-Ray Spectrometry, Vol. 7, Issue 1, pp. 23-25

Žitnik, M.; Kavčič, M.; Bučar, K.; Mihelič, A. (2008), X-ray resonant Raman scattering from noble gas atoms and beyond, Nuclear Instruments and Methods in Physics Research B, Vol. 267, Issue 2, pp. 221-225

# Part 2

# Characterization and Analytical Applications of X-Rays

# 6

# Chemical Quantification of Mo-S, W-Si and Ti-V by Energy Dispersive X-Ray Spectroscopy

Carlos Angeles-Chavez*,
Jose Antonio Toledo-Antonio and Maria Antonia Cortes-Jacome
*Instituto Mexicano del Petróleo, Programa de Ingeniería Molecular,*
*Eje Central Lázaro, D.F. México*
*Mexico*

## 1. Introduction

Elemental chemical identification of a specimen and its quantification is fundamental to obtain information in the characterization of the materials (Angeles et al., 2000; Cortes-Jacome et al., 2005). Energy dispersive X-ray spectroscopy (EDXS) is the technique that allows obtaining information concerning the elemental chemical composition using the EDX spectrometer. Generally is attached to a scanning electron microscope (SEM) (Goldstein & Newbury, 2003) and/or in a transmission electron microscope (Williams & Barry-Carter, 1996). The technique is very versatile because the spectrometer gives results in few minutes. The instrument is compact, stable, robust and easy to use and its results can be quickly interpreted. The analysis is based in the detection of the characteristic X-rays produced by the electron beam-specimen interaction. The information can be collected in very specific local points or on the whole sample. So, both electron microscopy and EDXS, give valuable information about the morphology and chemical composition of the sample.

In order to give an accurate interpretation of the data collected by the instrument is important to know the fundaments of the technique. The characteristic X-rays are produced by the atoms of the sample in a process called inner-shell ionization (Jenkins & De Dries, 1967). This process is carried out when an electron of inner-shell is removed by an electron of the beam generating a vacancy in the shell. At this moment the atom remain ionized during $10^{-14}$ second and then an electron of outer-shell fills the vacancy of the inner-shell. During this transition a photon is emitted with a characteristic energy of the chemical element and its shell ionized. The emitted photons are named by the shell-ionized type as K, L, M lines.... and α, β, γ... by the outer-shell corresponding to the electron that filled the inner-shell-ionized. For atoms with high atomic numbers, is important to note that some transitions are forbidden. Permissible transitions can be followed by the quantum selection rules and the notation can be followed by Manne Siegbahn and/or, IUPAC rules (Herglof & Birks, 1978). During the beam-sample interaction, another X-ray source is produced and it is known as Bremsstrahlung radiation or continuum X-rays which are generated for the deceleration of the electron beam in the Coulombic field of the specimen atoms. When the electrons are braked, they emit photons with any energy value giving rise to a continuous electromagnetic spectrum appearing in the EDX spectrum as

---

*Corresponding Author

background. In the case of high overvoltage, more than one electron may be ejected simultaneously from an atom and then, X-rays known as satellites peaks are generated (Deutcsh, et al., 1996). This simultaneous ejection of electrons causes a change in the overall structure of the energy levels resulting in the production of X-rays with slightly lower energies than those produced during single electron ionization appearing near to the characteristic peaks. Another important source of satellite peaks is the Auger process. Finally, in the X-ray spectrum can be detected shifts of peaks produced from a pure element and that produced by the same element contained in a compound. This variation occurs because the electronic configuration of the inner-shells of an atom is strongly influenced by the outer valence electrons. These shifts are most apparent when comparing metals with their corresponding oxides and halides. Consequently, it is not advisable to use a metallic standard material for analysis of oxide compounds. In the case of energy-dispersive X-ray analysis, the shifts of the peaks are undetectable and metals can be used as reference standards for quantification (Liebhafsky, 1976). All X-rays produced in the sample are detected and displayed in the EDX spectrum. Their identification in the spectrum is important because help to do an accurate identification of the chemical components remaining in the specimen and subsequently a more accurate quantification can be obtained. From all X-rays displayed in an EDX spectrum the most important are characteristic X-rays.

Another important parameter to consider in the EDXS is the acquisition of the data (Kenik, 2011; Scholossmacher, et al., 2010). The X-ray processing produced in the specimen is performed in three parts: detector, electronic processor and multichannel analyzer display. The overall process occurs as follow: the detector generates a charge pulse proportional to the X-ray energy. The produced pulse is first converted to a voltage and then the voltage signal is amplified through a field effect transistor, isolated from other pulses, further amplified, then identified electronically as resulting from an X-ray of specific energy. Finally, a digital signal is stored in a channel assigned to that energy in the multichannel analyzer. The speed of this process is such that the spectrum seems to be generated in parallel with the full range of X-ray energies detected simultaneously. Currently, the new software's generation delivers a spectrum ready to analyze and the previous process is not seen. However, there are many variables that must be taken into account to make a more accurate identification and subsequently their quantification. The most important variables are the time constant (Tc), acceleration voltage (AV), dead time (DT), acquisition time (AT), magnification and work distance (WD) which have direct effect on the energy resolution, peak intensity and natural width of characteristic X-ray lines. It is important to note that the calculation of the chemical composition is carried out considering the intensity and peak broadening in conjunction with the atomic number effect (Z), absorption correction (A) and characteristic fluorescence correction (F) (Newbury et al., 1986). For every group of samples with similar chemical components is recommendable to do a review of some of the previously mentioned operation variables to assure the truthfulness of the results. The chemical quantification has been very well studied for metals; however for powder samples of metallic and non-metallic oxides deposited on carbon tape, little work has been realized especially when the energy lines of two elements overlap. Generally, the chemical analysis in the scanning electron microscope can be obtained at different voltage, magnification, etc. depending on the information that it wants to reveal of a sample (Chung, et al., 1974). But if the overall chemical composition of the specimen analyzed is the primary requirement then a review of operation parameters of the instrument should be performed to ensure the accuracy of the results. In this study, the influence of operation conditions is presented to obtain overall elemental chemical composition representative of the sample by EDXS. Specifically of elements with overlapping characteristic X-ray lines.

In order to carry out this study the Mo-S, W-Si and Ti-V systems were choice because they are widely used in the catalysis field and their accurate chemical composition are required. Additionally, their more intense characteristic X-ray lines used to calculate the composition are overlapped. In order to know if the instrument is able to solve this problematic giving reliable composition results, samples of known Mo, S, W and V composition were prepared.

## 2. Experimental section

Samples of $Al_2O_3$, $MoO_3$ and S-$MoO_3$, $V_2O_5$-$TiO_2$ and $SiO_2$-$WO_3$ were chemically analyzed in an EDX spectroscope EDAX which is attached to the environmental scanning electron microscope PHILIPS XL30. The EDAX instrument has a detector type UTW-Sapphire, an energy resolution of 0.129 keV, tilt of 0.0 and take-off of 35.90. As the microscope is designed to operate at different conditions and to obtain useful information of the samples, a review of the variables of time constant, acceleration voltage and acquisition time in order to obtain more accurate overall quantification results was performed. In this study, the dead time, the magnification and the work distance were kept constant. The DT was adjusted with the spot size and it kept around of 30%. The deconvolutions of the experimental results were followed using the Phoenix software version 3.3 included with the instrument. The software uses the so-called segment Kramer´s fit method to model the background which include the collection efficiency of the detector, the processing efficiency of the detector and the absorption of X-ray within the specimen. This method is useful when the spectrum contains overlapping peaks. The chemical composition was calculated using the ZAF correction method. In order to establish the operation conditions of the instrument, samples of $Al_2O_3$ and $MoO_3$ were used. The experimental results were compared with their theoretical compositions and a statistical study of the exactitude and precision of the instrument was determined (Ellison, et al., 1995). Subsequently, under these conditions samples containing S-$MoO_3$, $V_2O_5$-$TiO_2$ and $SiO_2$-$WO_3$ were analyzed and their average compositions were reported. $Al_2O_3$, $MoO_3$, S, $V_2O_5$, $TiO_2$, $SiO_2$ and $WO_3$ analytic grade were used to prepare the samples. Samples with 1, 5, 10 and 15 wt% of Mo were prepared for the $MoO_3$-$Al_2O_3$ system. Same S concentrations in the S-$MoO_3$ system were prepared. For the $V_2O_5$-$TiO_2$ system, the samples were prepared at 0.5, 1, 3, 5 wt% of V while the $SiO_2$-$WO_3$ system the tungsten concentration in the samples were 20, 40, 50 and 60 wt% of W. All samples were mechanically mixed in an agate mortar. $V_2O_5$-$TiO_2$ and $SiO_2$-$WO_3$ systems were mixed and annealed at 500 °C. In order to calculate the chemical composition of the elements, more intense characteristic energy lines were used. Five measurements in different region were performed in each sample. The concentration values reported of each component for each sample was the average value. A monolayer of sample was deposited and dispersed on carbon tape to have a flat surface.

## 3. Results and discussion

### 3.1 Review of the operation conditions of the instrument

The best operation conditions in the instrument to calculate the more accurate elemental chemical composition were established using $Al_2O_3$ and $MoO_3$ samples. These samples were selected because they are components of the hydrodesulfuration catalysts. $Al_2O_3$ and $MoO_3$ are very stable under environmental conditions of temperature, humidity and pressure. Their chemical composition remains constant with the time.

The effect of the Tc, AV and AT variables on the chemical composition was studied using MoO₃. Figure 1 shows the typical EDX spectra of the MoO₃ displaying their characteristic energy lines located at 0.523 keV for OKα, 2.015, 2.293, 2.394 keV, for MoLl, MoLα1 and MoLβ1 and 17.376, 17.481, 19.609 keV for MoKα2, MoKα1 and MoKβ1. All lines were displayed in five Gaussian peaks. The more intense peaks correspond to the OKα and MoLα1 lines. The software calculates the composition using the energy lines corresponding to the shell ionized, for instance the OK and MoL lines for MoO₃. The first calculus was made with the EDX spectrum obtained to a Tc of 35.0µs, 25kV AV, 100s of lecture, around 30% of DT, 11 mm of WD and 100X of magnification using the most intense peaks, OK and MoL lines. The results obtained were 32.30 wt% O and 67.70 wt% Mo, composition close to the theoretical composition (33.35 wt% O and 66.65 wt% Mo). This result differs of the results obtained with MoK line. The chemical composition obtained considering MoK line was 23.40 wt% O and 76.60 wt% Mo, composition very far of the theoretical composition. Here is important to note the importance of consider the more intense characteristic energy lines to obtain more accurate composition results  in samples containing an element with two line series as K and L or M and L.

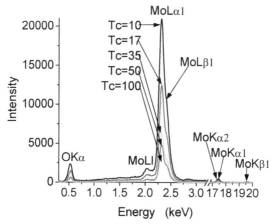

Fig. 1. EDX spectra showing the O and Mo experimental Gaussian peaks at different Tc. The spectra were obtained under operation conditions of DT around 30%, 25kV and 100s of acquisition time.

Knowing that with the MoL line the calculus of chemical composition gives a better approximation to the theoretical composition of MoO₃, the study of the effect of Tc in the chemical composition can be performed. Tc variable is the time allowed for the pulse processor to evaluate the magnitude time of pulse. For longer Tc, the ability of the system to assign an energy to the incoming pulse is better, but a few counts can be processed in a given analysis time. Therefore, longer Tc will give better spectrum resolution but the count rate will be lower. Shortest Tc will allow to process more counts per second but with a large error in the assignment of a specific energy to the pulse, and so the energy resolution will be poor. This can be observed in the Gaussian curves of MoO₃ EDX spectra displayed in Figure 1. A careful analysis of Figure 1 indicates that the resolution between the MoLα1 and MoLβ1 lines were better resolved at 100.0µs than at 10.0µs of Tc. In general, at shorter Tc larger intensity, wider broadening and minor resolution, while at higher Tc, minor intensity, minor broadening and better resolution is obtained. Nevertheless, the peak intensity and its broadening are important in the calculus of the composition and the resolution is also important to discern the characteristic energy of the chemical elements and as it can be seen in the Gaussian curves of Figure 1, the intensity, broadening and resolution changes for each Tc measured. Its effect in the chemical composition is displayed in table 1. A relative error was calculated between the theoretical and experimental composition data to analyze the results. The relative error given in the last column is lesser than 2.5% and it does not show some tendency. This means that the Tc has little effect in the chemical composition. The larger and smaller error was obtained at Tc=50.0µs and Tc=100.0µs, respectively. Therefore, a Tc optimum must be choice without affect both, intensity and resolution, too much.    In the instrument used in this work the Tc value suggested by the provider was 35.0µs that combine resolution and intensity giving a relative error of 1.6%.

| Tc(µs) | OK(Wt%) | MoL(Wt%) | Mo %Error |
|--------|---------|----------|-----------|
| 10     | 32.02   | 67.98    | 2.0       |
| 17     | 32.71   | 67.29    | 1.0       |
| 35     | 32.30   | 67.70    | 1.6       |
| 50     | 31.69   | 68.31    | 2.5       |
| 100    | 33.03   | 66.97    | 0.5       |

Table 1. Chemical composition of $MoO_3$ obtained with the OK and MoL lines at different time constant.

With a Tc=35.0µs and 100s of acquisition time, the effect of accelerating voltage on molybdenum oxide composition was studied. The DT was kept around of 30% manipulating the spot size. Figure 2 displays the EDX spectra of the OK and MoL characteristic energy lines. The Figure 2 shows a decreasing in the peak intensity with the increase of the accelerating voltage, see inset in Figure 2. However, the peak width fits with the others Gaussian peaks. This overlapping of curves was more evident for MoL line than OK line. The decreasing of the intensity produced by increase of the accelerating voltage without affect the broadening in the Gaussian curves suggests that the accelerating voltage has strong effect in the determination of the composition. This was revealed in the calculus of the composition, see table 2. At low accelerating voltage high error was obtained and it decreased with the increase of accelerating voltage obtaining the lowest error at 25 kV. Therefore, in order to obtain more accurate overall chemical composition is very important acquire experimental data at 25kV in the instrument used in this work for this sample.

| Acc. Voltage (kV) | OK(Wt%) | MoL(Wt%) | Mo %Error |
|-------------------|---------|----------|-----------|
| 15                | 27.20   | 72.80    | 9.2       |
| 20                | 29.55   | 70.45    | 5.7       |
| 25                | 32.47   | 67.53    | 1.3       |
| 30                | 34.31   | 65.69    | 1.4       |

Table 2. Chemical composition of $MoO_3$ obtained with OK and MoL lines at different accelerating voltage.

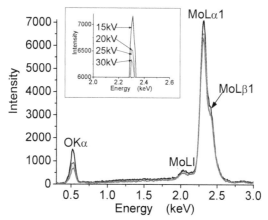

Fig. 2. EDX spectra showing the O and Mo experimental Gaussian peaks at different accelerating voltage. The spectra were obtained under operation conditions of Tc=35µs, DT around 30% and 100s of acquisition time.

The acquisition time is directly related with the intensity and it is expected to affect the calculus of the composition. Therefore, it is important to know how AT affects the $MoO_3$ composition and determine if there is an optimum acquisition time for more precise experimental results. Measurements carried out at Tc 35µs and 25kV and different acquisition time are displayed in the EDX spectra of Figure 3. The Gaussian peaks of the OK and MoL lines show an increase in the intensity and broadening with an increase on the acquisition time. This behavior suggests that the acquisition time has little effect in the chemical composition as it was revealed in the calculus displayed in table 3. A relative error of 1.7% was obtained at 100 and 200s of lecture while at 300 and 400s the error obtained was 1.4%. Therefore, the time acquisition only minimizes the integration error of the experimental results, and it does not affect the chemical composition. An acquisition time of 100s is adequate for performance the chemical analysis.

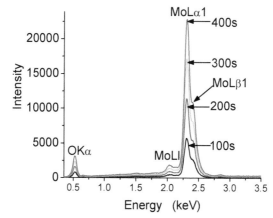

Fig. 3. EDX spectra showing the O and Mo experimental Gaussian peaks at different acquisition time. The spectra were obtained under operation conditions of Tc=35µs, DT around 30% and 25kV.

| Acquisition Time(s) | OK(Wt%) | MoL(Wt%) | Mo %Error |
|---|---|---|---|
| 100 | 34.48 | 65.52 | 1.7 |
| 200 | 34.50 | 65.50 | 1.7 |
| 300 | 34.25 | 65.75 | 1.4 |
| 400 | 34.30 | 65.70 | 1.4 |

Table 3. Chemical composition of $MoO_3$ obtained with OK and MoL lines at different acquisition time.

Knowing the values of the variables as Tc, accelerating voltage and acquisition time, the instrumental error was estimated performing measurements in a same region for five times. The measurements were performed at 35.0μs of Tc, 25kV of accelerating voltage 100s of acquisition time, around of 30% of DT, 100X of magnification and 11mm of WD. The results obtained are displayed in EDX spectra of Figure 4. As expected, the Gaussian peaks of the OK and MoL energy lines overlap with each measurement in the EDX spectra; however, the variations in the intensity of the peaks without affecting their broadening could affect the results of the chemical composition as was observed previously in the voltage study. These variations are related with the ability of the instrument to reproduce a measurement. The composition results and the relative errors calculated for every measurement is illustrated in table 4. As can be seen, the composition in every measurement is different indicating the instrumental error in the determination of the chemical composition. The variations are small without show any tendency. Under the measurement conditions used, the average error for the O was 1.9% while for the Mo it was 0.9%. From these results, it is possible to calculate the accuracy and precision of the instrument. Therefore, the instrument has an accuracy around 0.6wt% and a precision around 0.2wt%.

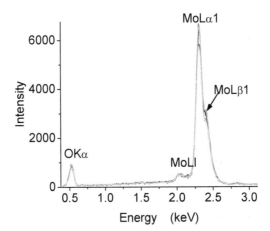

Fig. 4. EDX spectra showing the O and Mo experimental Gaussian peaks obtained in a same region under operation conditions of Tc=35μs, 25kV, DT around 30% and 100s of acquisition time.

| Measurement | OK(Wt%) | O %Error | MoL(Wt%) | Mo %Error |
|:-----------:|:-------:|:--------:|:--------:|:---------:|
| 1 | 33.44 | 0.3 | 66.56 | 0.1 |
| 2 | 34.04 | 2.1 | 65.96 | 1.0 |
| 3 | 34.16 | 2.4 | 65.84 | 1.2 |
| 4 | 33.75 | 1.2 | 66.25 | 0.6 |
| 5 | 34.48 | 3.4 | 65.52 | 1.7 |
| Average | 33.97 | 1.9 | 66.03 | 0.9 |

Table 4. Chemical composition of $MoO_3$ obtained with OK and MoL lines in the same analysis region.

Once the precision and exactitude of the instrument are known, the homogeneity of the O, Al and Mo composition in the $Al_2O_3$ and $MoO_3$ samples was evaluated. As the samples analyzed were analytical grade is expected a homogenous chemical composition and then the measurement error will be equal or minor to the instrumental error. To carry out the study, ten measurements in different regions of the $Al_2O_3$ sample deposited on the carbon tape were obtained using the same operation conditions of the instrument to determine the instrumental error. The results obtained are displayed in table 5. Characteristic energy corresponding to OK and AlK lines located at 0.523 and 1.486 keV were used to calculate the composition. Chemical composition expected is 47.07 wt% O and 52.93 wt% Al. This theoretical chemical composition was used in the statistical analysis. The average experimental composition obtained was around 46.46 wt% O and 53.54 wt% Al. The average calculated error was around of 1.3% for O and 1.2% for Al. The precision calculated for this sample was around of 0.2 wt% and the exactitude was around 0.6 wt%. The results are similar to the obtained in the study of the instrumental error. This similitude is indicating the high homogeneous distribution of the O and Al atoms in the alumina sample. Then, the measurement error is attributed to the instrument.

| Measurement | OK(Wt%) | O %Error | AlK(Wt%) | Al %Error |
|:-----------:|:-------:|:--------:|:--------:|:---------:|
| 1 | 46.85 | 0.5 | 53.15 | 0.4 |
| 2 | 46.42 | 1.4 | 53.58 | 1.2 |
| 3 | 46.70 | 0.8 | 53.30 | 0.7 |
| 4 | 46.07 | 2.1 | 53.93 | 1.9 |
| 5 | 46.47 | 1.3 | 53.53 | 1.1 |
| 6 | 46.92 | 0.3 | 53.08 | 0.3 |
| 7 | 46.80 | 0.6 | 53.20 | 0.5 |
| 8 | 46.74 | 0.7 | 53.26 | 0.6 |
| 9 | 46.49 | 1.2 | 53.51 | 1.1 |
| 10 | 45.14 | 4.1 | 54.86 | 3.6 |
| Average | 46.46 | 1.3 | 53.54 | 1.2 |

Table 5. Chemical composition of $Al_2O_3$ obtained with OK and AlK lines in different regions of the sample.

Experimental results corresponding to the molybdenum oxide is displayed in table 6. OK and MoL characteristic lines were used to calculate the chemical composition. In this sample, the average experimental composition obtained was 32.86 wt% O and 67.14 wt% Mo. The average measurement error calculated was 1.3% for O and 1.2% for Mo with an accuracy and precision of 0.5wt% and 0.3wt%, respectively. The exactitude value is into range calculated for the instrument but not for the precision. Precision is the degree of dispersion of the data, and then the result is indicating a high dispersion of the values of oxygen and molybdenum composition in the sample compared with the dispersion produced by the instrument. This dispersion could be related with the homogeneity grade in the chemical composition of the sample.

| Measurement | OK(Wt%) | O %Error | MoL(Wt%) | Mo %Error |
|---|---|---|---|---|
| 1 | 32.83 | 1.6 | 67.17 | 0.8 |
| 2 | 32.17 | 3.5 | 67.83 | 1.8 |
| 3 | 33.19 | 0.5 | 66.81 | 0.2 |
| 4 | 32.17 | 3.5 | 67.83 | 1.8 |
| 5 | 34.48 | -3.4 | 65.52 | -1.7 |
| 6 | 33.44 | -0.3 | 66.56 | -0.1 |
| 7 | 32.47 | 2.6 | 67.53 | 1.3 |
| 8 | 33.53 | -0.5 | 66.47 | -0.3 |
| 9 | 32.31 | 3.1 | 67.69 | 1.6 |
| 10 | 32.02 | 4.0 | 67.98 | 2.0 |
| Average | 32.86 | 1.5 | 67.14 | 0.7 |

Table 6. Chemical composition of $MoO_3$ obtained with OK and MoL lines in different regions of the sample.

## 3.2 $MoO_3$-$Al_2O_3$ system

With the operation conditions established on the instrument and knowing the exactitude and precision grade, a study on the determination of molybdenum in alumina was carried out. The purpose was to evaluate the response capability of the instrument to quantify molybdenum at different concentration in the alumina matrix. Theoretically, the instrument can detect concentrations around of 0.1wt% under special operation conditions. In this study, the samples analyzed were prepared with nominal molybdenum concentration of 1, 5, 10 and 15wt%. The results obtained are displayed in the EDX spectra of Figure 5. The more intense Gaussian peaks corresponding to the OK, AlK and MoL characteristic energy lines are observed. Figure 5 shows the EDX spectra of $MoO_3/Al_2O_3$ system with low and high Mo concentration. As expected, at low Mo concentration the MoL Gaussian peak is very small, while at high Mo concentration the MoL Gaussian peak is more intense indicating the qualitative increase of Mo. Additionally, in the Gaussian peaks can be observed the calculated curve, black line, overlapping the experimental curve, red line. Observing the difference between both curves, a very well fit can be determined. Latter indicates that all necessary considerations were made for a good interpretation of the experimental results. This analysis is very important when in the samples are present elements with very close characteristic energy lines that cannot be resolved by the instrument and/or when the concentration of one of the components is very low that cannot be detected its corresponding characteristic energy lines. Generally, the next energy line occurs at high energy value; for molybdenum the next energy line is the MoK and it is not detected at low concentration. Then, a careful analysis of calculated and experimental curves helps to resolve the presence of elements with characteristic energy lines very close and consequently their concentration can be calculated. The following systems analyzed in this work correspond to some examples to resolve the problematic presented when elements with similar energy lines are containing in a specimen. As can seen in the Figure 5, both EDX spectra the experimental and calculated curves follow the same trajectory indicating that chemical elements identified in the sample were correct. The average chemical composition calculated for each sample is displayed in table 7. Each molybdenum value obtained was into measurement error. Therefore, the results obtained operating the instrument under the conditions previously mentioned were enough reliable.

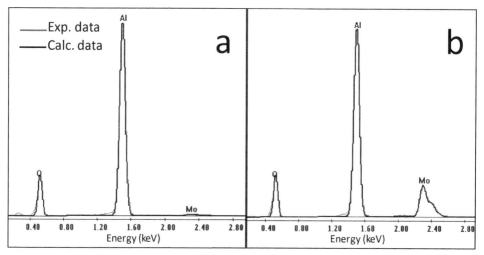

Fig. 5. EDX spectra showing the peak intensities corresponding to OK, AlK and MoL lines a)1.0 Wt% Mo and b)15Wt% Mo

| Nominal Mo (Wt%) | O (Wt%) | Al (Wt%) | Mo (Wt%) |
|---|---|---|---|
| 1 | 45.82 | 52.85 | 1.33 |
| 5 | 46.44 | 48.36 | 5.2 |
| 10 | 45.58 | 44.64 | 9.78 |
| 15 | 46.30 | 42.80 | 15.23 |

Table 7. Average chemical composition of the O, Al and Mo obtained of $Al_2O_3$ and $MoO_3$ samples at different molybdenum compositions.

### 3.3 S-MO₃ system

According with the previous results, measurements into the reliability range marked by the instrumental error can be obtained. Now, the capabilities of the instrument to resolve the chemical composition of two elements that have characteristic energy values very close were analyzed. The first example was the determination of S in S-MoO$_3$ samples. Their more intense characteristic energy lines are very close and when both components are present in the sample only one experimental Gaussian peak is generated overlapping the presence of both elements. The case was studied using an S-MoO$_3$ system where sulfur concentration was varied from 1 to 15 wt%. Experimentally, one Gaussian peak was detected in the EDX spectrum. Figure 6 shows experimental Gaussian peak around of SK and MoL lines in the sample with the lower sulfur concentration. This experimental curve can be explained if we analyzed the values of the characteristic energies of both elements. The characteristic energies of the sulfur are SKα and SKβ with values of 2.307 and 2.464 keV while for molybdenum are MoLl, MoLα and MoLβ with values of 2.015, 2.293 and 2.394 keV, respectively. The energy difference between more intense lines, SKα and MoLα, is 0.014 keV and the energy resolution of the instrument is 0.129 keV, which is larger than the separation between both energies. Therefore, only one experimental Gaussian peak is generated, see Figure 6. A careful analysis the Gaussian peaks evidence the presence of Mo by the presence of MoLβ line that it is separated 0.101 keV from MoLα line and 0.087 keV from SKα line. However, the SKα line is contained in it. To evidence the presence of sulfur component the overlapping of the calculated Gaussian peak with the experimental was made considering first individual elements and after both of them. The results obtained are displayed in Figure 6 for the sample with 1wt% S. Figure 6a show the calculated Gaussian curve considering only SK line, the fit with the experimental Gaussian curve was not good, see Figure 6a. Considering only the MoL line the calculated and experimental Gaussian curves are very close, Figure 6b, however, in the maximum of the calculated peak does not match with the maximum of the experimental one. The calculated curve is shift to the left of the experimental curve. In Figure 6c, both SK and MoL lines were considered in the calculus and the result obtained is lightly better than that obtained with individual MoL line in Figure 6b. As both results of Figure 6b and c almost are likeness, the presence of S in the sample can be doubtful. Then, the deconvolution and composition results must be reviewed. An integration error is given by the Phoenix software after chemical composition calculation. This error must be below 10% and for the analysis of sulfur in this sample was around 2.5% indicating good fit and confirming the presence of sulfur in the sample.

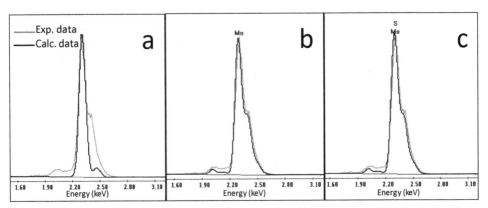

Fig. 6. EDX spectra of the sample with 1 wt% S showing the experimental and calculated Gaussian curves around of the SK and MoL lines. Calculated Gaussian curve considering a) SK line b)MoL line and  c) SK and MoL lines.

When the sulfur content is increased, the shape of the experimental curve is modified, see Figure 7. The shoulder generated by the MoLβ line was disappearing with the increase of the S concentration. In this case is easy to follow the presence of sulfur in the experimental curve. Calculated curves with individual SK or MoL line do not match the experimental Gaussian curve, Figure 7a and 7b. However, when both SK and MoL lines are considered, the calculated Gaussian curve match very well the experimental curve indicating the presence of both elements with high reliability grade, see Figure 7c. The integration error of the SK was below of 1% which is good.

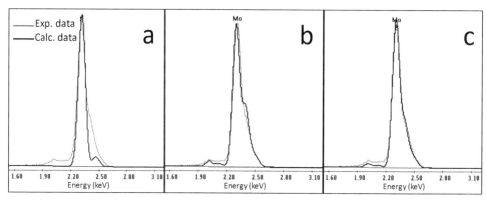

Fig. 7. EDX spectra of the sample with 15 wt% S showing the experimental and calculated Gaussian curves around of the SK and MoL lines. Calculated Gaussian curve considering a) SK line b)MoL line and c) SK and MoL lines

After the accurate identification of the chemical components in the EDX spectrum, the chemical composition of each sample was obtained. The average chemical composition of every component is displayed in table 8. The sulfur concentration values displayed in the last column are near to the nominal composition. The variations observed can be due to manipulation error during the preparation of the samples. The results indicate that not homogeneous samples were prepared by physical mixing. Another preparation route must be proposed for standard samples preparation. Nevertheless, for the purpose of measure the capability of the instrument to calculate the chemical composition of two elements with close characteristic energy lines that generate just one experimental Gaussian peak, the experimental values obtained were good.

| Nominal S (Wt%) | O (Wt%) | Mo (Wt%) | S (Wt%) |
|---|---|---|---|
| 1 | 31.87 | 66.47 | 1.56 |
| 5 | 30.81 | 64.76 | 4.33 |
| 10 | 28.47 | 60.91 | 10.62 |
| 15 | 27.95 | 58.11 | 13.89 |

Table 8. Average chemical composition of the O, Mo and S obtained of S and $MoO_3$ samples at different sulphur composition.

### 3.4 WO₃-SiO₂ system

The WO₃-SiO₂ system presents the same problematic with their corresponding more intense SiKα and WMα1 lines, they overlap giving rise only a Gaussian peak in the EDX spectrum. Therefore, it is not easy to evidence Si at low concentrations when the W is present at high concentration and vice-versa, because all energies of the SiK and WM are very close and at low tungsten concentration, the WLα line is not detected. Therefore, a carefully analysis of the experimental Gaussian peak generated by the presence of both elements must be carried out. In order to know the capability of the instrument to calculate the Si and W compositions using only one experimental curve containing the characteristic energy lines of both elements, SiO₂ and WO₃ mixed samples with concentration of 20, 40, 50 and 60 wt% of W were prepared. Just one experimental Gaussian peak was obtained for SiK and WM lines for the sample with 20 wt% W, Figure 8, making very difficult to evidence the presence of both Si and W components. The values of the characteristic energies of the SiK are 1.740 keV for SiKα and 1.829 keV for SiKβ. For WM are 1.776 keV for WMα1, 1.835 keV for WMβ, 2.035 for WMγ and 1.379 keV for WMz2. And the energy difference between SiKα and WMα1 lines is 0.036 keV which remain also below limit of energy resolution of the instrument (0.129 keV). For the analysis, first the presence just SiK line was considered in the calculus and it do not match very well with the experimental curve, Figure 8a. The calculated curve is shifted lightly to the left of the experimental Gaussian peak. In figure 8b shows the results obtained considering only WM line. The calculated Gaussian curve deviates strongly from the experimental one. In this case, the calculated curve is shifted to the right of the experimental curve and its calculated peak intensity stayed below the intensity of the experimental peak. Considering the presence of both elements, the calculated Gaussian peak with SiK and WL lines match very well the experimental Gaussian peak, Figure 8c. The WM and SiK integration error obtained for this sample was below 1%, and then the presence of both elements was confirmed.

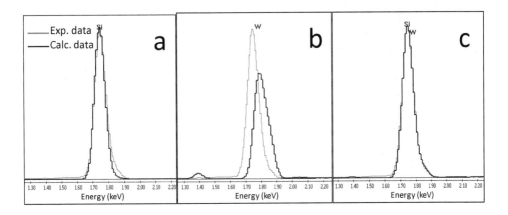

Fig. 8. EDX spectra of the sample with 20 wt% W showing the experimental and calculated Gaussian curves around of the SiK and WM lines. Calculated Gaussian curve considering a) SiK line b) WM line and c) SiK and WM lines

At higher tungsten concentrations (60 wt% W), the calculated Gaussian curve obtained considering only the presence of SiK line do not match with the experimental curve as is illustrated in Figure 9a. The calculated curve is shifted to the left of the experimental curve, whereas when just WM line is considered, the calculated Gaussian curve is shifted to the right of the experimental curve Figure 9b. Figure 9c shows the results obtained considering the SiK and WM lines in the calculus, the experimental Gaussian curve was fitted very well with the presence of both characteristic energy lines in the calculus. The integration error for SiK and WM remain below 1%. The error of integration is useful when the concentration of a component is low. When the concentration of the component is high, the integration error will be always less than 1%. Then, for these samples with the analysis of the Gaussian curve is enough.

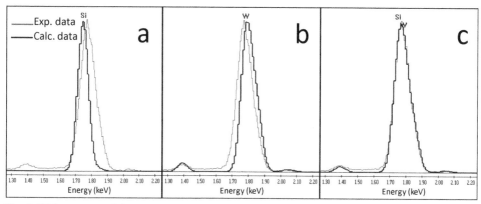

Fig. 9. EDX spectra of the sample with 60 wt% W showing the experimental and calculated Gaussian curves around of the SiK and WM lines. Calculated Gaussian curve considering a) SiK line b) WM line and c) SiK and WM lines

After determination of the presence of SiK and WL in the experimental Gaussian curve, the chemical composition on samples with different tungsten concentration was calculated. The obtained results are displayed in the table 9. The W compositions are close to the nominal composition. The deviation between the nominal and experimental results was around 1 to 1.7 wt % which can be attributed again to the manipulation process during the sample preparation. However, the results obtained are reliable and the fit of the experimental curve considering two elements has the ability to discern the composition of each element.

| Nominal W (Wt%) | O (Wt%) | Si(Wt%) | W (Wt%) |
|---|---|---|---|
| 20 | 40.78 | 40.74 | 18.48 |
| 40 | 33.47 | 27.23 | 39.30 |
| 50 | 28.67 | 19.64 | 51.69 |
| 60 | 24.63 | 14.40 | 60.97 |

Table 9. Average chemical composition of the O, Si and W obtained of $SiO_2$ and $WO_3$ samples at different tungsten composition.

## 3.5 $V_2O_5$-$TiO_2$ system

In this section will be analyzed the ability of the instrument to calculate the chemical composition in samples where the more intense characteristic energy line of one element is overlapped to one characteristic energy line of low intensity of the other component, specifically Kα line with Kβ line. This case is generally present in transition metals, and the $V_2O_5$-$TiO_2$ system was selected for this study. This problem occurs when one component is present at low concentrations, usually less than 5 wt% as in the case of vanadium. At V concentration below 5 wt% VKα line is too low and overlapped with TiKβ line of Ti. For this propose mechanical mixture of $V_2O_5$ and $TiO_2$ at vanadium concentration of 0.5, 1, 3 and 5 wt% were prepared. Figure 10 shows experimental Gaussian curve obtained corresponding to the TiKα and TiKβ lines of the sample with 0.5 wt% of V. Only one experimental Gaussian peak around of the VKα and TiKβ lines was observed, then, a careful analysis of the experimental curve to evidence the presence of V is necessary. The characteristic energies of Ti are located at 4.508 keV for TiKα and at 4.931 keV for TiKβ1 while for the V they are at 4.948 keV for VKα and at 5.426 keV for VKβ1. The more intense energies for Ti and V are TiKα and VKα which are resolved by the instrument at high concentrations; however at low vanadium concentration the intensity of VKα line is overlapped with the TiKβ1 line. Between these lines there is a difference of 0.017 keV which is below limit of energy resolution of the instrument, so only an experimental Gaussian curve is showed in Figure 10. The calculated Gaussian peak considering only the presence of TiK is displayed on the experimental Gaussian curve of Figure 10a. The calculated curve does not match with the experimental curve at VKα and TiKβ1 energy values. The calculated Gaussian curve is shifted to the left from the experimental Gaussian curve. For comparison, an EDX spectrum of a pure titanium oxide sample was obtained in which the calculated Gaussian curve for TiK line matched with the experimental one, see Figure 11. Therefore, the deviation observed in Figure 10a indicates qualitatively the presence of V in the sample. Considering both VK and TiK lines, the calculated Gaussian curve matched very well the experimental Gaussian curve. In this case, it is important to use the integration error to obtain better evidence of V presence in the sample. The integration error obtained with the VK line was around of 7% remaining lesser than 10%, confirming the presence of V in the sample.

At 5 wt% V concentration, a small Gaussian peak located at 5.426 keV corresponding to the VKβ1 line was detected in Figure 12, and then the presence of vanadium is easily identified. However, if only TiK line is considered the calculated curve do not match with the experimental curve at TiKβ line, Figure 12a. The position and intensity of calculated curve is different and is shifted to the left. This is less intense than the experimental curve. Therefore, the experimental Gaussian peak indicates the presence of VKα characteristic energy line. The calculated Gaussian curve considering the TiK and VK lines is showed in Figure 12b. This result matches very well the experimental Gaussian curve. Additionally, the small experimental Gaussian curve corresponding to the VKβ1 line is also fitted.

After analysis of the vanadium presence in the Gaussian peaks in the EDX spectrum, the chemical composition of samples with different V loading was obtained. The results are displayed in the table 10. As it can be observed, the vanadium concentrations are close to the nominal composition, confirming the capability of the instrument to give the composition of the chemical elements when their corresponding energy lines are close and their presence cannot be resolved by the instrument generating only one experimental Gaussian peak for both elements.

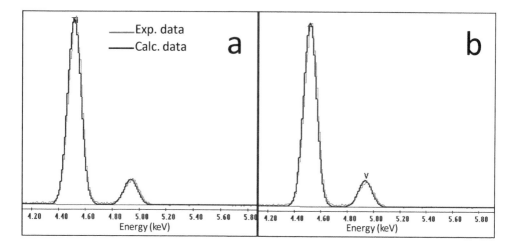

Fig. 10. EDX spectra of the sample with 0.5 wt% V showing the experimental and calculated Gaussian curves around of the TiK and VK lines. Calculated Gaussian curve considering a) TiK line and b) VK and TiK lines

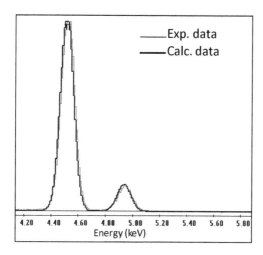

Fig. 11. EDX spectrum of a titanium oxide sample showing the calculated Gaussian curves with the TiK line. The calculated Gaussian peaks fit very well the experimental curves.

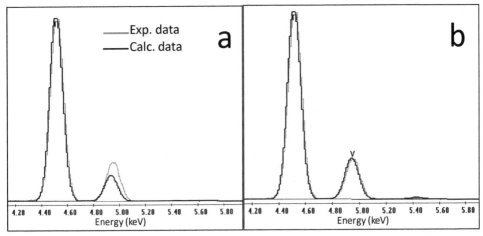

Fig. 12. EDX spectra of the sample with 5 wt% V showing the experimental and calculated Gaussian curves around of the TiK and VK lines. Calculated Gaussian curve considering a) TiK line and c) VK and TiK lines

| Nominal V (Wt%) | O (Wt%) | Ti (Wt%) | V (Wt%) |
|---|---|---|---|
| 0.5 | 44.97 | 54.36 | 0.67 |
| 1 | 42.47 | 56.75 | 0.78 |
| 3 | 43.40 | 54.20 | 2.39 |
| 5 | 41.69 | 53.79 | 4.52 |

Table 10. Average chemical composition of the O, Ti and V obtained from samples of $V_2O_5$ and $TiO_2$ at different vanadium composition.

## 4. Conclusion

The optimization of the operation conditions of the instrument to obtain more reliable elemental chemical composition for alumina and molybdenum oxide was reviewed. For the environmental scanning electron microscope PHILIPS XL30 the better operation conditions were a constant time of 35μs, 25kV of accelerating voltage and 100s of acquisition time keeping constant the dead time, the magnification, and the work distance. Under these operation conditions the exactitude and precision of the instrument to calculate the O, Al and Mo compositions for the alumina and molybdenum oxide was 0.6wt% and 0.2wt%, respectively. The more intense characteristic energy lines must be used to calculate the composition. The instrument is able to resolve very well the chemical composition in samples having elements with their more intense characteristic energy lines overlapped and they cannot be resolved by the limit of energy resolution of the instrument. Additionally, the preparation of the sample to be analyzed by SEM plays an important role for the determination of the chemical composition, then; a study of powder sample preparation is suggested. The results reported in this work correspond only to the samples analyzed with the instrument used.

## 5. Acknowledgment

This work was financially supported by IMP project D.00447.

## 6. References

Angeles, C.; Rosas, G. & Perez, R. (2000). Preparation of Al₃Ti and L1₂ Al₃Ti-Base Alloys Microalloyed with Fe By a Melting/Casting Rapid-Solidification Technique. *Materials and Manufacturing Processes*, Vol.15, No.2, (published 2000), pp.207-209, ISSN 1042-6914

Cortes-Jacome, M.A.; Toledo, J.A., Angeles-Chavez, C., Aguilar, M. & Wang, J.A. (2005). Influence of Synthesis Methods on Tungsten Dispersion, Structural Deformation, and Surface Acidity in Binary WO₃-ZrO₂, *Journal Physical Chemistry B*, Vol.109, No.48, (December 2005), pp. 22730-22739, ISSN 1520-6106

Goldstein, J.I. & Newbury, D.E. (2003). *Scanning Electron Microscopy and X-Ray Microanalysis*, Kluwer Academic, ISBN 0-306-47292-9, New York, USA

Williams, D.B . & Barry-Carter, C. (1996). *Transmission Electron Microscopy*, Plenum Press, ISBN 0-306-45247-2, New York, USA

Jenkins, K. & De Drier, J.L. (1967). *Practical X-Ray Spectrometry*, Springer-Verlag, BCIN No. 61267

Herglof, H. & Birk, L.S. (1978). *X-ray Spectrometry*, Dekker, ISBN-8247-6625-3, New York, USA

Deutsch, M.; Gang, O., Hamalainen, K. & Kao, C.C. (1996). Onset and Near Threshold Evolution of the Cu Kα X-Ray Satellites, *Physical Review Letters*, Vol.76, No.14, (April 1996), pp. 2424-2427, ISSN 0031-9007

Liebhafsky, H. A. (1976). *X-ray, Electrons, and Analytical Chemistry: Spectrochemical Analysis with X-rays*, Wiley- Interscience, ISBN 0471534285, New York, USA

Kenik, E.A. (2011). Evaluating the performance of a commercial Silicon Drift Detector for X-Ray Microanalysis, *Microscopy Today*, Vol.19, No.3, (May 2011), pp. 40-46, ISSN 1551-9295

Schlossmacher, P.; Kenov, D.O., Fritag, B. & Von Harrach, H.S. (2010). Enhanced Detection Sensitivity with a New Windowless XEDS System for AEM Based on Silicon Drift Detector Technology, *Microscopy today*, Vol.18, No.4, (July 2010), pp.14-20, ISSN 1551-9295

Newbury, D.E.; Joy, D.C., Echlin, P., Fiori, C.E. & Goldstein, J.I. (1986). *Advanced Scanning Electron Microscopy and X-Ray Microanalysis*, Plenum Press, ISBN 0-0306-42140-2, New York, USA

Chung, F.H.; Lentz, A.J. & Scott, R.W. (1974). Thin Film for Quantitative X-ray Emission Analysis, *X-ray Spectrometry*, Vol.3, No.4, (April 2005), pp. 172-175, ISSN 1097-4539

Ellison, S.L.R.; Roselein, M. & William, A. (1995). *Quantifying Uncertainty in Analytical Measurement*. EURACHEM/CITAC, ISBN 0-948926-08-2, London.

# 7

# The Interaction of High Brightness X-Rays with Clusters or Bio-Molecules

Kengo Moribayashi

*Japan Atomic Energy Institute*

*Japan*

## 1. Introduction

The three dimensional (3D) structures of bio-molecules such as proteins are important for the development of new drugs. At present, the structures are analyzed as follows. (i) The diffraction patterns of bio-molecules are produced by irradiation of synchrotron radiation x-rays onto crystallized bio-molecules. (ii) The structures are reproduced using computer simulations. However, there are a lot of bio-molecules which are very difficult to be crystallized. For such bio-molecules, Neutze et al. proposed that the diffraction patterns are produced from the irradiation of x-rays onto single bio-molecules. Then, the intensity of x-rays is required to be very bright. They also suggested that free electron x-ray laser (XFEL) light pulses, which have been developed by US, EU, and Japan, can have enough brightness x-rays (Neutze et al., 2000). When we use single molecules for the analysis of 3D structures, the study of the damage and the destruction of bio-molecules due to the irradiation of XFEL light pulses is indispensable (Hau-Riege et al., 2004, 2007, Jurek et al., 2004, Kai & Moribayashi, 2009, Kai, 2010, Moribayashi & Kai, 2009, Moribayashi, 2008, 2009, 2010, Neutze et al., 2000, Ziaja et al., 2006). We define the damage and the destruction as the ionization and the movement of atoms in a target, respectively (Hau-Riege et al., 2004, Moribayashi, 2008). This comes from the fact that places of the atoms are and are not changed due to the movement and the ionization, respectively. The change of the places means that the reconstruction of the 3D structure cannot be executed. The damage and the destruction mainly occur through the following occurrences: (i) the atoms in the target are ionized through the x-ray absorption or Compton scattering. (ii) From these ionization processes, free electrons, quasi-free electrons and ions are produced and move, where we define 'a free electron' and 'a quasi-free electron' as an electron, which is ionized from an atom, outside and inside the target, respectively (Hau-Riege et al., 2004, Moribayashi, 2008). (iii) Quasi-free electrons promote the ionization of the other atoms through electron impact ionization processes. (iv) Other ionization processes, such as Auger, also occur.

Before experiments of diffraction patterns start, simulations of the damage and the destruction play an important role. The simulations have been executed using various methods, such as molecular dynamics (MD) (Neutze et al., 2000, Jurek et al., 2004), rate equations (Hau-Riege et al., 2004, Hau-Riege et al., 2007, Kai & Moribayashi, 2009, Kai, 2010, Moribayashi, 2008, Moribayashi & Kai, 2009), and kinetic Boltzmann equations (Ziaja et al., 2006). All of these methods have both advantages and disadvantages. In the MD, accurate simulation can be executed for bio-molecules of small size, having up to 10,000 atoms.

However, MD is unsuitable for larger sizes because it takes too much time to calculate the damage and the movement of electrons and ions. The rate equations and the kinetic Boltzmann equations can treat bio-molecules larger size using spherically symmetry models, that is, one dimensional models. We have developed the Monte Carlo and the Newton equation (MCN) model (Moribayashi, 2010), which is almost the same as the MD except for the treatment of the movement of ions, as well as the rate equation (Moribayashi, et al., 1998, 2004, 2005, Moribayashi, 2007a, 2008, 2009, Moribayashi & Kai, 2009).

The 3D structures are investigated from diffraction patterns, which come from the irradiation of XFEL light pulses onto the bio-molecules. The shape of the diffraction patterns changes according to the x-ray flux which irradiates the bio-molecules. Generally speaking, larger numbers of x-rays produce better diffraction patterns. However, as the number of x-rays increases, bio-molecules are damaged, that is, the atoms in bio-molecules such as C, N, O are more often ionized and more highly charged ions are produced. The highly charged ions cause a Coulomb explosion, as a result, the bio-molecules are destroyed (Neutze et al., 2000). The damage and destruction appear as noise for the analysis of the three-dimensional structures. Namely, it is very important to know the x-ray flux irradiating the bio-molecules for the study of the 3D structures of bio-molecules. We have proposed the measurement of the x-ray flux using the x-ray emission from the hollow atoms ((Moribayashi et al., 2004, Moribayashi, 2008) and the energy loss of the photo-electrons (Moribayashi, 2009).

In this chapter, we discuss (i) the damage of bio-molecules or clusters (see Sec.3) and (ii)the measurement methods of x-ray fluxes (see Sec.4) by the irradiation of XFEL light pulses onto bio-molecules or clusters through the simulations using rate equations and the MCN models (see Sec.2). In our previous paper (Moribayashi and Kai, 2009, Moribayashi, 2008, 2009, 2010), we have only treated carbons in targets. On the other hand, in this chapter, we treat mixtures which have carbon, nitrogen and oxygen atoms. Then, the densities of C, N, and O are $1.8 \times 10^{22}/cm^3$, $6 \times 10^{21}/cm^3$, and $6 \times 10^{21}/cm^3$, respectively. We decided these densities according to the similar value to the existent ratio among these elements in bio-molecules.

## 2. Simulation methods

The analysis of 3D structures of bio-molecules is executed based on diffraction patterns, which come from x-rays scattered by electrons bounded in atoms. The intensity of the diffraction patterns ($I_o$) is given by

$$I_o(\vec{k}) \propto I_i \mid F(\vec{k}) \mid^2, \tag{1}$$

where $I_i$ is the intensity of XFEL light pulses and $F(k)$ defined by

$$F(\vec{k}) = \int \rho(\vec{r})e^{i\vec{k}\cdot\vec{r}} \, d\vec{r} = \sum_i e^{i\vec{k}\cdot\vec{r}_i} \int \rho_{atom}(\vec{r})e^{i\vec{k}\cdot\vec{r}} \, d\vec{r} + \int \rho_{fe}(\vec{r})e^{i\vec{k}\cdot\vec{r}} \, d\vec{r} \tag{2}$$

is the structure factor as a function of wave number vectors ($k$) with $k = K_i - K_f$. Here $r$, $\rho(r)$, $\rho_{atom}(r)$, and $\rho_{fe}(r)$ are the place of an atom in the target, the electron density in the bio-molecule, the electron density in the atom, and the density of free and quasi-free electrons, respectively and $K_i$ and $K_f$ are the wave number vectors of the incident and scattered x-rays, respectively. The change of $\rho_{atom}(r)$ and the places of atoms during the irradiation of x-rays are also conventionally ignored because of the small amount of the damage and destruction.

On the other hand, in XFEL light pulses, we may need to consider this second term on the right side of Eq.(2), the change of $\rho_{atom}(r)$ and the places of atoms because of the larger damage. Namely, the damage and the destruction change the diffraction patterns due to (i) ionization processes, which reduce $I_o$. [Then, we should change $\rho_{atom}(r)$ to the electron density according to ionized states [$\rho_{ion}(r)$] (Hau-Riege et al., 2007).], (ii) the interference of x-rays scattered by electrons bounded in the atoms with those by quasi-free and free electrons, which changes $I_o$ [see the second term of the right side of Eq.(2)], and (iii) the movement of atoms, which changes $r_i$. For the movement of the atoms, the distances over which the atoms move become one of the factors for the decision of the highest resolving power of the 3D structures obtained from the experiments. Therefore, we need to control these distances to be smaller than the desired resolving power during the irradiation of XFEL light pulses on the target. However, the movement of atoms may be able to be controlled using a short pulse of XFEL light pulses (Neutze et al., 2000) or a tamper target (Hau-Riege et al., 2007), where a tamper target has been defined as a bio-molecule surrounded by multi-layers of water. Hau-Riege et al. (Hau-Riege et al., 2007) showed from their simulation that the movement of atoms can be controlled using a tamper target and a pulse of 50 fs as diffraction patterns change little. On the other hand, it is almost impossible to control the movement of electrons. This means that the effect of the movement of atoms on the analysis of 3D structures is much smaller than that of the movement of electrons. In our simulations, we calculate the change of the electronic states of the atoms and the movement of free and quasi-free electrons. However, we ignore the movement of atoms.

We have developed two models, that is, the rate equation and the MCN models. Using the rate equations, we can roughly estimate the damage for the large size of spherical targets. The rate equations are suitable to research the most suitable experimental conditions. On the other hand, in the MCN model, we can treat atoms and electrons individually in various shapes with smaller number of atoms. The MCN model is suitable to reproduce experimental results of diffraction patterns or photo-electron spectroscopy.

## 2.1 Atomic processes

Atomic processes treated here are the x-ray absorption (e.g., $C + h\nu \rightarrow C^+ + e^-$), the Compton scattering ($C + h\nu \rightarrow C^+ + e^- + h\nu'$), the electron impact ionization ($C + e^- \rightarrow C^+ + 2\,e^-$), and the Auger ($C^{+*} \rightarrow C^{2+} + e^-$), where $h\nu$ and $h\nu'$ are the x-ray energies before and after the process occurs, respectively. We calculate the change of both of ionized and excited states of the atoms and the production of free and quasi-free electrons using rates or cross sections of these ionization processes as a function of times. We use the same rates or cross sections as those given in several papers (Bell et al., 1983, Henke et al., 1993, Kai & Moribayashi, 2009, Moribayashi, 2008). The x-ray absorption cross sections ($\sigma_{xa}$) are roughly calculated by

$$\sigma_{xa} \propto |< f\,|\,\bar{r}\,|\,i >|^2, \tag{3}$$

where $|\,i >$ and $|\,f >$ are the wave functions for the initial and final states, respectively (Cowan, 1968). On the other hand, the cross sections of Compton scattering ($\sigma_{CS}$) are determined by the Klein-Nishina formula (Klein & Nishina, 1929), that is,

$$\frac{d\sigma_{CS}}{d\Omega} = \frac{1}{2}r_c^{\,2}\frac{(h\nu')^2}{(h\nu)^2}(\frac{h\nu}{h\nu'} + \frac{h\nu'}{h\nu} - sin^2\theta), \tag{4}$$

where $r_c$, $\theta$, and, $\Omega$ are the radius of a classical electron, the scattering angle, and the solid angle, respectively and $h\nu'$ is given by

$$h\nu' = \frac{h\nu}{1 + \frac{h\nu}{m_e c^2}(1 - \cos\theta)}, \tag{5}$$

where $m_e$ and $c$ are the mass of an electron and the light speed, respectively. Then, the rates of the x-ray absorption ($R_{xa}$) and Compton scattering ($R_{CS}$) are given by

$$R_{xa} = \frac{I\sigma_{xa}}{h\nu} \quad and \quad R_{CS} = \frac{I\sigma_{CS}}{h\nu}, \tag{6}$$

respectively (Moribayashi et al., 1998, 1999, 2004, 2005, Moribayashi, 2008), where $I$ is the intensity of the x-rays. On the other hand, Auger rates are roughly given by

$$A_a \propto |<f|\frac{1}{r_{12}}|i>|^2, \tag{7}$$

where $r_{12}$ is the length between an electron transferred from an excited state to the ground state and that ionized from an ion (Cowan, 1968). We have used the Auger rates given in our previous paper (Moribayashi, 2008). For the cross sections of the electron impact ionization processes ($\sigma_e$), we employ the data given by Bell et al. (Bell et al., 1983).

Though we treat isolated atoms, that is, an isolated system, atoms in bio-molecules form a condensed matter system. Therefore, we have compared some atomic data of isolated atoms with those of molecules or solids. Photo-absorption measurements, which correspond to photo-ionization cross sections, in the foil targets of C show good agreement with those in the isolated C atom (Henke et al., 1993) at x-ray energies larger than 350 eV where inner-shell ionization dominates. Further, it was reported that photo-ionization cross sections of several molecules such as $CO_2$ and $C_3H_6$ almost equal to the sum of the cross sections for the constituent atoms in this x-ray energy region (Henke et al., 1993, Hatano, 1999). Coville and Thomas calculated the lifetimes due to Auger processes ($\sim 1/A_a$) of singly inner-shell ionized atoms of 14 molecules containing C, N, O and compared them with those of the isolated atoms (Coville & Thomas, 1991). They showed that the lifetimes of singly inner-shell ionized atoms of the molecules are 0.6 to 0.85 times shorter than that of the isolated atom. Their results agreed with the measured lifetimes to within 25 % except for $CO_2$. However, for $CO_2$, a newer experimental result showed good agreement with their lifetime (Neeb et al., 1991).

We encountered numerical difficulties in treating a large number of coupled rate equations associated with multiple energy levels in the singly inner-shell ionized atoms and hollow atoms in obtaining x-ray spectra due to the decay from these excited states. We have employed the approximation as follows (Moribayashi et al., 2004, 2005, Moribayashi, 2007a, 2008). Namely, the atomic data are averaged over the quantum numbers of spin angular momentum ($S$), orbital angular momentum ($L$), and total angular momentum ($J$). The averaged transition energy ($E_{av}$) and the averaged atomic data of $A_a$ and $A_r$ are given by

$$E_{avn}(2p \rightarrow 1s) = \frac{\sum_{S,L,J} g_{SLJ} \sum_{S'L'J'} A_{rSLJS'L'J'} E_{nSLJS'L'J'}}{\sum_{S,L,J} g_{SLJ} \sum_{S'L'J'} A_{rSLJ,S'L'J7}}, \tag{8}$$

and

$$A_{av} = \frac{\displaystyle\sum_{S,L,J} g_{SLJ} A_{SLJ}}{\displaystyle\sum_{S,L,J} g_{SLJ}}, \quad (9)$$

respectively, where $g_{SLJ}$ expresses the statistical weight and $A_{rSLJS'L'J'}$ and $E_{n\ SLJS'L'J'}(2p \rightarrow 1s)$ are the radiative transition probability and the energy difference between the states of $1s\ 2s^2 2p^n$ $^SL_J$ and $1s^2 2s^2 2p^{n-1}\ ^{S'}L'_{J'}$, respectively.

## 2.2 Rate equations

We calculate the damage, the x-ray emission from the hollow atoms, photo-electron spectroscopy for the various parameters of XFEL light pulses using rate equations. With these atomic rates, the population dynamics of the various atomic may be investigated by the following rate equations:

$$\frac{dN_0}{dt} = -\beta_0 N_0,$$
$$\frac{dN_k}{dt} = \alpha_{k-1} N_{k-1} - \beta_k N_k \quad (k = 1, 2, ..., n), \quad (10)$$

where $N_0, N_1, N_2, ..., N_n$ are the populations of the ground state of the atom and the ions, singly inner-shell ionized atoms and hollow atoms of the ions, $\alpha_{m,k}$ is the transition rate via the transition processes such as photo-ionization, electron impact ionization from the $m'$th to the $k'$th state and $\beta_k$ is the decay rate via transition processes in the $k'$th state. The number of fluorescent x-ray photons per volume ($P_{ek}$) from singly inner-shell ionized states of the atoms or the hollow atoms is given by

$$P_{ek} = \int_0^\infty N_k A_{rav} dt, \quad (11)$$

where $N_k$ is the population of singly inner-shell ionized atoms or hollow atoms.

## 2.3 Monte Carlo and Newton equation (MCN) model

The Monte Carlo and Newton equation (MCN) model employed here is almost the same method as that treated in the MD (Jurek et al., 2004) except for the movement of atoms or ions as mentioned in Sec.1. In the MCN model, we can treat the change of the electric states of atoms and electrons individually in various shapes such as bio-molecules. The MCN model is applied to reproduce the experimental results of the diffraction pattern and photo electron spectroscopy.

For the reconstruction of the 3D structures of bio-molecules, a lot of pulses are required. It should be noted that the production and the movement of electrons depend on the initial values of the random numbers (seeds) and that we can demonstrate the calculations of the damage and the electron distributions for different pulses using different initial seeds for the random number generated. We will show the results averaged by a few hundred pulses.

### 2.3.1 Initial electron energies

The initial energies and velocities of electrons produced from these ionization processes should be mentioned because they contribute significantly not only to the movement of free and quasi-free electrons but also to the treatment of electron impact ionization processes. (i) The x-ray absorption processes: The initial electron energy corresponds to the value that subtracted a bound energy ($E_B$) of atoms or ions from the x-ray energy. Since the x-ray energy treated here is much larger than $E_B$ of H, C, N, and O, which are main elements of bio-molecules, the initial electron energy is almost the same as the x-ray energy. (ii) The Compton scattering: The value of $\theta$ is determined randomly by treating the right side of Eq.(4) multiplied by $d\Omega$ as a weighting factor and the initial electron energy is $h\nu - h\nu' - E_B$. (iii) Auger: We employed the initial electron energy calculated by Cowan's code (Cowan, 1968). (iv) Electron impact ionization processes: We calculate the initial electron energy from the binary encounter dipole (BED) theory (Kim et al., 2000) or use the data given by Nakazaki et al. (Nakazaki et al., 1991). After the initial electron energy is determined, the initial direction of the electron velocity is given randomly except for that due to Compton scattering. In Compton scattering, the initial direction is determined from the electron energy, $\theta$, and the momentum conservation law.

### 2.3.2 Monte Carlo

The x-ray absorption, Auger, and Compton scattering processes are treated using the Monte Carlo method as follows (Moribayashi, 2007b, 2009, 2010): (i) just when an XFEL light pulse begins to irradiate a target, we start the calculation and set the time of $t = 0$. We also set the neutral and the ground states for ionized and excited states of all atoms in the target, respectively. (ii) We calculate the transition rates [$R_{ifp}$ ($m$)] of all the possible ionization processes according to the ionized and excited states of all the atoms and random numbers [$N_R$ ($m$)]. One random number is given to each atom at the time interval between $t$ and $t + \Delta t$, where $R_{ifp}$ ($m$) and $N_R$ ($m$) are the transition rate from the $i'$th state to the $f'$th one of the $m'$th atom due to the $p'$th ionization process and the random number given to the $m'$th atom, respectively. (iii) Only when

$$\sum_p \sum_f R_{ifp}(m)\Delta t > N_R(m), \tag{12}$$

one process for the $m'$th atom occurs. When this equation is satisfied, the state where the ionization occurs is chosen randomly among all the possible transitions using the respective $R_{ijp}$ ($m$) as weighting factors. (iv) The value of $t$ increases by $\Delta t$ and procedures (ii) and (iii) are executed. (v) We reiterate procedures (ii) - (iv) until the XFEL light pulse passes through the target.

### 2.3.3 Electron movements

As for the electron impact ionization process, it is judged that the process occurs only when a quasi-free electron crosses the area of a cross section according to an ionized state of an atom. The center of the cross section is located at the place of the atomic nucleus and the cross section is perpendicular to the direction of the electron velocity (Jurek et al., 2004, Moribayashi, 2009, 2010, 2011). Specifically, we use the relationship of cross sections with impact parameters ($b$) where $b$ is defined as the perpendicular distance between the path of an incident ion and the center of the atom. The electron impact cross section ($\sigma$) is given by

$$\sigma = \pi \int_0^{b_{max}} P(b)bdb, \tag{13}$$

where $P(b)$ is the probability that the corresponding processes occur as a function of $b$ and $b_{max}$ is the maximum $b$ where the process occurs. When we assume $P(b)$ to be a step function with value 0 outside of $b_{max}$, $\sigma = \pi b_{max}^2$. Only when $b$ becomes smaller than $(\sigma/\pi)^{1/2}$, we judge that the particle impact process occurs (Moribayashi, 2011)

The Coulomb forces due to ions and electrons act on free and quasi-free electrons. The movement of these electrons is solved by the Newton's equations, that is,

$$\vec{F} = m_e \frac{d\vec{v}_{ei}}{dt} = -\sum_{j \neq i} \frac{e^2 \vec{r}_{ij}}{4\pi\varepsilon_0 r_{ij}^3} + \sum_l \frac{q_l e \vec{r}_{il}}{4\pi\varepsilon_0 r_{il}^3}, \tag{14}$$

where $\varepsilon_0$, $m_e$, $\vec{v}_{ei}$, $q_l$, and $\vec{r}_{ij(l)}$ are the dielectric constant in vacuum, the mass of an electron, the velocity of the $i$'th electron, the charge of the $l$'th ion, and the distances between the $i$'th electron and the $j$'th free and quasi-free electron (the $l$'th ion), respectively. In order to avoid the divergence near $r_{ij(l)} = 0$, we use an approximation where $r_{ij(l)}$ is approximately replaced by $(r_{ij(l)}^2 + a_s^2)^{1/2}$ (Jurek et al. 2004, Moribayashi, 2010, 2011).

### 2.3.4 Spherically symmetric models

In the case of a spherical target with a radius of 100 nm, the number of atoms is larger than $10^7$. In our calculation using the MCN developed here, it takes about 12 hours to calculate the damage and the movement of free and quasi-free electrons for the number of atoms of only 8000 and the x-ray flux of $3 \times 10^{20}$ photons/pulse/mm$^2$. Therefore, it takes too much time to execute the 3D calculation for the damage of bio-molecules when we treat a target with a radius around 100 nm. Then, spherically symmetric models become useful.

When we study the irradiation of XFEL light pulses with the clusters or bio-molecules, the uniform space charge, $Q_e(r)$ is produced from electrons escaped from the target, that is.

$$Q_e(r) = 4/3\pi r^3 D_{pe}e, \qquad (r < r_0)$$
$$Q_e(r) = 4/3\pi r_0^3 D_{pe}e, \qquad (r \geq r_0) \tag{15}$$

with $D_{pe} = N_e/V_t$, where $r_0$, $e$, $N_{ee}$, and $V_t$ are the radius of the target, the charge of an electron, the number of the electrons which escape from the target, and the volume of the target, respectively. This comes from the Gauss law for the sphere. Then, in our first approximation (which we call SSM1), we use Eq.(15) for the space charge, where the uniform charge distribution in the spherical targets is assumed. In the case of ellipsoids, we define an escaped electron as an electron, which has a value of $r$ larger than that of the atom furthest from the center of the target ($r_{alm}$), that is, $r_0 = r_{alm}$. Then, the force acting on an electron becomes

$$F(r) = \frac{Q_e(r)e}{4\pi\varepsilon_0 r^2} = \frac{1}{3\varepsilon_0} r D_{pe}e^2, \qquad (r < r_0)$$
$$F(r) = \frac{Q_e(r_0)e}{4\pi\varepsilon_0 r^2}. \qquad (r \geq r_0) \tag{16}$$

Here, the force is directed toward the center. It should be noted that Eq. (16) follows the Gauss law in the case of the uniform charge distribution in spherical targets. Namely, electric fields $(F(r)/e)$ are produced from the charge, which exists inside the place of interest, and the charge outside it can be ignored because of cancellation. We treat Eq. (16) instead of Eq. (14) for the movement of electrons in the SSM1. In Eq. (16), there is no divergence near $r = 0$, which often appears for the point charge, because $F(0) = 0$. This approximation is useable only when the number of quasi-free electrons is too small to effect on the charge distribution. Since the SSM1 is useful for the saving of calculation time, we examine the limits of application of the SSM1. When we consider that quasi-free electrons effect on the charge distribution, the charge distribution for $r$ becomes non-uniform. Then, in our second approximation (SSM2), we estimate the charge $Q_{es}(r)$ by counting the total charge inside the place where the electron of interest exists and we use $Q_{es}(r)$ instead of $Q_e(r)$ in Eq. (16).

## 3. X-ray damage

Here, we study (i) the relationship between the damage and the parameters of XFEL light pulses such as pulse widths, wavelenths, and x-ray fluxes using rate equations in order to research the most suitable experimental conditions and (ii) the free and qausi-free electron movement in the target using MCN model in order to aim at the reproduction of the experimental data.

### 3.1 The most suitable XFEL parameters

We have been studying the role of atomic processes such as photo-ionization, Auger, radiative transition, and electron impact ionization processes for the damage of bio-molecules irradiated by an XFEL light pulse. By considering these roles, we have constructed the models mentioned in Sec.2..

We have calculated the changes of the electronic states of atoms or ions in a bio-molecule as a function of time. However, the number of the electronic states is too large. Then, for clearer figures, we show the changes of charge determined from the electronic states.

Figures 1 (a – i) show the populations of the charge of C, N, and O as a function of time for the pulse width ($\tau$) and wavelength ($\lambda$) of (a – c) an XFEL light pulse of 100 fs and 0.1 nm, (d – f) 10 fs and 0.1 nm, and (g – i) 10 fs and 0.06 nm, respectively. The x-ray flux of the XFEL light pulses and the radius of bio-molecules are $10^{22}$/pulse/mm$^2$ and 10 nm, respectively. A gauss type time function is employed for the fluxes of the XFEL light pulses (see upper sides of Fig.1) and the time of 0 is set when the peak intensity of x-rays is located in the bio-molecule. We have found that (i) the damage becomes larger as the atomic number increases, (ii) a shorter value of $\tau$ produces smaller damage (see Figs.1 (a – c ) and (d – f)), and (iii) a shorter value of $\lambda$ also produces smaller damage (see Figs.1 (d - f) and (g – i)). The reason why a shorter pulse produces smaller damage is due to the fact that time scale of Auger processes is about 10 fs. Namely, photo-absorption ionization and Auger processes occur only once in the case of $\tau = 10$ fs and several times in the case of 100 fs, respectively. Auger-electrons also give more significant contribution to the damage in the case of $\tau = 100$ fs. On the other hand, the dependence of wavelength on the damage comes from the inner-shell photo-ionization cross sections. The photo absorption ionization cross sections for $\lambda = 0.1$ nm is about 10 times larger than that for $\lambda = 0.06$ nm (Henke et al, 1993).

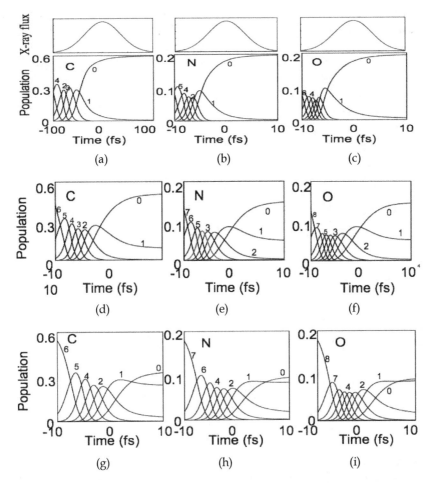

Fig. 1. Population of the charge of C (a, d,g), N (b, e, h), and O (c, f, i) atoms as a function of time for the x-ray flux of $10^{22}$/pulse/mm$^2$, the radius of bio-molecules of 10 nm, and (a - c) $\tau$ = 100 fs, $\lambda$ = 0.1 nm, (d - f) $\tau$ = 10 fs, $\lambda$ = 0.1 nm, (g - i) $\tau$ = 10 fs, $\lambda$ = 0.06 nm, respectively. The numbers written in the figures express charge (Lower figures). Upper figures: the x-ray intensity of an XFEL light pulse as a function of time. The time of 0 is set when the peak intensity of x-rays passes through the bio-molecule.

Figures 2 (a - f) show the population of C with different charge states as a function of time for x-ray fluxes of $10^{19}$ - $10^{21}$/pulse/mm$^2$ for the wavelength of an XFEL light pulse of (a - c) 0.1 nm and (d - f) 0.06 nm, respectively. X-ray fluxes correspond to the time scale of the photo absorption ionization processes because the rates of photo absorption ionization processes ($R_{ap}$) are in proportion to the x-ray flux. Smaller x-ray fluxes produce smaller damage. However, the resolution powers or the intensities for the diffraction pattern correspond to x-ray fluxes. Intensities for the diffraction patterns ($I_O$) are given by Eq.(1) using the structure factor defined in Eq.(2). Therefore, we need to study the intensities of the diffraction patterns including the damage in order to propose the best parameter for the experiments in future.

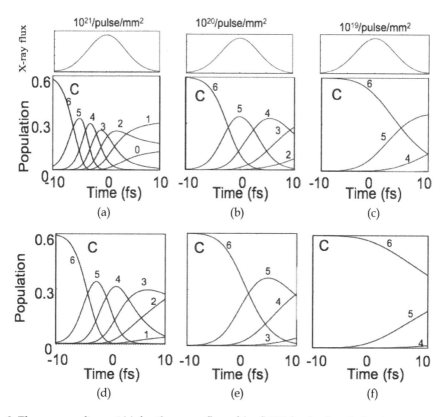

Fig. 2. The same as figure 1(a) for the x-ray flux of (a, d)$10^{21}$/pulse/mm$^2$, (b, e) $10^{20}$/pulse/mm$^2$, (c, f) $10^{19}$/pulse/mm$^2$ and the radius is 10 nm. The wavelengths are are (a - c) 0.1 nm and (d - f) 0.06 nm.

### 3.2 Electron movement

We treat model clusters with spheres at a solid density ($3 \times 10^{22}$/cm$^3$). We decide places inside and outside the target from the number and the density of atoms. Then, the places of the atoms are assigned randomly on the condition that they are located inside the target and that lengths among the atoms are larger than 2.7 Å, which is almost the same as the length between carbons in proteins. Then, we attempt to apply our MCN models to one bio-molecule, that is, a lysozyme which has elements of H, C, N, O and, S. We use the place coordinate data of a lysozyme in the protein data bank (PDB) (http://www.pdb.org/pdb/home/home.do), in which we employ 2LZM as PDB ID.

For the parameters of XFEL light pulses, it is estimated that x-ray fluxes around $10^{20}$ photons/pulse/mm$^2$ and wavelength around 1 Å are required (Neutze et al., 2000). In this paper, we treat x-ray fluxes of $10^{20}$ to $5 \times 10^{20}$ photons/pulse/mm$^2$, a wavelength of 1 Å, a pulse of 10 fs, and the number of atoms of 1000 - 8000.

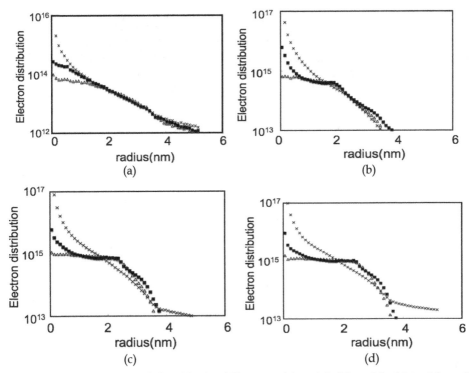

Fig. 3. Electron distribution defined by Eq. (17) vs. $r$ at (a) $t = 1$ fs (b) $t = 3$ fs, (c) $t = 6$ fs, and (d) $t = 9$ fs for a spherical target. The calculation methods are the MCN ($\triangle$), the SSM1 ($\times$), and the SSM2 ($\blacksquare$). The x-ray fluxes, the number of atoms in a target, a wavelength, and a pulse of XFEL are $3 \times 10^{20}$ photons/pulse/mm$^2$, 2000, 1 Å, and 10 fs, respectively.

Figure 3 shows the free and quasi-free electron distribution as a function of $r$ for a spherical target at $t = 1, 3, 6$, and 9 fs calculated by the MCN, SSM1, and the SSM2, where $r$ is the length from the center of the target. We use the constant x-ray flux for the XFEL light pulses as a function of time and set the time of $t = 0$ just when an XFEL light pulse begins to irradiate the target. The electron distribution treated here is defined as follows: (i) we count the number of electrons [$N_e(r)$] at the interval between $r$ and $r + \Delta r$, where we take to be 0.1 nm for $\Delta r$. (ii) The electron distribution $F_{ed}(r)$ is given by

$$F_{ed}(r) = \frac{N_e(r)}{4\pi(r + \Delta r/2)^2}. \tag{17}$$

The results calculated by the SSM1 and the SSM2 show good agreement with those of the MCN at $t = 1$ fs [see Figs.3 (a)]. Then, a lot of quasi-free electrons can escape from the target. Then, the distribution becomes smaller as $r$ increases. We predict from Fig.3 (a) as follows: (i) since quasi-free electrons are accelerated toward $r = 0$, the electrons become concentrated near $r = 0$. (ii) As the charge becomes smaller near $r = 0$, the acceleration becomes weaker as time progresses. (iii) This reduces the invasion of quasi-free electrons into $r = 0$. As a result, the distribution near $r = 0$ becomes almost a constant value as a function of $r$. This trend agrees well with that given by Hau-Riege et al.( Hau-Riege et al., 2004). For $t \geq 3$ fs [see Figs.3 (b - d)], the SSM1 seems to become useless because the number of quasi-free electrons is enough large to effect on the charge distribution. The electron distribution calculated by the SSM2 still shows good agreement with that of the MCN except for the places near $r = 0$. We give up the use of the SSM1 because we have judged from Figs.3 (b – d) that it is danger to apply the SSM1 to the calculation of the electron distribution.

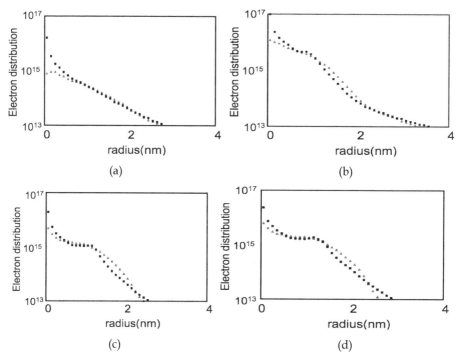

Fig. 4. The same as Fig.3 for a lysozyme target. Electron distribution defined by Eq. (17) vs. $r$ at (a) $t = 1$ fs (b) $t = 3$ fs, (c) $t = 6$ fs, and (d) $t = 9$ fs. The calculation methods are the MCN ($\triangle$) and the SSM2 ($\blacksquare$). The x-ray fluxes, a wavelength, and a pulse of XFEL are 3 × $10^{20}$ photons/pulse/mm², 1 Å, and 10 fs, respectively.

Figure 4 shows the same as Fig.3 for the application to a lysozyme. A lysozyme has elements of H, N, O, and S, as well as C. We treat the ionization processes using cross sections or rates corresponding to each element except for S because the number of S is much smaller than that of the other elements. We have found the same trends as those in Fig.3, that is, the electron distribution near $r = 0$ remains almost a constant value and good agreement between the electron distributions calculated by the SSM2 and the MCN is shown. This may result from the fact that a lysozyme has a shape close to a sphere. We derive the relationship between the radius of the sphere into which the bio-molecule could be transformed ($r_{0l}$) and the average value among the lengths of the places of atoms from the center ($r_{av}$). The relationship between $r_{av}$ and $r_{0l}$ is given by

$$r_{av} = \frac{\int_0^{r_{0t}} r 4\pi r^2 \, dr}{4/3\pi r^3} = \frac{3}{4} r_{0t}, \qquad (18)$$

that is, we assume $r_{0t} = 4/3 \, r_{av}$. From this equation, we estimate that $r_{0t}$ for a lysozyme is approximately 0.21 nm. We conclude that we may apply the SSM2 to the calculation of the electron movement on bio-molecules with the shape close to a sphere.

## 4. The measurement of x-ray flux

In the reconstruction of the 3D structure of bio-molecules from diffraction patterns using XFEL light pulses, a lot of patterns, that is, a lot of pulses are required. Then, the x-ray fluxes should be almost the same for each shot. The method which measures the x-ray flux is required. Therefore, we have proposed the measurement of x-ray fluxes using the x-ray emission from hollow atoms and photo-electron specrtroscopy.

For the measurement of the x-ray flux, there are other methods such as using scattered x-rays, the number and degree of ionization of all the ions as well as photo-electron spectrums and x-ray emission from hollow atoms. Since we believe that all of the methods have both advantage and disadvantage for the measurement. For example, for the scattered x-rays, x-rays scatted by electrons are measured. In this measurement, high-energy electrons escaped from the target reduce the intensity of scatted x-rays. Further, an interface between x-rays scattered through electrons bounded in and ionized from the atoms changes the intensity randomly. The interface comes from the fact that XFEL light pulses have full coherence. Therefore, since we forecast that the relationships of the x-ray fluxes with intensities of scattered x-rays become non-monotonic, it is not simple to use the scattered x-rays for the measurement. We will not intend to say that the x-ray emission from hollow atoms or photo-electron spectroscopy are the best for the measurement of the x-ray flux. We should use all the methods mentioned here after we understand the mechanism of them. Fortunately, we can measure them at the same time.

### 4.1 X-ray emission from hollow atoms

Moribayashi et al. have proposed a new method for the measurement of the x-ray intensity or x-ray flux using the x-ray emission from hollow atoms produced by high intensity x-rays (Moribayashi et al., 2004, Moribayashi, 2008). As the x-ray intensity increases, the rates of inner-shell ionization processes also increase, while the rates of other atomic processes such as Auger and radiation transition processes remain constant. As a result, the ratio of production of hollow atoms to that of singly inner-shell ionized atoms increases with the x-ray flux (Moribayashi et al., 1998, 2004, 2005, Moribayashi, 2007a, 2008). From this ratio, we may measure the x-ray flux (Moribayashi et al., 2004). We showed concrete calculation results of the application of this method to the measurement of the x-ray flux irradiating bio-molecules or clusters (Moribayashi, 2008) where we treated targets which have one element among carbon, nitrogen, oxygen atoms and electron impact ionization processes were ignored. Here, we treat mixtures which have carbon, nitrogen and oxygen atoms. Then, the populations of C, N, and O are $1.8 \times 10^{22}/cm^3$, $6 \times 10^{21}/cm^3$, and $6 \times 10^{21}/cm^3$, respectively and consider electron impact ionization processes.

Figure 5 shows atomic processes relevant to hollow atoms due to the interaction of x-rays with carbon atoms. We calculate (i) the population of inner-shell excited states and hollow atoms and (ii) the x-ray emission intensity from inner-shell ionization states (Ar1, Ar3, ---) and hollow atoms (Ar2, Ar4, ---).

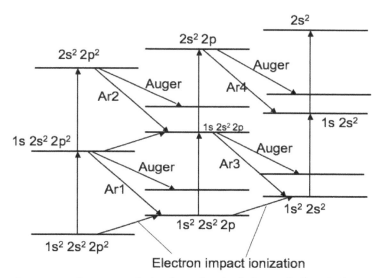

Fig. 5. Atomic processes in x-ray emission from the singly inner-shell excited states and hollow atoms.

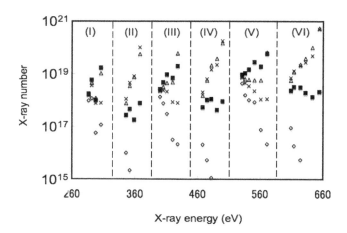

Fig. 6. Spectroscopy emitted from the inner-shell excited states (IES) and hollow atoms (HA): The electronic states and their energies are listed in Table 1 in our previous paper (Moribayashi, 2008). The values of the x-ray flux are given by the symbols of $\diamondsuit$: $10^{19}$/pulse/mm$^2$, $\blacklozenge$:$10^{20}$/pulse/mm$^2$, $\triangle$: $10^{21}$/pulse/mm$^2$, and X: $10^{22}$/pulse/mm$^2$. In the regions (I – VI), the x-rays are emitted from IES of C, HA of C, IES of N, HA of N, IES of O, and HA of O, respectively.

Figure 6 shows the x-ray emission from inner-shell excited states and hollow atoms of carbon, nitrogen, and oxygen atoms as a function of energies of the x-ray emission. We employ various values of x-ray flux and take 0.1 nm for wavelengths of x-ray sources. The electronic states and their x-ray emission energies are listed in Table 1 in our previous paper (Moribayashi,2008). The x-ray emission for a x-ray photo-energy is decided by the the number of the outer-shell electrons, which correspond to $2s$ and $2p$ electrons. The left and right sides of the spectroscopy correspond to the x-ray emissions from the inner-shell excited states and hollow atoms, respectively. As the energies increase, the charge becomes larger.

Figure 7 shows the ratio of the number of fluorescent x-ray photons emitted from the first hollow atoms to that from the first singly inner-shell ionized atoms as a function of x-ray fluxes ($F_x$). As $F_x$ increases, x-ray emissions from highly charged ions and hollow atoms become larger. Namely, this may inform us of x-ray flux from the spectroscopy. This is consisted with the trend given in our previous paper (Moribayashi,2008). The ratio increases in proportion to $F_x$ when $F_x$ is smaller than $10^{20}$/pulse/$mm^2$ and $10^{21}$/pulse/$mm^2$ for the x-ray wavelength of 0.1 and 0.06 nm, respectively. The difference between the wavelengths of 0.1 nm and 0.06 nm comes from the fact that the production of hollow atoms increases according to the square of the inner-shell ionization cross sections. Namely, the cross sections of the wavelength of 0.1 nm are about 10 times as large as those of 0.06 nm as mentioned before. We can see almost the same trends among the elements of C, N, and O. This may mean that this method of the measurement of the x-ray flux can apply to the bio-molecules, which are mainly constructed by these three elements.

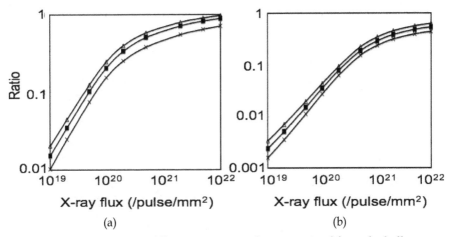

(a)                                                              (b)

Fig. 7. The ratio of the number of fluorescent x-ray photons emitted from the hollow atoms to that from the singly inner-shell excited states as a function of x-ray fluxes. The target materials are C ($\times$), N ($\blacksquare$), and O($\triangle$). The wavelength of x-ray sources treated here are (a) 0.1 nm and (b) 0.06 nm.

## 4.2 Photo-electron spectroscopy

The scenario adopted here for the energy losses of photo-electrons is as follows. The photo-electrons are produced through x-ray absorption processes, and their initial energies are almost the same as that of the x-rays just when the XFEL light pulse enters the target. The photo-electrons can escape from the targets. When one electron escapes from them, one charge is added. The charge increases according to the number of electrons escaped from the target. The Coulomb force due to this charge reduces the photo-electron energies. The total energy losses ($E_p$ - $E_{min}$) of the photo-electrons depend on the total charge, which is decided by the x-ray flux, the size and density of the targets and the energies of x-rays.

As mentioned before, energy losses of photo-electrons are caused by a space charge ($Q$) in the target, which is produced by the escape of electrons from the target. We consider not only photo-electrons but also Auger electrons for the calculation of $Q$ values. Since we assume that the space distribution of each ion state becomes almost uniform in a target, $Q$ may be considered to be concentrated at the center of the target. The charge affecting an electron ($Q_e(r)$) which is located at a distance of $r$ from this center is given by

$$Q_e(r) = \frac{4}{3}\pi r^3 D_{pe}e.$$  (19)

Then, the force acting on the electron is

$$F = \frac{Q_e(r)e^2}{4\pi\varepsilon_0 r^2} = \frac{1}{3\varepsilon_0}rD_{pe}e^2.$$  (20)

Suppose that the radius and volume of the target are $r_0$ and $V$, respectively. Just when the photo-electron produced at $r$ moves from the target to the surface of the target, the energy loss of the photo-electron is given by

$$\Delta E(r) = \int_r^{r_0} F dr = = \frac{1}{6\varepsilon_0}D_{pe}e^2(r_0^2 - r^2).$$  (21)

Here, we assume that the $Q_e(r)$ remains constant from the production of an electron to its escape, because the photo-electron is too fast for the value of $D_{pe}$ to change during the escape. Then, the averaged energy loss is

$$\Delta E_a = \frac{\int_0^V \Delta E(r)dV}{V} = D_{pe}e^2\frac{1}{15\varepsilon_0}r_0^2.$$  (22)

We assume that $\Delta E = \Delta E_a$, when $r = r_a$. Using Eqs. (21) and (22), we obtain

$$D_{pe}e^2\frac{1}{15\varepsilon_0}r_0^2 = D_{pe}e^2\frac{1}{6\varepsilon_0}(r_0^2 - r_a^2),$$  (23)

that is, $r_a = \sqrt{3/5}r_0$. Therefore, we assume that photo-ionization processes occur at $r = r_a$ in our calculation. The energy loss of the electrons after the escape until they reach the detector is given by

$$\frac{Q_e(r_0)e^2}{4\pi\varepsilon_0 r_0} = \frac{1}{3\varepsilon_0}r_0^2 D_{pe}e^2.$$  (24)

Then, adding Eq.(22) to this equation, the total energy loss is given by

$$\Delta E_{tot} = \frac{2}{5\varepsilon_0} r_0^2 D_{pe} e^2. \tag{25}$$

The electrons are produced through ionization processes of atoms or ions such as photo-absorption, Compton Scattering, Auger electron emission, and electron impact ionization. In order to count the number of the electrons, firstly, we calculate the population of several electronic states due to these ionization processes of atoms or ions using rate equations (Kai, 2010, Moribayashi et al. 1998, 2004, 2005, Morbayashi, 2007a, 2008). Supposed that the density of photo-electrons from a target is $D_{pe}$ which is calculated by

$$\frac{dD_{pe}}{dt} = \sum_j (R_{pj} + R_{Augerj})N_j, \tag{26}$$

with

$$R_{pj} \sim \frac{I\sigma_{pj}}{E_p}, \tag{27}$$

where $I$, $\sigma_{pj}$, and $E_p$ are the intensity of x-rays, a photo-absorption cross section from the $j$ state, and the energy of x-rays, respectively, and $R_{Augrj}$ is the Auger rate from the $j$ state.

Since an easy estimation of the total energy loss is useful for experiments, we have derived an easy approximation equation. For $N_0 \gg N_1, N_2, ---, N_n$ where the small x-ray flux irradiates a target, we may approximate $D_{pe}$ by using the following equations

$$\frac{dN_0}{dt} = -R_{p0}N_0, \quad \frac{dD_{pe}}{dt} = R_{p0}, \tag{28}$$

so that $N_0$ and $D_{pe}$ become

$$N_0 \sim exp(-R_{p0}t)N_{00}, D_{pe} \sim (1-exp(-R_{p0}t))N_{00}, \tag{29}$$

where $N_{00}$ is the initial density of atoms in the target. By inserting this equation into Eq. (25), the total energy loss is rewritten as

$$\Delta E_{tot} \sim \frac{2}{5\varepsilon_0} r_0^2 e^2 N_{00}(1-\exp(-\frac{I\sigma_{p0}}{E_p}t)). \tag{30}$$

Furthermore, $I$ is estimated as

$$I \sim \frac{F_X E_p}{\tau}, \tag{31}$$

where $\tau$, $F_x$, and $E_p$ are the pulse length, the x-ray flux, and the energy of XFEL light pulses, respectively and $\sigma_{p0}$ is the photo-ionization cross section from the ground state of an atom. Therefore, we can derive the approximation equation of $\Delta E_{tot}$ as a function of $F_X$ as follows:

$$\Delta E_{tot} \sim \frac{2}{5\varepsilon_0} r_0^2 e^2 N_{00}(1-\exp(-\frac{F_X\sigma_{p0}}{\tau}t)). \tag{32}$$

At $t = \tau$, $\Delta E_{tot}$ becomes maximum, that is,

$$\Delta E_{tot,max} \sim \frac{2}{5\varepsilon_0}r_0^2 e^2 N_{00}(1 - exp(-F_X\sigma_{p0})).$$                                    (33)

This equation is applied for the case where one element of atoms exists in the target (Moribayashi, 2009). In the case where three elements of atoms exist in the target, Eq.(33) is changed to

$$\Delta E_{tot,max} \sim \frac{2}{5\varepsilon_0}r_0^2 e^2 N_{00}(1 - \sum_{i=1}^{3}P_i exp(-F_X\sigma_{ip0})),$$                        (34)

where $i$, $P_i$ and and $\sigma_{ip0}$ are the element number of atoms, the ratio of the initial density, and the photo-ionization cross section from the ground state of the atom $i$, respectively. Here, the element numbers $i$= 1, 2, and 3 correspond to the elements of C, N, and O, respectively, and $P_1 = 0.6$, $P_2 = P_3 = 0.2$.

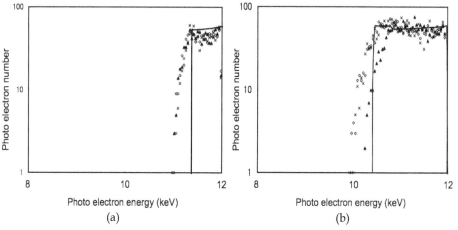

Fig. 8. The comparison of the photo-electron spectrums calculated by the rate equation method (solid lines) with those by Monte Carlo and Newton equation (MCN) (symbols). In the MCN, spectra given by three pulses are shown separately. The target radius treated here is 2.5 nm. The x-ray fluxes are (a) $10^{20}$ photons/pulse/mm$^2$ and (b) $3 \times 10^{20}$ photons/pulse/mm$^2$.

In order to verify our model, we also calculate photo-electron spectrums for a small size of a cluster of a few 1000 atoms using Monte Carlo and Newton equations (MCN) (Moribayashi, 2009, 2010) and compare them with those calculated by the rate equations (Moribayashi et al., 1998, 1999, 2004, 2005, Moribayashi, 2007a, 2008). It should be noted that they depend on the initial values of the random number, which are employ in the Monte Carlo method and that we can demonstrate the calculations of photo-electron spectrums for different pulses by using a different initial value of a random number. We show some examples of photo-electron spectra by each shot. For the rate equation method, we employ using Eqs.(20 – 24) where only energies of the electrons are treated.

Figure 8 shows the comparison of the photo-electron spectrums calculated by these two methods. The target radius treated here is 2.5 nm and the x-ray fluxes are (a) $10^{20}$ photons/pulse/mm$^2$ and (b) $3 \times 10^{20}$ photons/pulse/mm$^2$. For the MCN method, individual spectrums for three pulses are shown. Good agreement among them are shown for the minimum energies though the absolute values of the number of electrons do not remain at a constant value for the results of the MCN with one pulse. This means that the plasmas in the target give little contribution to the spectroscopy and furthermore, photo-electron spectrums can be treated by the simple model. This may come from the fact that the energy losses after the escape from the target are much larger than those before the escape. Since it takes too much time to calculate photo-electron spectrums for much larger sizes of targets (Moribayashi, 2009, 2010), we employ the rate equation method to calculate the size dependence on the spectroscopy. Furthermore, we compared these calculation results with the approximations given by Eq. (34).

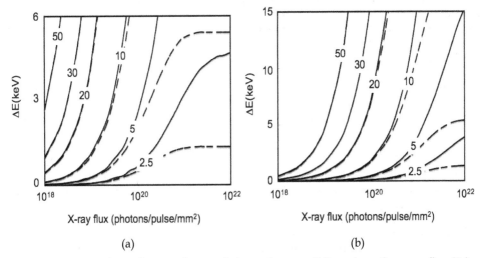

Fig. 9. Maximum values of energy losses of photo-electrons ($\Delta E_{tot,max}$) vs. the x-ray flux ($F_X$) for various radius of a target, the pulse of 10 fs, and x-ray energies ($E_p$) of (a) 12 keV and (b) 20 keV: Approximation solutions of Eq.(34) are also shown. solid lines: the calculation results, dotted lines: approximations. Radius values of the target are taken to be 2.5, 5, 10, 20, 30, and 50 nm which are shown in the lines.

Figure 9 shows the maximum values of $\Delta E_{tot}$ ($\Delta E_{tot,max}$) as a function of $F_X$ for various values of $r_0$, a pulse of 10 fs, and (a) $E_p$ = 12 keV and (b) $E_p$ = 20 keV. Approximation solutions given by Eq.(34) are also shown. The values of $\Delta E_{tot,max}$ increase with increase in $F_X$. As $r_0$ increases, the upper limit of values of $F_X$ which can be measured, where $\Delta E_{tot\ max} < E_p$, becomes lower because $\Delta E_{tot,max}$ increases in direct proportion to $r_0^2$ and $\Delta E_{tot,max}$ is independent of the x-ray flux. Much larger values of x-ray flux can be measured where $E_p$ = 20 keV than where $E_p$ = 12 keV. This comes from the fact that the photo-absorption cross section at $E_p$ = 20 keV is much smaller than that at $E_p$ = 12 keV (Henke et al., 1993) and $\Delta E_{tot,max}$ depends on $F_X \sigma_{p0}$ as seen in Eq. (34). Eq.(34) is accurate to within 20 % with x-ray flux smaller than $10^{20}$ photons/pulse/mm$^2$ at $E_p$ = 12 keV and smaller than $10^{21}$

photons/pulse/mm² at $E_p$ = 20 keV, independent of $r_0$ where $F_X \, \sigma_{p0} \ll 1$. This means that the approximations are useful for larger values of $E_p$, as seen in Figs.9. Then, as $F_X$ increases, the approximations become worse because a larger number of ions is produced. The discrepancy between the calculation and the approximation results mainly comes from the fact that ionization starting from inner-shell excited states and ions produced through Auger processes are ignored in the approximation. However, approximations with simple equations may be useful for the analysis of experiments. It was found that enough values of $\Delta E_{tot,max}$ could be measured that differences in the x-ray flux within a factor of 2 could be detected in the case where $\Delta E_{tot,max}$ is larger than the resolution power of the detector, because $\Delta E_{tot,max}$ increases with increase in $F_X$.

## 5. Summary

We theoretically study (i) the most suitable experimental conditions for the reconstruction of three dimensional structure of bio-molecules, (ii) free and quais-free electrons movement and (ii) the measurement methods of x-ray fluxes by the irradiation of XFEL light pulses onto bio-molecules or clusters using x-ray emission from hollow atoms and phot-electron spectroscopy. We employ rate equations and the MCN models as a simulation method. In our previous paper (Moribayashi and Kai, 2009, Moribayashi, 2008, 2009, 2010), we only treat carbon atoms in the targets. On the other hand, here, we treat mixtures which have carbon, nitrogen and oxygen atoms. Then, the densities of C, N, and O are $1.8 \times 10^{22}$/cm³, $6 \times 10^{21}$/cm³, and $6 \times 10^{21}$/cm³, respectively. These populations come from the ratio of the elements in proteins.

We have shown the relationship of the damage with the parameters of pulse length, wavelength, and x-ray flux of XFEL light pulses. We have found that the shorter pulse widths, shorter wavelengths, and smaller x-ray fluxes reduce the damage. We believe that these results become important for the experiment of the three-dimensional structure of a single bio-molecule.

We discuss the space distribution of free and quasi-free electrons. The electron distribution calculated by our spherically symmetric model agrees well with that calculated by our more accurate model except for the place near the center of the targets. Our spherically symmetric model can be applied to a lysozyme. We may apply our spherically symmetric models developed here to the calculation of the movement of free and quasi-free electrons in bio-molecules with a shape close to a sphere.

We study hollow atom production processes by high brightness x-rays and propose the application of fluorescent x-rays emitted from singly inner-shell ionized atoms and hollow atoms to the measurement of x-ray flux irradiating bio-molecules. We have found that the ratio of the number of fluorescent x-ray photons from the hollow atoms to that from the singly inner-shell ionized atoms increase according to x-ray flux irradiating C, N, and O atoms. This ratio may be employed for this measurement.

We propose measuring the x-ray flux irradiating a single cluster or a bio-molecule using photo-electron spectroscopy. As the size of the targets increases, the x-ray flux which can be measured become smaller. Much larger values of the x-ray flux can be measured at the x-ray energy of 20 keV than that of 12 keV. We derived an easy approximation equation. The equation is valid for x-ray flux smaller than $10^{20}$ photons/pulse/mm² at $E_p$ = 12 keV and $10^{21}$ photons/pulse/mm² at $E_p$ = 20 keV, and independent of the size of the cluster or the bio-molecule.

## 6. Acknowledgements

We wish to thank Profs. Go N., Tajima T., Shinohara K., and Namikawa K.,  Dr. Kono H, Dr. Kai T, Dr. Tokushisa A, Dr. Koga J, Dr. Yamagiwa M., and Dr. Kimura T. for their useful discussions. This study has been supported by the 'X-ray Free Electron Laser Utilization Research Project' of the Ministry of Education, Culture, Sports, Science and Technology of Japan (MEXT). We employ the data in the protein data bank (PDB) for the place coordinate of a lysozyme. For the calculation of atomic data, we have employed Cowan's code.

## 7. References

Bell, K. L.; Gillbody, H. B.; Hughes, J. G.; Kingston, A. E. & Smith, F. J. (1983). Recommended data on the electron impact ionization of light atoms and ions. *J. Phys. Chem. Ref. Data,* Vol. 12, No.4, pp.891-916.

Cowan, R. D. (1968). Theoretical calculation of atomic spectra using digital computers. *J. Opt. Soc. Am.,* Vol.58, No.6, pp.808-818.

Coville, M. & Thomas, T.D. (1991). Molecular effects on inner-shell lifetimes: Possible test of the one-center model of Auger decay. *Phys.Rev.A,* Vol.43, No.11, pp.6053 – 6056.

Hatano, Y. (1999). Interaction of vacuum ultraviolet photons with molecules. Formation and dissociation dynamics of molecular superexcited states. *Phys.Rep.,* Vol.313, No.3, pp.109 – 169.

Henke, B.L.; Gullikson, E.M. & Davis J.C. (1993). X-Ray interactions: photoabsorption, scattering, transmission, and reflection at $E = 50$-30,000 eV, $Z = 1$-92. *Atomic data & Nuc. Data Tables,* Vol.54, No.2, pp.181-342.

Gaffney, K. J. & Chapman, H. N. (2007). Imaging Atomic Structure and Dynamics with Ultrafast X-ray Scattering. *Science,* Vol.316, No.5830, pp.1444 - 1448.

Hau-Riege, S.P.; London, R.A. & Szoke, A. (2004). Dynamics of biological molecules irradiated by short X-Ray pulses. *Phys.Rev.E,* Vol.69, No.5, pp.051906.

Hau-Riege, S.P.; London, R.A.; Chapman, H. N.; Szoke, A. & Timneanu, N. (2007). Encapsulation and diffraction-pattern- correction methods to reduce the effect of damage in X-Ray diffraction imaging of single biological molecules. *Phys. Rev. Lett.,* Vol.98, No.19, pp.198302.

Jurek, Z.; Faigel, G. & Tegze, M.  (2004). Dynamics in a cluster under the influence of intense femtosecond hard X-Ray pulses. *Eur. Phys. J. D,* Vol.29, No.2, pp.217 - 229.

Kai,T. & Moribayashi, K. (2009). Effect of electron-impact ionization in damage of Bio-Molecules irradiated by XFEL. *J. Phys.: Conf. Ser.,* Vol.163, No.1, pp.012035.

Kai, T. (2010). Single-differential and integral cross sections for electron-impact ionization for the damage of carbon clusters irradiated with X-Ray free-electron lasers. *Phys. Rev. A,* Vol.81, No.2, pp.023201.

Kim, Y.K.; Santos, J.P. & Parente, F. (2000). Cross sections for singly differential and total ionization of helium by electron impact. *Phys.Rev.A,* Vol.62, No.5, pp.052710.

Klein, V. O. & Nishina, Y. (1929). Über die Streuung von Strahlung durch freie elektronen nach der neuen relativistischen quntendynamik von Dirac. *Z. Phys.,* Vol.52, 853-868.

Moribayashi, K.; Sasaki, A. & Tajima, T. (1998). Ultrafast X-Ray processes wit hollow atoms. *Phys. Rev. A,* Vol.58, No.3, pp.2007-2015.

Moribayashi, K.; Sasaki, A. & Tajima, T. (1999). X-ray emission by ultrafast inner-shell ionization from vapors of Na, Mg, and Al. *Phys.Rev.A*, Vol.59, No.4, pp.2732-2737.

Moribayashi, K.; Kagawa, T. & Kim, D. E. (2004). Theoretical study of the application of hollow atom production to the intensity measurement of short-pulse high-intensity x-ray sources. *J. Phys. B*, Vol.37, No.20, pp.4119 – 4126.

Moribayashi, K.; Kagawa, T. & Kim, D. E. (2005). Application of x-ray non-linear processes to the measurement of 10 fs to sub-ps of x-ray pulses. *J. Phys. B*, Vol.38, No.13, pp.2187-2194.

Moribayashi, K. (2007a). Multiply inner-shell excited states produced through multiple X-Ray absorption relevant to X-Ray pulses. *Phys. Rev. A*, Vol.76, No.4, pp.042705.

Moribayashi, K. (2007b). Comparison of the stopping powers calculated by using rate equation with those by Monte Carlo method. *J. Phys.: Conf. Ser.*, Vol.58, No.1, pp.192-194.

Moribayashi, K. (2008). Application of XFEL to the measurement of X-Ray flux irradiating Bio-Molecules by using X-Ray emission from hollow atoms produced from multiple X-Ray absorptions. *J. Phys. B*, Vol.41, No.8, pp.085602.

Moribayashi, K. & Kai, T. (2009). Atomic processes for the damage on bio-molecules irradiated by XFEL. *J. Phys.: Conf. Ser.*, Vol.163, No.1, pp.012097.

Moribayashi, K. (2009). Application of photoelectron spectroscopy to the measurement of the flux of X-Ray free-electron lasers irradiating clusters or Bio-Molecules. *Phys. Rev. A*, Vol.80, No.2, pp.025403.

Moribayashi, K. (2010). Spherically symmetric the models for the X-Ray damage and the movement of electrons produced in non-spherically symmetric targets such as Bio-Molecules. *J. Phys. B*, Vol.43, No,16, pp.165602.

Moribayashi, K. (2011). Incorporation of the effect of the composite electric fields of molecular ions as a simulation tool for biological damage due to heavy ion irradiation. *Phys. Rev. A*, vol.84, No.1, pp.012702.

Nakazaki, S.; Nakashima, M.; Takebe, H. & Takayanagi, K. (1991). Energy distribution of secondary electrons in electron- impact ionization of hydrogen-like ions. *J. Phys. Soc. Japan*, Vol.60, No.5, pp.1565-1571.

Neeb, M.; Kempgens, B.; Kivimäki, A.; Köppe, H.M.; Maier, K.; Hergenhahn, U.; Piancastelli, M.N.; Rüdel, A. & Bradshaw, A.M. (1998). Vibrational fine structure on the core level photoelectron lines of small polyatomic molecules. *J. Electron Spectr. Relat. Phenom.*, Vol. 88-91, pp.19 – 27.

Neutze, R.; Wouts, R.; Spoel, D.; Weckert. E. & Hajdu J. (2000). Potential for biomolecular imaging with femtosecond X-Ray pulses. *Nature*, Vol.406, No. 6797, pp.752-757.

Ziaja, B.; de Castro, A.R.B.; Weckert, E. & Möller, T. (2006). Modelling dynamics of samples exposed to free-electron-laser radiation with Boltzmann equations. *Eur. Phys. J. D*, Vol.40, No.3, pp.465 - 480.

# Quantification in X-Ray Fluorescence Spectrometry

Rafał Sitko and Beata Zawisza
*Department of Analytical Chemistry,*
*Institute of Chemistry, University of Silesia*
*Poland*

## 1. Introduction

X-ray fluorescence spectrometry (XRF) is a versatile tool in many analytical problems. Major, minor and trace elements can be qualitatively and quantitatively determined in various kinds of samples: metals, alloys, glasses, cements, minerals, rocks, ores, polymers as well as environmental and biological materials. Elements from Na to U are routinely determined using energy-dispersive X-ray fluorescence spectrometry (EDXRF) whereas application of wavelength-dispersive spectrometers (WDXRF) allows efficient determination of low-Z elements down to even Be. Although the samples can be analyzed without treatment, high quality results can be ensured if appropriate sample preparation is applied. This may vary from simple cleaning and polishing of the sample (metals, alloys), powdering and pelletizing with or without binder (ceramics, minerals, ores, soils, etc.), fusing the sample with appropriate flux (ceramics, rocks, ores, etc.) to digestion with acids (metals, alloys). This way errors resulting from surface roughness, particle size effect or inhomogeneity of the material can be eliminated or minimized. Due to the nondestructive character of X-ray measurement, the XRF spectrometry is widely applied in analysis of art, museum and archeological objects such as manuscripts, paintings, icons, pottery, ancient glasses, ceramics, coins. Moreover, XRF spectrometry is utilized for simultaneous determination of thickness and composition of various materials such as semiconductors, electrooptic and solar cell devices, etc., in electronic industry and other branches of technology. Typical detection limits for medium- and high-Z elements are in the ppm range, which is satisfactory for several applications. However, in some cases, the elemental concentrations are too low for a direct analysis. Then, the analytes must be preconcentrated prior to analysis using physical or chemical preconcentration or separation methods.

Quantitative analysis of all types of aforementioned samples requires applying adequate empirical or theoretical methods. In quantitative XRF analysis, the measured fluorescent intensities are converted into the concentration of the analytes. This issue is rather complicated because the measured intensities depend not only on the analyte concentration but also on accompanying elements (matrix), sample type (solid, liquid or powder sample, etc.), method of sample preparation, shape and thickness of the analyzed sample and measurement conditions such as geometrical setup of the spectrometer, irradiated size, flux and spectral distribution of the exciting radiation and the efficiency of detection systems. The simplest equation relating radiation intensity $I_i$ to weight fraction of analyte $W_i$ can be expressed as follows:

$$I_i = k_i W_i \tag{1}$$

Where $k_i$ is a constant. The radiation intensity $I_i$ in Eq. (1) is corrected for background, line overlap, and so forth. In practice, the subtraction of background is not perfectly performed. Thus, Eq. (1) can be expressed in a more general form:

$$I_i = k_i W_i + b_i \tag{2}$$

Where $b_i$ is the radiation intensity when analyte concentration equals zero. The constant $k_i$ is called the sensitivity and is expressed in counts per second per unit of concentration. The constants $k_i$ and $b_i$ are determined by least-squares fit on the basis of measured reference samples from the following formulas:

$$k_i = \frac{n\sum_{j=1}^{n} W_{ij} I_{ij} - \sum_{i=1}^{n} W_{ij} \sum_{i=1}^{n} I_{ij}}{n\sum_{i=1}^{n} W_{ij}^2 - \left(\sum_{i=1}^{n} W_{ij}\right)^2} \tag{3}$$

$$b_i = \frac{\sum_{i=1}^{n} I_{ij} - K_i' \sum_{i=1}^{n} W_{ij}}{n}$$

Where $n$ is the number of standard samples used for analyte $i$, $W_{ij}$ is the weight fraction of analyte $i$ in standard $j$, $I_{ij}$ is the radiation intensity of analyte $i$ in standard $j$.
In practice, the concentrations of the standard samples have to cover the concentration in unknown sample. Moreover, the calculated concentration is more accurate at the center of the calibration line than at the extremities. The Eq. (2) can be rewritten as follows:

$$W_i = K_i I_i + B_i \tag{4}$$

If the calibration based on Eq. (2) or Eq. (4) and standard samples similar to the unknown are carefully applied, several parameters such as sample type, method of sample preparation, and measurement conditions, i.e. the geometrical setup of the spectrometer, irradiated size, flux and the efficiency of detection systems are included in the slope $K_i$ and can be omitted in further stages of quantification. However, a simple linear calibration is not the rule in the XRF analysis. In general, applying the linear Eq. (2) or Eq. (4) requires not only that all standards are similar to the unknown and but also the set of standards with a very limited range of concentrations must be applied for calibration. Only then, the matrix effects in all samples are similar and linear relationship between radiation intensity and analyte concentration can be obtained. On the other hand, the use of standards with a very limited range of concentrations will lead to a calibration graph with large uncertainty on the slope and intercept. Because the matrix effects play an important role in XRF analysis, a more general equation should be applied:

$$W_i = K_i I_i M_i + B_i \tag{5}$$

Where $M_i$ is the total matrix effects term. The $M_i$ differs from one if matrix effects cannot be neglected. When the analyte radiation is absorbed by the matrix or when the absorption effects are dominating over enhancement effects, $M_i$ is larger than 1. On the other hand, when enhancement effects are dominant over absorption, $M_i$ is smaller than 1.

In matrix correction methods, the radiation intensity $I_i$ is usually replaced with relative radiation intensity $R_i$ defined as fluorescent radiation intensity of analyte in binary, ternary or in multielement specimen $I_{specimen,i}$ to fluorescent radiation intensity of pure element or compound $I_{pure-element,i}$, e.g. oxide:

$$R_i = \frac{I_{specimen,i}}{I_{pure-element,i}} \qquad (6)$$

If matrix effects can be neglected, the relative radiation intensity equals weight fraction of analyte:

$$W_i = R_i \qquad (7)$$

In practice, the matrix effects play an important role in XRF analysis. Therefore, the relative radiation intensity has to be corrected using total matrix effects term:

$$W_i = R_i M_i \qquad (8)$$

Because the matrix effects are the major source of errors in X-ray fluorescence analysis, this chapter is devoted to matrix correction methods applied in quantitative XRF analysis. The matrix effects (absorption and secondary fluorescence) and necessary background information on theoretical relationship between radiation intensity and sample composition will be provided first. In the next part of the chapter, the quantification methods applied in XRF will be discussed.

## 2. General relationship between radiation intensity and concentration

In 1955, Sherman proposed a mathematical formula to calculate radiation intensity of analyte in a specimen of a known composition (Sherman, 1955). Later, Shiraiwa and Fujino corrected the enhancement part of this formula by introducing a missing factor of ½ (Shiraiwa and Fujino, 1966). The general equation to calculate X-ray fluorescence intensity $I_i$ emitted by an analyte in the specimen of thickness $t$ when it is irradiated by a polychromatic X-ray beam can be expressed as follows:

$$I_i = \frac{d\Omega}{4\pi \sin\phi_1} Q_i q_i W_i \int_{\lambda\min}^{\lambda edge} \tau_i(\lambda) I_0(\lambda) \frac{1 - \exp[-\chi(\lambda,\lambda_i)\rho t]}{\chi(\lambda,\lambda_i)} \left(1 + \sum_j W_j S_{ij}\right) d\lambda \qquad (9)$$

Where $d\Omega$ is the differential solid angle for the characteristic radiation; $i, j$ are the subscripts for the analyte and matrix element, respectively; $Q_i$ is the sensitivity of the spectrometer for characteristic radiation of analyte $i$; $W_i$, $W_j$ are weight fractions of the analyte $i$ and matrix element $j$, respectively; $\lambda_{min}$ and $\lambda_{edge}$ are short-wavelength limit and wavelength of analyte absorption edge, respectively; $\tau_i(\lambda)$ is the photoelectric absorption coefficient for analyte $i$ and primary radiation of wavelength $\lambda$; $I_0(\lambda)$ is intensity of the primary radiation, $\rho$ is the density of the sample; $t$ is the sample thickness; $q_i$ is sensitivity of the analyte $i$ (if the $K_\alpha$ line is chosen then $q_i = \omega_{K,i} f_{i,K\alpha}(1-1/J_{i,K})$, where $\omega_{K,i}$ is fluorescence yield of K radiation; $f_{i,K\alpha}$ is weight of $K_\alpha$ line within K series; $J_{i,K}$ is absorption edge jump ratio. If the $L_\alpha$ or $L_\beta$ is chosen as the analytical line, then the Cöster–Kronig transition probabilities have to be additionally taken into consideration;); $\chi(\lambda,\lambda_i)$ is total mass-attenuation coefficient of the sample for the incident and fluorescent radiation:

$$\chi(\lambda,\lambda_i) = \frac{\mu(\lambda)}{\sin\phi_1} + \frac{\mu(\lambda_i)}{\sin\phi_2}$$

$$\mu(\lambda) = W_i\mu_i(\lambda) + \sum_j W_j\mu_j(\lambda) \tag{10}$$

$$\mu(\lambda_i) = W_i\mu_i(\lambda_i) + \sum_j W_j\mu_j(\lambda_i)$$

Where $\mu(\lambda)$ and $\mu(\lambda_i)$ are the total mass-attenuation coefficients of the specimen for the incident radiation $\lambda$ and characteristic radiation $\lambda_i$, respectively; $\phi_1$ and $\phi_2$ are the incidence and take-off angles, respectively; $\mu_i(\lambda)$, $\mu_i(\lambda_i)$, $\mu_j(\lambda)$, $\mu_j(\lambda_i)$ are the mass-attenuation coefficients of the analyte $i$ and matrix element $j$ present in the specimen for the incident radiation $\lambda$ and characteristic radiation $\lambda_i$. The $S_{ij}$ (in Eq. 9) is the enhancement term for the matrix element $j$, which can enhance the analyte $i$ (if analyte is not enhanced by matrix element $j$ then $S_{ij} = 0$):

$$S_{ij} = \frac{1}{2} q_j \tau_j(\lambda) \frac{\tau_i(\lambda_j)}{\tau_i(\lambda)} \frac{\chi(\lambda,\lambda_i)}{1-\exp[-\chi(\lambda,\lambda_i)\rho t]} D_{ij}$$

$$D_{ij} = \int_0^{\pi/2} \tan(\theta) \left[ \frac{1-\exp[-\chi_1(\lambda_i,\lambda_j)\rho t]}{\chi_1(\lambda_i,\lambda_j)\chi_2(\lambda,\lambda_j)} - \frac{1-\exp[-\chi(\lambda,\lambda_i)\rho t]}{\chi(\lambda,\lambda_i)\chi_2(\lambda,\lambda_j)} \right] d\theta +$$

$$\int_{\pi/2}^{\pi} \tan(\theta) \left[ \frac{\exp[-\chi_2(\lambda,\lambda_j)\rho t]-\exp[-\chi(\lambda,\lambda_i)\rho t]}{\chi_1(\lambda_i,\lambda_j)\chi_2(\lambda,\lambda_j)} - \frac{1-\exp[-\chi(\lambda,\lambda_i)\rho t]}{\chi(\lambda,\lambda_i)\chi_2(\lambda,\lambda_j)} \right] d\theta \tag{11}$$

$$\chi_1(\lambda_i,\lambda_j) = \frac{\mu(\lambda_i)}{\sin\varphi_1} + \frac{\mu(\lambda_j)}{\cos\theta}$$

$$\chi_2(\lambda,\lambda_j) = \frac{\mu(\lambda)}{\sin\phi_1} - \frac{\mu(\lambda_j)}{\cos\theta}$$

It is beyond the scope of this chapter to derive the Eq. (9). Details on the derivation of this equation can be found elsewhere (Mantler, 1986; Van Dyck et al., 1986; He and Van Espen, 1991; Węgrzynek et al. 1993). As seen from Eq. (9), the intensity of characteristic radiation is the complex function of sample composition and sample thickness. The primary and fluorescent radiation are attenuated by atoms of the analyte and by any other atoms present in the matrix (see Eq. (10)). If matrix element emits a characteristic line that has sufficient energy to excite the analyte, the fluorescent intensity is higher than expected from primary excitation only (the enhancement term $S_{ij}$ in Eq. (9)). The so-called matrix effects (absorption and enhancement) will be discussed in the next section of this chapter.

It should be emphasized that the enhancement term $S_{ij}$ in Eq. (9) cannot be expressed as analytical function and numerical integration is required. Therefore, the matrix correction methods require complex mathematical treatment. Nevertheless, if the thickness of the sample is greater than the so-called saturation thickness ($t \to \infty$), Eq. (9) simplifies to:

$$I_i = \frac{d\Omega}{4\pi \sin\phi_1} Q_i q_i W_i \int_{\lambda\min}^{\lambda edge} \frac{\tau_i(\lambda)I_0(\lambda)}{\chi(\lambda,\lambda_i)}\left(1+\sum_j W_j S_{ij}\right)d\lambda$$

(12)

$$S_{ij} = \frac{1}{2}q_j\tau_j(\lambda)\frac{\tau_i(\lambda_j)}{\tau_i(\lambda)}\left[\ln\left(1+\frac{\mu(\lambda)}{\mu(\lambda_j)\sin\phi_1}\right)\frac{\sin\phi_1}{\mu(\lambda)} + \ln\left(1+\frac{\mu(\lambda_i)}{\mu(\lambda_j)\sin\phi_2}\right)\frac{\sin\phi_2}{\mu(\lambda_i)}\right]$$

Eq. (12) is applied in analysis of the so-called infinitely thick specimens. In practice, the sample should satisfy Eq. (13). Then, the relative error resulting from applying Eq. (10) instead of Eq. (7) does not exceed 1%.

$$t \geq \frac{4.61}{\chi(\lambda,\lambda_i)\rho}$$

(13)

As seen from Eq. (12), the intensity of characteristic radiation of analyte present in infinitely thick sample depends not only on analyte concentration but also on full matrix composition. Because sample thickness is greater than the saturation thickness, the intensity of characteristic radiation of analyte does not depend on sample thickness, which considerably simplifies mathematical treatment.

If the sample is infinitely thin ($t \to 0$), then the enhancement effects can be neglected ($S_{ij} \to 0$) and the approximation $\exp(-x) \approx 1-x$ can be applied and the Eq. (9) simplifies to:

$$I_i = \frac{d\Omega}{4\pi \sin\phi_1} Q_i q_i W_i \rho t \int_{\lambda\min}^{\lambda edge} \tau_i(\lambda)I_0(\lambda)d\lambda$$

(14)

In practice, the sample is not infinitely thin, therefore it should satisfy Eq. (15). Then, the relative error resulting from applying Eq. (14) instead of Eq. (9) does not exceed 0.5 %.

$$\rho t \leq \frac{0.1}{\chi(\lambda,\lambda_i)}$$

(15)

For thin samples, the intensity of characteristic radiation of analyte does not depend on matrix composition – matrix effects can be neglected. In consequence, the linear relationship between radiation intensity and mass per unit area of the analyte is observed (mass per unit area of the sample: $m = \rho t$ [g cm$^{-2}$], mass per unit area of the analyte: $W_i m$).

The samples of less than critical thickness for which matrix effects cannot be neglected are called intermediate-thickness samples. Review of quantitative analysis of these samples including many references can be found in Ref. (Markowicz and Van Grieken, 2002; Sitko, 2009). The general division of the sample in X-ray fluorescence analysis is presented in Fig. 1.

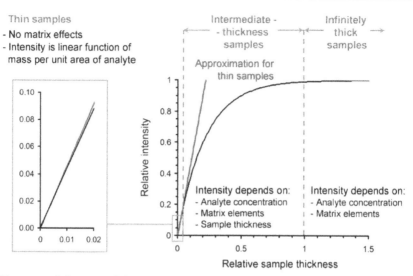

Fig. 1. The general division of the sample in XRF analysis.

## 3. Matrix effects in XRF analysis

Matrix effects in XRF spectrometry are caused by absorption and enhancement of X-ray radiation in the specimen. The primary and secondary absorption occur as the elements in the specimen absorb the primary and characteristic radiation, respectively. The strong absorption is observed if the specimen contains an element with absorption edge of slightly lower energy than the energy of the characteristic line of the analyte. When matrix elements emit characteristic radiation of slightly higher energy than the energy of analyte absorption edge, the analyte is excited to emit characteristic radiation in addition to that excited directly by the X-ray source. This is called secondary fluorescence or enhancement. The absorption and enhancement effects are shown in Fig. 2 using binaries (FeMn, FeCr and FeNi) as examples. When matrix effects are either negligible or constant, the linear relationship

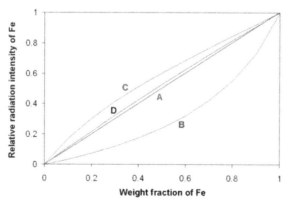

Fig. 2. Relationship between radiation intensity of Fe and weight fraction of Fe: Curve A – matrix effects are negligible, Curve B – FeCr, Curve C – FeNi, Curve D – FeMn.

between radiation intensity and weight fraction of analyte is obtained (Curve A). Curve B is obtained when the absorption by the matrix elements in the specimen is greater than the absorption by the analyte alone (the so-called positive absorption, e.g. determination of Fe in FeCr binaries). Curve C illustrates an enhancement effect, e.g. in the case of determination of Fe in FeNi binaries. Curve D is observed when the matrix element in the specimen absorbs the analyte radiation to a lesser degree than the analyte alone (the so-called negative absorption, e.g. determination of Fe in FeMn binaries).

## 4. Quantification in XRF analysis

Numerous methods, both empirical and theoretical, have been proposed for quantitative XRF analysis. They are divided into two major groups: compensation and matrix correction methods (Fig. 3). Moreover, only one method allows minimizing matrix effects. This method is based on preparation of thin samples. For these samples, matrix effects are not observed under measurement conditions and linear relationship between radiation intensity and analyte concentration is observed. In other quantitative methods, the matrix effects are still present but they are corrected or compensated.

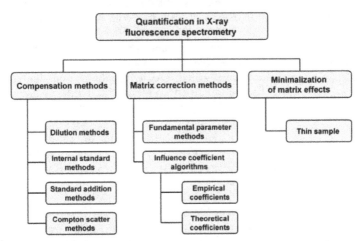

Fig. 3. General division of methods applied in quantitative XRF analysis.

The compensation methods (variations in matrix effects resulting from various specimen compositions are minimized), except Compton scatter, are all well-known in other analytical techniques, e.g. atomic absorption or emission spectrometry. In these methods, special sample preparation is required and only one or few elements can be quantitatively determined. Therefore, the compensation methods are less popular than matrix correction methods. It should be emphasized that radiation intensity of analyte can be calculated from theory. This feature is unique to X-ray spectrometry techniques. No other analytical technique allows such a combination of theoretical calculations and experimental results. Due to the increasing power of computers during the past few years, the theoretical methods, both fundamental parameters and theoretical influence coefficients, became the most popular in routine XRF analysis. Therefore, these methods are discussed in this chapter. Review of the compensation methods including many references can be found in

"Handbook of X-Ray Spectrometry," edited by Van Grieken and Markowicz (de Vries and Vrebos, 2002) and "Handbook of Practical X-Ray Fluorescence Analysis" edited by Beckhoff, Kanngießer, Langhoff, Wedell and Wolff (Vrebos, 2006).

## 5. Fundamental parameter methods

The fundamental parameter methods are based on Sherman equation (Eq. (9)) considering both primary and secondary fluorescence. The tertiary fluorescence and the effects caused by scattered radiation are usually neglected. Eq. (9) enables to calculate the intensity of fluorescent radiation of analyte in the specimen of known composition. It is possible if all physical constants are known: photoelectric absorption coefficients, mass-attenuation coefficients, Cöster–Kronig transition probabilities, fluorescence yields, weight of analytical line within the series, absorption jump ratios, whose values can be found in updated databases (Elam et al., 2001; Ebel et al., 2003). X-ray tube spectrum, required in calculation, can be obtained from published experimental data. Nevertheless, the experimental spectral distributions are published only for selected voltages and types of X-ray tube (take-off angle and thickness of the Be window although absorption in Be window can be easily corrected). Therefore, X-ray tube spectrum (both characteristic line intensities and continuum intensity) can be calculated from theoretical or semi-empirical algorithms (Ebel, 1999; Pella et al., 1985 and 1991; Finkelshtein and Pavlova, 1999).

In practice, the application of fundamental parameter method consists of two steps: calibration and analysis of unknown sample.

Calibration is a crucial issue in assuring high quality results of quantitative analysis. In fundamental parameter methods, calibration can be performed in different ways:

1.  Calibration can be performed using pure-element standards (thin or bulk) for each element or multielement standard similar to unknown sample. Then, the calibration constant $(d\Omega/4\pi\sin\phi_1)Q_iq_i$ in Eq. (9) is calculated from the measured intensity and theoretical intensity of analyte is calculated for a given composition of standard specimen. If multi-element standards similar to unknown sample are applied, then the best analysis results can be expected.

2.  The product $(d\Omega/4\pi\sin\phi_1)Q_i$ can be determined for a few pure-element standards, the element sensitivity $q_i$ is calculated from theory. Then, the relationship between $(d\Omega/4\pi\sin\phi_1)Q_i$ and wavelength can be established. This procedure allows the determination of all elements using only a few pure-element standards. The determination of the relationship between $(d\Omega/4\pi\sin\phi_1)Q_i$ and wavelength of analyte in WDXRF is described in Ref. (Sitko and Zawisza, 2005).

3.  The fundamental parameter method allows performing standardless analysis. Both $Q_i$ and $q_i$ are calculated from theory. Nevertheless, the element concentrations have to be normalized because absolute radiation intensity of X-ray tube is very difficult to calculate from theory. Such a procedure can be strictly performed in EDXRF spectrometry where $Q_i$ depends on the applied semiconductor detector, filters and measurement atmosphere. In the WDXRF spectrometry, $Q_i$ is a complex function of detection efficiency of proportional and scintillation detectors, measurement atmosphere, as well as the reflectivity of the analyzing crystal (very difficult or even impossible to calculate from theory with efficient accuracy). Therefore, standardless methods included in commercial software packages are not strictly standardless because they are based on spectrometer sensitivities determined experimentally by manufacturer.

Sherman equation enables to calculate radiation intensity of analyte in a sample of known composition. In practice, the aim of the analysis is to calculate analyte concentration from the actual measurements. Unfortunately, the Sherman equation cannot be transformed in order to calculate analyte concentration directly (analyte concentration is included in total mass-attenuation coefficient $\chi(\lambda, \lambda_i)$ and in enhancement term $S_{ij}$). Therefore, the analysis of unknown sample is performed using iteration. At the beginning, the first estimate of the composition is made, which can be done in several ways. For example, in this step, the matrix effects are neglected and weight fraction of analyte $W_i$ equals relative intensity $R_i$ determined using the calibration data. In the next step, the theoretical intensity $R_i'$ is calculated for the first estimate of composition. Then, the next estimate of composition can be calculated from the difference between measured and theoretical intensities e.g. using linear interpolation. For example, if measured intensity $R_i$ is 10% higher than theoretical intensity $R_i'$, the weight fraction of analyte is increased by 10%. The process is repeated until convergence is obtained, i.e. the weight fraction element does not change from one step to another by more than a present proportion e.g. 0.0001. In the analysis of specimen containing $n$ elements (or stoichiometric compounds, e.g. oxides), a set of $n$ Eq. (9) has to be solved for the unknown weight fractions by iteration.

The fundamental parameter methods have several advantages. First of all, these methods can be applied in analysis of thick samples, thin films and multilayers (simultaneous determination of composition and thickness is possible). The comprehensive algorithms dedicated to analysis of multiple layer films assume such effects as primary fluorescence, secondary fluorescence within the same layer, secondary enhancement between different layers and also secondary enhancement from the bulk substrate (Mantler, 1986; Willis, 1989; De Boer, 1990). A serious advantage of fundamental parameter methods is the possibility of using any standard specimen for calibration: pure-element thick or thin standard, one standard similar to unknown sample, series of standards similar to unknown sample, etc. Moreover, standardless analysis can be performed. The overview of fundamental parameter methods applied in analysis of thin films and multilayers can be found in Ref. (Sitko, 2009).

The fundamental parameter methods have some limitations. They do not usually consider all physical processes in the sample such as: tertiary fluorescence, scatter of both the primary and fluorescence radiation and photoelectrons (important in the case of low-Z elements). Moreover, the accuracy of fundamental parameters methods strongly depends on uncertainty of atomic parameters (mass-attenuation coefficients, fluorescent yield, etc.), measurement geometry and spectral distribution of X-ray tube. Nevertheless, the use of standards similar to unknown will compensate these effects and will lead to more accurate results. The accuracy of fundamental parameter methods is discussed in previously published papers (Mantler and Kawahara, 2004; De Boer et al., 1993; Elam et al., 2004; Sitko, 2007 and 2008a). The "classical" fundamental parameters method can be applied only if all elements in the specimen are detectable. Then, the total mass-attenuation coefficient $\chi(\lambda, \lambda_i)$ and enhancement term $S_{ij}$ can be calculated during iteration. The quantitative analysis is hampered when undetectable low-Z elements (e.g. H, C, N, O) are present in the material, e.g. geological, environmental and biological samples. Then, the methods utilizing scattered primary radiation or emission-transmission method can be successfully applied.

## 6. Methods based on scattered radiation

The intensity of the Compton scattered radiation can be used to obtain estimate of the attenuation coefficient of the specimen at the wavelength of the scattered photons. Because the intensity of the scattered radiation is inversely proportional to the mass attenuation coefficient, the Compton scattered radiation $I_{Comp}$ can be used as internal standard:

$$W_i = K_i \frac{I_i}{I_{Comp}} + B_i \tag{16}$$

The most common application of Eq. (16) is in the determination of trace elements, e.g. Sr, in geological samples. The method is limited to those cases where only trace elements have absorption edges between the wavelength of the analyte characteristic radiation and the wavelength of scattered radiation. Otherwise, the characteristic radiation of a major matrix element, e.g. Fe, can be used for matrix correction together with scattered radiation (Nesbitt, 1976). If matrix diversification is significant or heavy absorbers are present in large amounts, the Compton scattered radiation is preferred using a power function rather than a simple inverse proportion (Bao, 1997).

The scattered primary radiation is also used in more sophisticated strategies, i.e. in backscattered fundamental parameter methods. Then, scattered primary radiation is applied for the evaluation of the so-called "dark matrix" which consists of undetectable low-Z elements (e.g. H, C, N, O). Nielson (Nielson, 1977) proposed the backscattered fundamental parameter method which utilizes incoherently and coherently scattered radiations to choose and determine quantities of two light elements representative of the 'dark matrix.' To improve the accuracy of the analysis, Węgrzynek et al. applied differential scattering cross sections instead of total scattering cross sections (Węgrzynek et al., 2003a). Another strategy is the use of average atomic number (e.g. Szalóki et al., 1999). Theory and experiment show that the coherent/incoherent scatter ratio is sensitive to average atomic number of the sample $Z_M$:

$$\frac{I_{coh}}{I_{Comp}} = aZ_M^n \tag{17}$$

Where $I_{coh}$ is coherent (Rayleigh) scattered radiation intensity, $a$ and $n$ are constants. Average atomic number is defined as follows:

$$Z_M = W_i Z_i + \sum_j W_j Z_j \tag{18}$$

Where $Z_i$ and $Z_j$ are atomic numbers of analyte $i$ and matrix element $j$, respectively. The summation in Eq. (18) is over all elements present in the specimen. Because the detectable elements can be distinguished from undetectable elements, the average atomic number of the 'dark matrix' $Z_{low-Z}$ can be calculated from the Eq. (19).

$$Z_{low-Z} = \frac{Z_M - \sum_k W_k Z_k}{1 - \sum_k W_k} \tag{19}$$

Where $W_k$ and $Z_K$ are weight fraction and atomic number of detectable element, respectively. If $Z_{low-Z}$ is known, then the mass-attenuation coefficient of the 'dark matrix' for given radiation energy $E$ can be calculated from the empirical Eq. (20).

$$\mu_{low-Z}(E) = b(E) \times Z_{low-Z}^{c(E)} \tag{20}$$

Where $b(E)$ and $c(E)$ are calculated from the least-squares fits applied to the published values of mass-attenuation coefficients.

The backscattered fundamental parameter methods allow calculating contribution of undetectable elements to absorption effects. Therefore, they are usually applied in determination of heavy elements in light matrix samples, e.g. environmental samples (plants, soils, etc.), biological samples and plastics. A strong advantage of the backscattered fundamental parameter methods is the fact that full matrix composition of the sample does not need to be known. Unfortunately, additional calibration has to be performed with the use of the standard samples of known $Z_M$ (determination of constants $a$ and $n$). It should be noted that the scattered radiation can also be applied to estimate the sample thickness (Araujo et al., 1990; Giauque et al., 1994).

An empirical algorithm for correction of matrix effects in light matrix samples was proposed in Ref. (Sitko, 2006 a). The algorithm was derived for the analysis of samples collected onto membrane filters:

$$W_i m = K_i I_i \left[ a_i m \left( \frac{I_{coh}}{I_{Comp}} \right)^b + 1 \right] + B_i \tag{21}$$

Where $K_i$, $B_i$ and $a_i$ are the constants calculated by the least-squares fit on the basis of experimental results for standard samples, $m$ is the mass per unit area of the sample. The coefficient $b$ depends on the filter applied and the mass per unit area of the collected sample. This coefficient can be described by least-squares fit polynomials of second order in ln–ln scale:

$$b = \exp \left[ p_0 + p_1 \ln m + p_2 (\ln m)^2 \right] \tag{22}$$

Where $p_0$, $p_1$, $p_2$ are the constants determined on the basis of experimental results for standard samples. If diversification of sample thickness is limited, this coefficient can be also treated as constant.

## 7. Emission–Transmission method

The emission–transmission (E–T) method is one of the most popular methods based on transmission measurement (Fig. 4). The method is frequently applied for correction of absorption effects in light matrix samples pressed into pellets or collected onto filters. Nevertheless, applicability of E–T method is limited to the samples that are partially transparent for X-ray beams, i.e. intermediate-thickness samples. The method consists of measuring the X-ray fluorescent radiation from the specimen alone $I_{i,S}$, from specimen with a target located at a position adjacent to the back of the specimen $I_{i,S+T}$ and from the target alone $I_{i,T}$. Taking into account these measurements, the total mass-attenuation coefficient $\chi(\lambda, \lambda_i)$ can be calculated from the following relationship:

$$\exp[-\chi(\lambda, \lambda_i)\rho t] = \frac{I_{i,S+T} - I_{i,S}}{I_{i,T}} \tag{23}$$

The most important advantage of the E–T method is the possibility of determining $\chi(\lambda, \lambda_i)$ without the knowledge of the sample composition. In consequence, the absorption correction can be performed very easily and enhancement term $S_{ij}$ can be calculated without using iterative approach (Węgrzynek et al., 1993). In multielement analysis, the measurements are usually performed for a few elements and then relationship between $\chi(\lambda, \lambda_i)$ and wavelength is established. If minor and/or major elements are present in a sample, discontinuities in the relationship between $\chi(\lambda, \lambda_i)$ and wavelength resulting from absorption edges are observed. Then, at least two measurements for each wavelength region are performed or appropriate iteration procedure can be applied (Markowicz and Haselberger, 1992). Quantification based on E-T method, including accuracy and calibration can be found in Ref. (Markowicz et al., 1992; Markowicz and Van Grieken, 2002; Węgrzynek et al., 2003b).

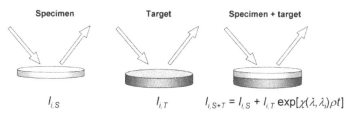

Fig. 4. Measurements in emission-transmission method.

## 8. Influence coefficient algorithms

Many influence coefficient algorithms have been developed. They were reviewed and discussed in one chapter of "Handbook of X-Ray Spectrometry," edited by Van Grieken and Markowicz (de Vries and Vrebos, 2002) and numerous papers (Rousseau, 2001, 2002, 2004, 2006), (Willis and Lachance, 2000, 2004). The algorithms can be divided in different ways. The influence coefficients can be calculated from theory (using Sherman equation) or from the measurements, therefore the algorithms are generally divided into two groups: theoretical and empirical influence coefficients algorithms (Fig. 5). The algorithms can use a single or more than one coefficient per matrix element. Moreover, the influence coefficients can be constant or can vary with composition of the sample.

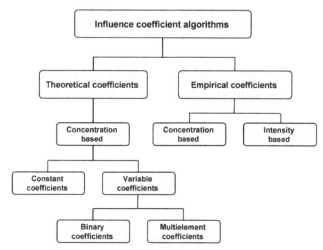

Fig. 5. General division of influence coefficient algorithms.

In general, the total matrix correction term $M_i$ is expressed as a linear combination of weight fractions of matrix elements $W_j$:

$$M_i = 1 + \sum_j \alpha_{ij} W_j \tag{24}$$

Where $\alpha_{ij}$ is influence coefficient describing the matrix effect of the interfering element $j$ on the analyte $i$. Combining Eq. (8) with Eq. (24) leads to the general form of concentration based algorithm:

$$W_i = R_i \left[ 1 + \sum_j \alpha_{ij} W_j \right] \tag{25}$$

Eq. (25) is usually applied if influence coefficients are determined using $R_i$ calculated from theory. The total matrix correction term expressed by Eq. (24) can also be combined with Eq. (5). Then, we obtain the general equation that can be used during calibration if the matrix correction term is calculated from theory for each standard specimen:

$$W_i = K_i I_i \left[ 1 + \sum_j \alpha_{ij} W_j \right] + B_i \tag{26}$$

Eq. (26) is also used in empirical algorithms if influence coefficients $\alpha_{ij}$, slope $K_i$ and intercept $B_i$ are determined from multiple-regression analysis on a large suite of standards.

One of the simplest algorithms was proposed by Lachance and Traill in 1966 (Lachance and Traill, 1966). The algorithm can be easily derived from the Sherman equation if the following assumptions are made: the sample is infinitely thick, the monochromatic excitation is applied and enhancement effects are negligible. Then, the Sherman equation simplifies to:

$$I_i = \frac{d\Omega}{4\pi \sin \phi_1} Q_i q_i W_i \tau_i(\lambda) I_0(\lambda) \frac{1}{\chi(\lambda, \lambda_i)} \tag{27}$$

The total mass-attenuation coefficient can be expressed as follows:

$$\chi(\lambda, \lambda_i) = W_i \chi_i(\lambda, \lambda_i) + \sum_j W_j \chi_j(\lambda, \lambda_i) \tag{28}$$

Where:

$$\chi_i(\lambda, \lambda_i) = \frac{\mu_i(\lambda)}{\sin \phi_1} + \frac{\mu_i(\lambda_i)}{\sin \phi_2}$$

$$\chi_j(\lambda, \lambda_i) = \frac{\mu_j(\lambda)}{\sin \phi_1} + \frac{\mu_j(\lambda_i)}{\sin \phi_2} \tag{29}$$

The total mass attenuation coefficient $\chi(\lambda, \lambda_i)$ depends on both matrix elements $j$ and analyte element $i$. Assuming that the sum of the element weight fractions in the specimen equals one (then $W_i = 1 - \Sigma W_j$), the weight fraction of analyte element $i$ can be eliminated from Eq. (28) and then $\chi(\lambda,\lambda_i)$ is given by:

$$\chi(\lambda,\lambda_i) = \chi_i(\lambda,\lambda_i)\left[1 + \sum_j \alpha_{ij}W_j\right] \tag{30}$$

Where:

$$\alpha_{ij} = \frac{\chi_j(\lambda,\lambda_i)}{\chi_i(\lambda,\lambda_i)} - 1 \tag{31}$$

Finally, if the absolute intensity $I_i$ is replaced by the relative radiation intensity $R_i$ the Lachance-Traill equation can be obtained from the simplified Sherman formula (Eq. (27)):

$$W_i = R_i\left[1 + \sum_j \alpha_{ij}W_j\right] \tag{32}$$

In Lachance-Traill algorithm, the influence coefficient $\alpha_{ij}$ corrects for the absorption effects of the matrix element $j$ on the analyte $i$ in the case of monochromatic excitation of wavelength $\lambda$. The coefficient can be positive or negative. If the analyte is determined in presence of a heavier matrix element, then $\chi_i(\lambda,\lambda_i) < \chi_j(\lambda,\lambda_i)$ and $\alpha_{ij}$ is positive. If the analyte is determined in presence of a lighter matrix element, then $\chi_i(\lambda,\lambda_i) > \chi_j(\lambda,\lambda_i)$ and $\alpha_{ij}$ is negative. The influence coefficients can be calculated in different ways:

- It can be calculated directly from the Eq. (31).
- It can be determined from multiple-regression analysis on a large suite of standards.
- It can be calculated from the relative radiation intensity $R_i$.

The influence coefficients can be calculated directly from the Eq. (31) only if monochromatic excitation is applied (or the effective wavelength is used although it is composition dependent) and there are no enhancement effects. The determination of Ni in ternary system FeNiCr is shown here as an example. The analyte is excited by molybdenum radiation (Mo K$_\alpha$) and following measurement geometry is assumed: $\phi_1 = 60°$ and $\phi_2 = 40°$. In this case, the influence coefficient $\alpha_{NiFe}$ describing influence of Fe on Ni is calculated from the following equation:

$$\alpha_{NiFe} = \frac{\dfrac{\mu_{Fe}(\lambda_{Mo})}{\sin\phi_1} + \dfrac{\mu_{Fe}(\lambda_{Ni})}{\sin\phi_2}}{\dfrac{\mu_{Ni}(\lambda_{Mo})}{\sin\phi_1} + \dfrac{\mu_{Ni}(\lambda_{Ni})}{\sin\phi_2}} - 1 = \frac{\dfrac{37.4}{0.87} + \dfrac{363.3}{0.64}}{\dfrac{44.6}{0.87} + \dfrac{59.8}{0.64}} - 1 = 3.21 \tag{33}$$

The influence coefficient $\alpha_{NiCr}$ describing influence of Cr on Ni is calculated in the same way. Finally, the following equation is obtained for determination of Ni in FeNiCr using aforementioned measurement conditions:

$$W_{Ni} = R_{Ni}\left[1 + 3.21 \times W_{Fe} + 2.47 \times W_{Cr}\right] \tag{34}$$

The influence coefficients can also be calculated from relative radiation intensity of the analyte. In this case, the influence coefficient $\alpha_{ij}$ is calculated for binary systems. Then Eq. (32) simplifies to:

$$W_i = R_i \left[ 1 + \alpha_{ij} W_j \right] \tag{35}$$

Eq. (35) can be rewritten to obtain $\alpha_{ij}$:

$$\alpha_{ij} = \frac{W_i/R_i - 1}{W_j} \tag{36}$$

For example, to determine influence coefficient describing influence of Cr on Ni in FeNiCr, the binary system NiCr is taken into account, and $R_{Ni}$ for this system has to be determined. The relative intensity can be determined based on actual measurements. In this case, the pure element specimens and suitable binary specimens must be available. Therefore, $R_i$ is usually calculated from theory using fundamental parameter method. In the first step, the composition of hypothetical binary specimen is assumed. In next step, $R_i$ is calculated from theory for actual measurement conditions (voltage of X-ray tube, incident and take-off angles, medium: air, helium or vacuum). Finally, $\alpha_{ij}$ is calculated from Eq. (36). The influence coefficients can be determined for any quantitative composition of binary systems. Table 1 shows influence coefficients calculated for various compositions of binary systems. The example is given for determination of Ni in ternary system FeNiCr. Thus, in this case the secondary fluorescence does not exist. As can be observed, if monochromatic excitation is applied, the coefficients are constants, i.e. they do not vary with composition. A different situation is observed for polychromatic excitation, where both $\alpha_{NiFe}$ and $\alpha_{NiCr}$ vary with composition of the specimen.

| $W_{Ni}$ | $W_{Fe}$ | $W_{Cr}$ | Monochromatic excitation, Mo $K_\alpha$ | | | Polychromatic excitation, Mo target X-ray tube operated at 45kV | | |
|---|---|---|---|---|---|---|---|---|
| | | | $R_{Ni}$ | $\alpha_{NiFe}$ | $\alpha_{NiCr}$ | $R_{Ni}$ | $\alpha_{NiFe}$ | $\alpha_{NiCr}$ |
| 0.2 | 0.8 | | 0.0561 | 3.21 | | 0.0740 | 2.13 | |
| 0.5 | 0.5 | | 0.1920 | 3.21 | | 0.2396 | 2.17 | |
| 0.8 | 0.2 | | 0.4873 | 3.21 | | 0.5510 | 2.26 | |
| 0.2 | | 0.8 | 0.0672 | | 2.47 | 0.0883 | | 1.58 |
| 0.5 | | 0.5 | 0.2238 | | 2.47 | 0.2759 | | 1.63 |
| 0.8 | | 0.2 | 0.5355 | | 2.47 | 0.5972 | | 1.70 |

Table 1. Influence coefficient calculated from $R_i$ for determination of Ni in ternary system FeNiCr using monochromatic and polychromatic excitation.

A similar situation is observed in determination of Fe in FeNiCr if polychromatic excitation is applied (Table 2). Both $\alpha_{FeCr}$ and $\alpha_{FeNi}$ are variable. The coefficient describing influence of Cr on Fe is positive because only absorption effect is observed in this case. The coefficient describing influence of Ni on Fe is negative because the enhancement effect dominates. Moreover, it strongly depends on specimen composition.

Summarizing, the influence coefficients can be treated as constants only if monochromatic excitation is applied and there are no enhancement effects. Fig. 6 presents results for determination of Cr in stainless steel before and after matrix correction using constant coefficients (e.g. Lachance-Traill algorithm) and variable linear coefficients (e.g. Claisse-Quintin algorithm). Before matrix correction, the mean error of Cr determination equals 1.11 % Cr. Considerable improvement is obtained after matrix correction with constant coefficient - the mean error is equal to 0.35 % Cr. Even better results are obtained for linear coefficients - the mean error equals 0.14 % Cr.

| $W_{Fe}$ | $W_{Cr}$ | $W_{Ni}$ | $R_{Fe}$ | $\alpha_{FeCr}$ | $\alpha_{FeNi}$ |
|------|------|------|--------|------|-------|
| 0.2 | 0.8 |     | 0.0717 | 2.24 |       |
| 0.5 | 0.5 |     | 0.2329 | 2.29 |       |
| 0.8 | 0.2 |     | 0.5405 | 2.40 |       |
| 0.2 |     | 0.8 | 0.3017 |      | -0.42 |
| 0.5 |     | 0.5 | 0.5985 |      | -0.33 |
| 0.8 |     | 0.2 | 0.8431 |      | -0.26 |

Table 2. Influence coefficient calculated from $R_i$ for determination of Fe in ternary system FeNiCr using polychromatic excitation: Mo target X-ray tube operated at 45kV.

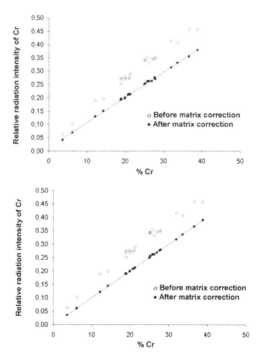

Fig. 6. Determination of Cr in stainless steel before and after matrix correction using (a) Lachance-Traill algorithm and (b) Claisse-Quintin algorithm (see subchapters 8.1 and 8.2).

The coefficients calculated from $R_i$ for binaries are the so-called binary coefficients, i.e. they describe influence of matrix element on analyte but they do not take into account other matrix elements. Therefore, applying binary coefficients to multielement samples leads to an incomplete matrix correction. It can be explained using simple example: the enhancement of Fe by Ni in two different matrices - Al and Cr matrix. In Cr matrix, Ni radiation is strongly absorbed by Cr. Therefore, Ni excites Fe to a lesser extent in Cr matrix than in Al matrix. In both cases, the same influence coefficient $\alpha_{FeNi}$ is used if binary coefficient algorithm is applied. Therefore, a better solution is the use of multielement coefficient algorithms, e.g. Lachance-Claisse, Broll-Tertain or Rousseau algorithm.

In general, the constant coefficient algorithms (e.g. Lachance-Traill algorithm) give satisfactory results but concentration range should be limited (0-10%). The algorithms using binary variable coefficients (e.g. Claisse-Quintin algorithm) give excellent results over medium concentration range of 0-40%. Multielement coefficient algorithms can be used in a full concentration range between 0 and 100%.

A number of influence coefficient algorithms have been proposed for correction of matrix effects. They can be divided in several ways: by the way of calculating influence coefficients (theoretical or empirical), by the variables used in matrix correction term (concentration or intensity based) or by the analytical context of determination of coefficients (constant or variable, binary or multielement). The general scheme of division and sub-divisions of influence coefficient algorithms is presented in Fig. 5.

In the next part of the chapter, various influence coefficient algorithms will be reviewed. The application of these algorithms for calibration and analysis of unknown specimen discussed here is valid for any of the algorithms. Some differences are observed in the use of theoretical and empirical algorithms during the calibration stage.

In the case of theoretical influence coefficient algorithms, the coefficients are determined from theory e.g. using $R_i$ calculated from Sherman equation for given measurement conditions. In the next stage, the matrix correction term (Eq. (24)) is calculated for all standard specimens and for a given analyte. Then, the calibration graph is plotted: the measured radiation intensity of the analyte multiplied by the corresponding matrix correction versus weight fraction of analyte. Then, slope $K_i$ and intercept $B_i$ are determined by least-squares fit using set of standard specimens.

In the case of empirical algorithms, the influence coefficients, slope $K_i$ and intercept $B_i$ are determined from multiple-regression analysis on a large suite of standards. If there are $n$ matrix elements, $n+2$ coefficients have to be determined ($n$ influence coefficients, slope $K_i$ and intercept $B_i$). For the calculation of $n+2$ coefficients, $n+2$ standards are required. In practice, a much larger number of standards is used: $2n$ or even $3n$. It should be noted, that the matrix element can be included in matrix correction term only if its influence on the analyte is significant. Otherwise, the matrix element should be omitted in matrix correction term to obtain correct values of other influence coefficients. Moreover, the concentration of all elements in the reference specimens should cover composition of unknown sample. Only then, accurate results can be obtained.

Analysis of unknown sample is similar for both theoretical and empirical algorithms. In this step, a set of equations has to be solved for the unknowns: $W_i, W_j, \ldots W_n$. If there are $n$ linear equations with $n$ unknowns, the set of equations can be solved algebraically. However, an iterative procedure is usually applied, especially in the case of a set of non-linear equations. General scheme of using theoretical influence algorithm is presented in Fig. 7.

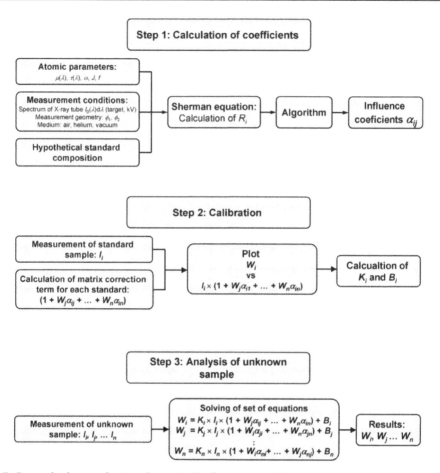

Fig. 7. General scheme of using theoretical influence algorithm.

## 8.1 Constant coefficient algorithms

The previous section shows that influence coefficients can be treated as constants only if: monochromatic excitation is applied and there are no enhancement effects. However, the monochromatic excitation is rarely used in practice and enhancement effects are usually observed in multielement specimens. Therefore, to obtain accurate results with constant coefficients, the concentration ranges of analyte and matrix elements should be limited. The constant coefficients are calculated for a given composition range or a series of reference samples that cover composition of unknown sample. If composition range is wide, then the sample can be diluted by pelletizing with a binder or by fusion.

**The Lachance-Traill algorithm (1966)**

$$W_i = R_i \left[ 1 + \sum_j \alpha_{ij} W_j \right] \qquad (37)$$

The algorithm was discussed in detail in the previous section. The influence coefficients can be determined from the relative radiation intensity or from multiple-regression analysis on a large suite of standards. They can also be calculated directly from the Eq. (31) if monochromatic excitation is applied and enhancement effects are not observed. If polychromatic beam is applied, then the effective wavelength can be used, however it is also composition dependent.

**The de Jongh algorithm (1973)**

$$W_i = E_i R_i \left[ 1 + \sum_{\substack{j \\ j \neq e}} \alpha_{ij} W_j \right]$$

(38)

Where $E_i$ is a constant determined during calibration. The algorithm proposed by de Jongh (de Jongh, 1973) uses theoretical coefficients calculated from fundamental parameters. This algorithm looks similar to the Lachance-Traill equation but the eliminated element $e$ is the same for all equations (in Lachance-Traill algorithm, the analyte is eliminated from each equation, i.e. $\alpha_{ii} = 0$). The influence coefficients are calculated by a Taylor series expansion around an average composition. They are multielement coefficients rather than binary coefficients, i.e. the influence coefficient describes influence of matrix element $j$ on analyte $i$ in the presence of all matrix elements. The advantage of the de Jongh algorithm is that one element can be arbitrarily eliminated from the correction procedure, so there is no need to measure it. For example, in ferrous alloys, iron is often the major element and it is usually determined by difference, and therefore, can be eliminated from the correction procedure.

**The Rasberry-Heinrich algorithm (1974)**

$$W_i = R_i \left[ 1 + \sum_{j} a_{ij} W_j + \sum_{k} \frac{b_{ik}}{1 + W_i} W_k \right]$$

(39)

In this algorithm, two different coefficients are used for correction of absorption effect (coefficient $a_{ij}$) and enhancement effect (coefficient $b_{ij}$) and only one coefficient is used for each matrix element. The coefficient $a_{ij}$ is used if absorption is the dominant effect (then $b_{ik} = 0$). The coefficients $b_{ik}$ are used if enhancement is the dominant effect (then $a_{ij} = 0$). The coefficients are determined experimentally on the basis of measured reference samples. The serious disadvantage of the algorithm is the fact that it is not clear which matrix element should be assigned an $a_{ij}$ and which one a $b_{ik}$. Therefore, the Rasberry-Heinrich algorithm is not considered to be generally applicable, but it gives satisfactory results for FeNiCr alloys. It is for this system that the algorithm was originally developed.

**The Lucas-Tooth and Price algorithm (1964)**

$$W_i = I_i \left[ k_i + \sum_{j} a_{ij} I_j \right] + B_i$$

(40)

Where $k_i$ and $a_{ij}$ are the correction coefficients, $B_i$ is a background term. The coefficients are determined by a least-squares fit on the basis of measured reference samples. The correction is done using intensities rather than concentrations. The algorithm assumes that the matrix effect of an element $j$ on the analyte $i$ is proportional to its intensity. Because this approximation is unreliable (the measured intensities are modified by matrix effects and are not directly proportional to concentrations), the algorithm has a limited range of applicability. However, determination of a single element is possible using Lucas-Tooth and Price algorithm.

## 8.2 Variable binary coefficient algorithms

The Claisse-Quintin algorithm (1967)

$$W_i = R_i \left[ 1 + \sum_j \alpha_{lin,ij} W_j + \sum_j \sum_k \alpha_{ijk} W_j W_k \right]$$

(41)

$$\alpha_{lin,ij} = \alpha_{ij} + \alpha_{ijj} W_M$$

In 1967, Claisse and Quintin proposed an improved version of Lachance-Traill algorithm (Claisse and Quintin, 1967). The algorithm uses linear coefficients $\alpha_{lin,ij}$ instead of constant coefficients and can be applied for polychromatic excitation in a medium range of concentration. The cross-product coefficient $\alpha_{ijk}$ included in matrix correction term corrects for the simultaneous presence of both $j$ and $k$ and compensates for the fact that the total matrix effect correction cannot be represented as a sum of binary matrix effect corrections. $W_M$ is matrix concentration ($W_M = 1 - W_i$) proposed by Tertian instead of $W_j$ in the original equation in order to obtain high quality results (Tertian, 1976). Equations for calculating accurate and valid theoretical binary influence coefficients in the Claisse–Quintin algorithm were first proposed by Rousseau and Claisse (1974) and later improved by Rousseau (1984 b). The influence coefficients are calculated from $R_i$ using fundamental parameters at two binaries ($W_i = 0.2$ and $0.8$). The algorithm introduces a theoretical mean relative error of 0.04% on the calculated concentrations in the case of cement samples prepared as pressed powder pellets (medium concentration range of 0 - 40%). For a large concentration range (0–100%), e.g. in the case of alloy analysis, the Claisse–Quintin algorithm introduces 0.3% relative error (Rousseau, 2001).

The COLA algorithm (1981)

$$W_i = R_i \left[ 1 + \sum_j \alpha_{hyp,ij} W_j + \sum_j \sum_k \alpha_{ijk} W_j W_k \right]$$

(42)

$$\alpha_{hyp,ij} = \alpha_{ij,1} + \frac{\alpha_{ij,2} W_M}{1 + \alpha_{ij,3}(1 - W_M)}$$

In 1981, Lachance proposed an algorithm called COLA (comprehensive Lachance) with a new approximation to the binary influence coefficients (Lachance, 1981). This time, hyperbolic influence coefficients were applied. The influence coefficients are calculated from $R_i$ using fundamental parameters at three binaries ($W_i = 0.001$, $0.5$ and $0.999$). Similarly to the Claisse-Quintin algorithm, the cross-product coefficients are calculated from a ternary system ($W_i = 0.30$, $W_j = 0.35$, $W_k = 0.35$). The applicability of COLA algorithm is similar to the Claisse-Quintin algorithm. The algorithm corrects for both absorption and enhancement effects over a broad range of concentration.

## 8.3 Variable multielement coefficient algorithms

Binary influence coefficients are based on an approximation: the total matrix effect on the analyte $i$ equals the sum of the effects of each matrix element j and each effect is calculated independently of each other. The binary coefficient algorithms give accurate results if composition range is limited, i.e. 0-10% for the constant coefficient algorithms and 0-40% for binary variable coefficient algorithms. The coefficients are calculated for a given composition range rather than for a given sample composition. The multielement coefficients are exact and are calculated for each individual reference and unknown specimens. Therefore, the multielement coefficient algorithms are applicable in full composition range from 0 to 100%. The most important algorithms were proposed by Broll and Tertian (1983), Rousseau (1984), and Lachance and Claisse (1995).

**The Broll-Tertain algorithm (1983)**

$$W_i = R_i \left[ 1 + \sum_j \left( \alpha_{ij} - \varepsilon_{ij} \frac{W_i}{R_i} \right) W_j \right] \tag{43}$$

**The Rousseau algorithm (1984)**

$$W_i = R_i \frac{1 + \sum_j \alpha_{ij} W_j}{1 + \sum_j \varepsilon_{ij} W_j} \tag{44}$$

**The Lachance-Claisse algorithm (1995)**

$$W_i = R_i \left[ 1 + \sum_j \frac{\alpha_{ij} - \varepsilon_{ij}}{1 + \sum_j \varepsilon_{ij} W_j} W_j \right] \tag{45}$$

Where $\alpha_{ij}$ and $\varepsilon_{ij}$ are influence coefficients correcting for absorption and enhancement effects, respectively. The multielement influence coefficients can be easily calculated for reference specimens of known composition. In the case of unknown specimens, the iteration is used. In Rousseau algorithm, a first estimate of the composition of the unknown specimen is calculated using the Claisse-Quintin algorithm, then the influence coefficients are calculated for this composition. An exact estimate of composition is finally obtained by applying the iterative process to Eq. (44). Willis and Lachance (2004) showed that the aforementioned multielement coefficient algorithms give the same high-quality results and no algorithm can outperform any of the others.

## 8.4 Influence coefficient algorithms for intermediate-thickness samples

The influence coefficients algorithms are widely applied for quantitative analysis of infinitely thick samples. Influence coefficients algorithms can be deduced from the Sherman equation because exponential term equals zero for thick samples. In consequence, the matrix correction term can be expressed as a linear combination of matrix elements concentrations. The analysis of intermediate-thickness samples is much more complicated because the exponential term in Eq. (9) cannot be neglected. However, Sitko (2005) proposed a certain approximation to eliminate the exponential term:

$$1 - \exp(-x) = \frac{1}{k_1 + k_0/x} \tag{46}$$

Where $k_1$ and $k_0$ are constants, $x = \chi(\lambda, \lambda_i)\rho t$. The proposed approximation allows not only for eliminating the exponential term but also transforming Sherman equation to a form in which the matrix correction term is expressed as a linear combination of all matrix elements concentrations. Thus, the proposed approximation was a base for the first empirical influence coefficient algorithm dedicated for the analysis of intermediate-thickness samples (Sitko, 2005):

$$W_i m = I_i \left[ a_i + c_i m + \sum_j \alpha_{ij} W_j m \right] + B_i \tag{47}$$

Where $W_i m$ and $W_j m$ are the masses per unit area of the analyte $i$ and the matrix element $j$, respectively; $\alpha_{ij}$ is the influence coefficient; $a_i$ and $c_i$ are the coefficients dependent on the range of thickness of calibration samples. The coefficients $a_i$, $c_i$, $\alpha_{ij}$ and $B_i$ are calculated by the least-squares fit on the basis of reference samples of various thickness and composition. The influence coefficients in this algorithm are treated as constants. Therefore, to obtain accurate results, sample thickness and composition range should be limited. To overcome this inconvenience, the theoretical influence coefficient algorithm for intermediate-thickness samples was proposed (Sitko, 2006 b):

$$W_i m = K_i I_i \left[ a_i + c_i m + \sum_j \alpha_{ij} W_j m \right] + B_i$$

$$\alpha_{ij} = \frac{\alpha_{ij} c_i - S_{ij}\left(a_i/m + c_i\right)}{1 + \sum_j S_{ij} W_j} \tag{48}$$

Where $K_i$, $B_i$ are the calibration constants calculated by the least-squares fit on the basis of experimental results for reference samples; $a_i$ and $c_i$ are the coefficients dependent on the thickness and total mass-attenuation coefficient of the sample; $\alpha_{ij}$ is the absorption influence coefficient; $S_{ij}$ is the enhancement term. The coefficients $a_i$, $c_i$, $\alpha_{ij}$ and $S_{ij}$ are multielement coefficients and are calculated directly from theory. The proposed theoretical influence algorithm can be applied in analysis of samples of any thickness and is more general in its form than algorithms for thick samples. It is worth emphasizing that the algorithm takes the well-known form for thick samples, e.g. the Lachance-Claisse algorithm (if sample thickness $\rightarrow \infty$ then $a_i \rightarrow 0$ and $c_i \rightarrow 1$). The theoretical influence algorithm is more flexible than the previously proposed empirical algorithm and can be applied for a wide range of thickness and composition.

In the next paper (Sitko, 2008b), two algorithms of constant and linear coefficients for simultaneous determination of composition and thickness of thin films were proposed. This time, the coefficients are not calculated directly from theory but from the relative radiation intensity (calculated from theory) of hypothetical pure element films and binary films. The potential of the algorithms was demonstrated with hypothetical ternary and binary systems: FeCrNi, FeCr, FeNi, CrNi and experimental data of FeNi and Cu films.

## 9. Conclusions

Many methods, both empirical and theoretical, have been proposed for quantitative XRF analysis. Method selection usually depends on sample type (thin or bulk, alloys or rocks, etc.), method of sample preparation (without treatment, fusion, etc.), expected results (quantitative or semi-quantitative analysis, determination of a single element or multielement analysis) and availability of standard samples. If the compensation methods are applied, then complicated sample preparation is required, only one or few elements can be quantitatively determined (standard addition, internal standard, Compton scatter) and additional matrix correction may be required (dilution). In influence coefficient algorithm, the matrix correction and calibration is clear, therefore, interpretation of data is very easy. The fundamental parameter methods are like "a black box". Nevertheless, the serious advantage is their versatility: the analysis of bulk, thin samples and multilayers is possible. Fundamental parameter methods are usually considered to be less accurate than the influence coefficient algorithms. This results from the fact that the fundamental parameter methods are usually used with only a few standards. The accuracy of fundamental parameter methods is very similar to that of influence coefficient algorithms when the same standards (many standards similar to the unknown) are used in both cases.

## 10. References

Araujo, M.F.; Van Espen, P.; Van Grieken, R. (1990). Determination of sample thickness via scattered radiation in X-ray fluorescence spectrometry with filtered continuum excitation, *X-Ray Spectrometry*, Vol. 19, pp. 29–33.

Bao, S.X. (1997). A power function relation between mass attenuation coefficient and Rh Kα Compton peak intensity and its application to XRF analysis, *X-Ray Spectrometry*, Vol. 26, pp. 23-27.

Broll, N.; Tertian, R. (1983). Quantitative X-ray fluorescence analysis by use of fundamental influence coefficients, *X-ray Spectrometry*, Vol. 12, pp. 30–37.

De Boer, D.K.G. (1990). Calculation of X-ray fluorescence intensities from bulk and multilayer samples, *X-Ray Spectrometry*, Vol. 19, pp. 145–154.

De Boer, D.K.G.; Borstrok, J.J.M.; Leenaers, A.J.G.; Van Sprang, H.A.; Brouwer, P.N. (1993). How accurate is the fundamental parameter approach? XRF analysis of bulk and multilayer samples, *X-Ray Spectrometry*, Vol. 22, pp. 33–38.

Claisse, F.; Quintin, M. (1967). Generalization of the Lachance–Traill method for the correction of the matrix effect in X-ray fluorescence analysis, *Canadian Journal of Spectroscopy*, Vol. 12, pp. 129–134.

de Jongh, W.K. (1973). X-ray fluorescence analysis applying theoretical matrix correction. Stainless steel, *X-ray Spectrometry*, Vol. 2, pp. 151–158.

de Vries, J.L.; Vrebos, B.A.R. (2002). Quantification of infinitely thick specimen by XRF analysis, in: R. Van Grieken, A. Markowicz (Eds.), Handbook of X-Ray Spectrometry, 2nd ed., Marcel Dekker, New York, 2002, pp. 341–405.

Ebel, H. (1999). X-ray tube spectra, *X-Ray Spectrometry*, Vol. 28, pp. 255–266.

Ebel, H.; Svagera, R.; Ebel, M.F.; Shaltout, A.; Hubbell, J.H. (2003). Numerical description of photoelectric absorption coefficients for fundamental parameter programs, *X-Ray Spectrometry*, Vol. 32, pp. 442–451.

Elam, W.T.; Ravel, B.D.; Sieber, J.R. (2001). A new atomic database for X-ray spectroscopic calculations, *Radiation Physics and Chemistry*, Vol. 63, pp. 121–128.

Elam, W.T.; Shen, R.B.; Scruggs, B.; Nicolosi, J. (2004). Accuracy of standardless FP analysis of bulk and thin film samples using a new atomic database, *Advances in X-Ray Analysis*, Vol. 47, pp. 104–109.

Finkelshtein, A.L.; Pavlova, T.O. (1999). Calculation of X-ray tube spectral distributions, *X-Ray Spectrometry*, Vol. 28, pp. 27–32.

Giauque, R.D. (1994). A novel method to ascertain sample mass thickness and matrix effects for X-ray fluorescence element determinations, *X-ray Spectrometry*, Vol. 23, pp. 160–168.

He, F.; Van Espen, P.J. (1991). General approach for quantitative energy dispersive X-ray fluorescence analysis based on fundamental parameters, *Analytical Chemistry*, Vol. 63, pp. 2237–2244.

Lachance, G.R. (1981). The role of alpha coefficients in X-ray Spectrom, Paper presented at the International Conference on Industrial Inorganic Elemental Analysis, Metz, France, 3 June 1981.

Lachance, G.R.; Claisse, F. (1995). Quantitative X-ray Fluorescence Analysis, Theory and Application, John Wiley & Sons.

Lachance, G.R.; Traill, R.J. (1966). A practical solution to the matrix problem in X-ray analysis, *Canadian Journal of Spectroscopy*, Vol. 11, pp. 43–48.

Lucas-Tooth, H.; Pyne, C. (1964). The accurate determination of major constituents by X-ray fluorescent analysis in the presence of large interelement effects, *Advances in X-ray Analysis*, Vol. 7, pp. 523–541.

Mantler, M. (1986). X-ray fluorescence analysis of multiple-layer films, *Analytica Chimica Acta*, Vol. 188, pp. 25–35.

Mantler, M.; Kawahara, N. (2004). How accurate are modern fundamental parameter methods? *The Rigaku Journal*, Vol. 21, pp. 17–25.

Markowicz, A.; Haselberger, N. (1992). A modification of the emission–transmission method for the determination of trace and minor elements by XRF, *Applied Radiation and Isotopes*, Vol. 43, pp. 777–779.

Markowicz, A.; Van Grieken, R.E. (2002). Quantification in XRF analysis of intermediate thickness samples, in: R. Van Grieken, A. Markowicz (Eds.), Handbook of X-Ray Spectrometry, 2nd ed., Marcel Dekker, New York, 2002, pp. 407–431.

Markowicz, A.; Haselberger, N.; Mulenga, P. (1992). Accuracy of calibration procedure for energy-dispersive X-ray fluorescence spectrometry, *X-Ray Spectrometry*, Vol. 21, pp. 271–276.

Nesbitt, R.W.; Mastins, H.; Stolz, G.W.; Bruce, D.R. (1976). Matrix corrections in trace-element analysis by X-ray fluorescence: An extension of the Compton scattering technique to long wavelengths, *Chemical Geology*, Vol. 18, pp. 203-213

Nielson, K.K. (1977). Matrix correction for energy dispersive X-ray fluorescence analysis of environmental samples with coherent/incoherent scattered X-rays, *Analytical Chemistry*, Vol. 49, pp. 641–648.

Pella, P.A.; Feng, L.; Small, J.A. (1985). An analytical algorithm for calculation of spectral distributions of X-ray tubes for quantitative X-ray fluorescence analysis, *X-Ray Spectrometry*, Vol. 14, pp. 125–135.

Pella, P.A.; Feng, L.; Small, J.A. (1991). Addition of M- and L-series lines to NIST algorithm for calculation of X-ray tube output spectral distributions, X-Ray Spectrometry, Vol 20, pp. 109–110.

Rasberry, S.D.; Heinrich, K.F.J. (1974). Calibration for interelement effects in X-ray fluorescence analysis, *Analytical Chemistry*, Vol. 46, pp. 81–89.

Rousseau, R.M. (1984 a). Fundamental algorithm between concentration and intensity in XRF analysis. 1: theory, *X-Ray Spectrometry*, Vol. 13, pp. 115–120.

Rousseau, R.M. (1984 b). Fundamental algorithm between concentration and intensity in XRF analysis: 2. Practical application, X-Ray Spectrometry, Vol. 13, pp. 121–125.

Rousseau, R.M. (2001). Concept of the influence coefficient, *The Rigaku Journal*, Vol. 18, pp. 8–21.

Rousseau, R.M. (2002). Debate on some algorithms relating concentration to intensity in XRF spectrometry, *The Rigaku Journal*, Vol. 19, pp. 25–34.

Rousseau, R.M. (2004). Some considerations on how to solve the Sherman equation in practice, *Spectrochimica Acta Part B*, Vol. 59, pp. 1491–1502.

Rousseau, R.M. (2006). Corrections for matrix effects in X-ray fluorescence analysis - A tutorial, *Spectrochimica Acta Part B*, Vol. 61, pp. 759–777

Rousseau, R.; Claisse, F. (1974). Theoretical alpha coefficients for the Claisse–Quintin relation for X-ray spectrochemical analysis, *X-Ray Spectrometry*, Vol. 3, pp. 31–36.

Sherman, J. (1955). The theoretical derivation of fluorescent X-ray intensities from mixtures, *Spectrochimica Acta*, Vol. 7, pp. 283–306.

Shiraiwa, T.; Fujino, N. (1966). Theoretical calculation of fluorescentX-ray intensities in fluorescent X-ray spectrochemical analysis, *Japan Journal of Applied Physics*, Vol. 5, pp. 886–899.

Sitko, R. (2005). Empirical coefficients models for X-ray fluorescence analysis of intermediate-thickness samples, *X-Ray Spectrometry*, Vol. 34, pp. 11–18.

Sitko, R.; Zawisza, B. (2005). Calibration of wavelength-dispersive X-ray spectrometer for standardless analysis, *Spectrochim. Acta Part B*, Vol. 60, pp. 95–100.

Sitko, R. (2006a). Correction of matrix effects via scattered radiation in X-ray fluorescence analysis of samples collected on membrane filters, *Journal of Analytical Atomic Spectrometry*, Vol. 21, pp. 1062–1067.

Sitko, R. (2006b). Theoretical influence coefficients for correction of matrix effects in X-ray fluorescence analysis of intermediate-thickness samples, *X-Ray Spectrometry*, Vol. 35, pp. 93–100.

Sitko, R. (2007). Influence of X-ray tube spectral distribution on uncertainty of calculated fluorescent radiation intensity, *Spectrochimica Acta Part B*, Vol. 62, pp 777–786.

Sitko, R. (2008a). Study on the influence of X-ray tube spectral distribution on the analysis of bulk samples and thin films: fundamental parameters method and theoretical coefficient algorithms, *Spectrochimica Acta Part B*, Vol. 63, pp. 1297–1302.

Sitko, R. (2008b). Determination of thickness and composition of thin films by X-ray fluorescence spectrometry using theoretical influence coefficient algorithms, *X-Ray Spectrometry*, Vol. 37, pp. 265–272.

Sitko, R. (2009). Quantitative X-ray fluorescence analysis of samples of less than 'infinite thickness': Difficulties and possibilities, *Spectrochimica Acta Part* B, Vol. 64, pp. 1161-1172

Szalóki, I.; Somogyi, A.; Braun, M.; Tóth, A. (1999). Investigation of geochemical composition of lake sediments using ED-XRF and ICP-AES techniques, *X-Ray Spectrometry*, Vol. 28, pp. 399–405.

Tertian, R. (1976). An accurate coefficient method for X-ray fluorescence analysis, *Advances in X-Ray Analysis*, Vol. 19, pp. 85–111.

Van Dyck, P.M.; Török, S.B.; Van Grieken, R.E. (1986). Enhancement effect in X-ray fluorescence analysis of environmental samples of medium thickness, *Analytical Chemistry*, Vol. 58, pp. 1761–1766.

Vrebos, B.A.R. (2006). Compensation methods, in: B. Beckhoff, B. Kanngießer, N. Langhoff, R. Wedell and H. Wolff (Eds.), Handbook of Practical X-Ray fluorescence Analysis, 1st ed., Springer, Berlin, 2006, pp. 358–369.

Węgrzynek, D.; Hołyńska, B.; Pilarski, T. (1993). The fundamental parameter method for energy-dispersive X-ray fluorescence analysis of intermediate thickness samples with the use of monochromatic excitation, *X-Ray Spectrometry*, Vol. 22, pp. 80–85.

Węgrzynek, D.; Markowicz, A.; Chinea-Cano, E. (2003a). Application of the backscatter fundamental parameter method for in situ element determination using a portable energy-dispersive X-ray fluorescence spectrometer, *X-Ray Spectrometry*, Vol. 32, pp. 119–128.

Węgrzynek, D.; Markowicz, A.; Chinea-Cano, E.; Bamford, S. (2003b). Evaluation of the uncertainty of element determination using the energy-dispersive X-ray fluorescence technique and the emission–transmission method, *X-Ray Spectrometry*, Vol. 32, pp. 317–335.

Willis, J.E. (1989). Enhancement between layers in multiple-layer thin-film samples, *X-Ray Spectrometry*, Vol. 18, pp. 143–149.

Willis, J.P.; Lachance, G.R. (2000). Resolving apparent differences in mathematical expressions relating intensity to concentration in X-ray fluorescence spectrometry, *The Rigaku Journal*, Vol. 17, pp. 23-33.

Willis, J.P.; Lachance, G.R. (2004). Comparison between some common influence coefficient algorithms, *X-Ray Spectrometry*, Vol. 33, pp. 181–188.

# 9

# The Use of Electron Probe MicroAnalysis to Determine the Thickness of Thin Films in Materials Science

Frédéric Christien, Edouard Ferchaud,
Pawel Nowakowski and Marion Allart
*Polytech'Nantes, University of Nantes*
*France*

## 1. Introduction

Electron Probe MicroAnalysis (EPMA) was born around 1950 when Raymond Castaing, a French graduate student working under the supervision of André Guinier, built his first microanalyser (Castaing & Guinier, 1950; Castaing, 1951; Grillon & Philibert, 2002). The principle of EPMA is to bombard the sample surface with a focused electron beam and to collect the X-rays emitted from the sample. The X-rays are dispersed using Bragg diffraction on a mobile monochromator, which enables to get the whole X-ray spectrum from zero to more than 10 keV. The technique of dispersion is called WDS (Wavelength Dispersive X-ray Spectroscopy). The spectrum is made of continuous background (Bremstrahlung emission) and characteristic peaks, which allow elemental qualitative and quantitative analysis of the material. EPMA-WDS can be carried out in dedicated instruments called "microprobes" (usually fitted with 4 WDS spectometers) or in high current Scanning Electron Microscopes (SEM) equipped with one single WDS spectrometer.

More recently, Energy Dispersive X-ray Spectroscopy has been developed and has nowadays become a cheap and widespread technique. Most SEM in materials science laboratories are equipped with an EDS spectrometer. EPMA-EDS performances are far below those of EPMA-WDS with regards to sensitivity and spectral resolution, although it can achieve very good qualitative and quantitative results in many cases.

EPMA quantification of the sample composition is usually possible by using a standard material of known composition. Unfortunately there is in general no direct proportionality between the concentration of an element and the X-ray emission intensity (peak height) coming from this element. Since the early years of EPMA, many models have been proposed in the literature to correlate the concentration of an element and the associated X-ray emission intensity: ZAF (Philibert & Tixier, 1968), MSG (Packwood & Brown, 1981), PAP and XPP (Pouchou & Pichoir, 1987; Pouchou et al., 1990)… The XPP model (implemented in several microanalysis software packages) has now reached a good level of maturity and reliability, even in difficult situations involving light elements and strong bulk absorption. It is based on an accurate calculation of the $\xi(\rho z)$ curve, which describes the depth-dependence of the X-ray emission intensity of a particular line in the sample. In EPMA analysis of homogeneous bulk samples, the correlation between X-ray emission intensity and the concentration of an element A in the sample is given by:

$$\frac{I}{I_{Std}} = \frac{X}{X_{Std}} \frac{\int\limits_{\rho z=0}^{\infty} \xi(\rho z)d\rho z}{\int\limits_{\rho z=0}^{\infty} \xi_{Std}(\rho z)d\rho z} \tag{1}$$

where $I$ is the X-ray intensity of a particular line of element A (for example $K\alpha$), $I_{Std}$ is the intensity of the same line measured on a standard material containing a known and homogeneous concentration of element A, $z$ is the depth (cm), $\rho$ is the material density (g cm$^{-3}$), $\rho z$ is the mass thickness (g cm$^{-2}$), i.e. the mass per unit area of a layer of thickness $z$, $X$ is the weight fraction of element A in the sample, $X_{Std}$ is the weight fraction of element A in the standard material, and $\xi(\rho z)$ and $\xi_{Std}(\rho z)$ are dimensionless functions that describe the depth distribution of the chosen X-ray line emerging intensity respectively for the sample and for the standard material. The $\xi$ functions depends on the X-ray line chosen for analysis, the matrix composition, the beam voltage and the orientation of the sample surface with respect to the primary beam and the spectrometer. For example, Fig. 1 shows the $\xi$ functions for the palladium $L\alpha$ line calculated using the XPP model for different matrix compositions (pure palladium and pure nickel) and different accelerating voltages (5, 10 and 20 kV). The calculation was made using the Stratagem™ software (http://www.samx.com) in which the XPP model is implemented (Pouchou, 1993). The influence of matrix composition and beam voltage on the $\xi$ function is demonstrated.

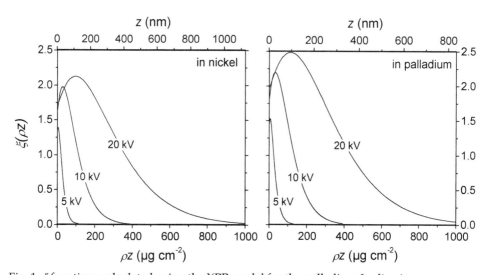

Fig. 1. $\xi$ functions calculated using the XPP model for the palladium $L\alpha$ line in pure palladium and in pure nickel (containing traces of palladium) at 5, 10 and 20 kV.

From Eq. (1), the correlation between elemental weight fraction and X-ray emission intensity can be determined. Let us consider for example the simple case of a nickel-palladium binary alloy (this system was considered here because nickel and palladium are totally miscible in the solid state and it is possible to get homogenous Ni-Pd alloys of any composition from pure nickel to pure palladium). Fig. 2 shows the composition dependence of the Pd Lα k-ratio ($I/I_{Std}$) at 20 kV. The standard material was assumed to be pure palladium here. It can be observed from Fig. 2 that there is no simple proportionality between the Pd Lα k-ratio and the Pd weight fraction. The reason for this is that the $\xi$ function in Eq. (1) depends on the composition.

In general, the composition of a sample analysed by EPMA is not known a priori. It is then not possible to accurately calculate the $\xi$ function, which is needed to correlate the elemental weight fraction and the k-ratio using Eq. (1). An iterative calculation is then needed: in a first step, the elemental weight fraction is assumed equal to the measured k-ratio (1st iteration), which enables to do a first estimation of the $\xi$ function. Then, the elemental weight fraction can be re-estimated using Eq. (1) (2nd iteration). This iterative process is carried out again until it converges to the correct value of the weight fraction (in most cases, a few iterations are sufficient).

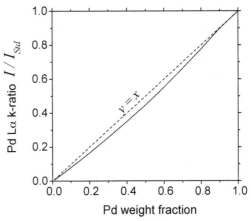

Fig. 2. Composition dependence of the Pd Lα k-ratio ($I/I_{Std}$) in the Ni-Pd system at 20 kV calculated from Eq. (1).

EPMA is usually presented as a "bulk" analysis technique, as the typical depth of analysis is about 1 µm. It should be mentioned that this 1 µm depth of analysis is only "typical" as it actually strongly depends on both the beam voltage and on the material density (see Fig. 1). Furthermore, EPMA is also very sensitive to "surface" composition and can be used to derive quantitative data such as the thickness and composition of very thin films, down to the monolayer range. The "difficulty" of EPMA applied to thin films is that the probed volume is larger than the film thickness. Since the late eighties, dedicated models have been proposed to overcome that difficulty and to derive thin film thicknesses and compositions from simple and non destructive EPMA measurements (Bastin et al., 1990; Pouchou & Pichoir, 1991; Pouchou, 1993; Pouchou & Pichoir, 1993; Staub et al., 1998; Llovet & Merlet, 2010).

It can be observed nevertheless that EPMA has remained rather underused for the study of thin films with regards to its good sensitivity and quantitative capabilities, real "surface" techniques such as Auger spectroscopy or XPS being often preferred to EPMA. The aim of this paper is to demonstrate the quantitative capabilities of EPMA (associated with either EDS or WDS) for the thickness determination of thin films in materials science.

## 2. EPMA-EDS: A fast, cheap and simple tool for the characterization of surface thin films from 10 to 1000 nm

### 2.1 Principle

Let us consider a simple example: a substrate of pure nickel coated with a pure copper layer whose thickness is smaller than the depth of analysis (Fig. 3). In that situation, both nickel and copper X-rays are detected on the EDS spectrum. Assuming that the standard material is pure copper, the Cu k-ratio is given by:

$$\frac{I}{I_{Std}} = \frac{\int_{\rho z=0}^{\mu} \xi(\rho z)d\rho z}{\int_{\rho z=0}^{\infty} \xi_{Std}(\rho z)d\rho z} \tag{2}$$

where $\mu$ is the copper film mass thickness (g cm$^{-2}$). $\xi(\rho z)$ and $\xi_{Std}(\rho z)$ describe the Cu K$\alpha$ emission respectively in the Cu/Ni stratified sample and in the standard material (pure copper). Fig. 4 shows the beam voltage dependence of the Cu K$\alpha$ line k-ratio for several copper mass thicknesses $\mu$. The calculation was carried out using the Stratagem™ software (http://www.samx.com) based on the XPP model (Pouchou, 1993). In such a simple situation (film of known composition on a substrate of known composition), a single voltage approach is sufficient, i.e. $\mu$ can be determined from the k-ratio measured at one voltage only. The voltage must be high enough so that the depth of analysis should be higher than the film thickness, otherwise the k-ratio is always equal (or close) to 1 no matter how thick the film is, and the determination of $\mu$ is not possible.

Once $\mu$ is determined, the copper thickness can be worked out from Eq. (3), assuming that the density $\rho$ (g cm$^{-3}$) of the film is known:

$$e = \frac{\mu}{\rho} \tag{3}$$

Examples of thin film thickness determination by EPMA-EDS are given in the following sections. Unless otherwise stated, the experimental data were acquired on a Leo S440 SEM equipped with an INCA EDS system from Oxford Instruments.

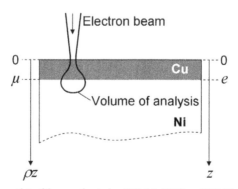

Fig. 3. Principle of surface thin film analysis by EPMA-EDS or EPMA-WDS.

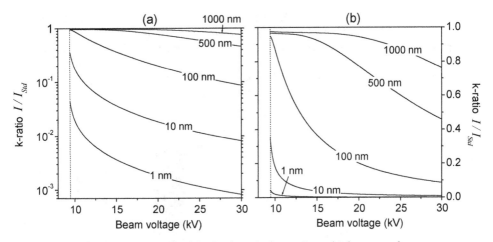

Fig. 4. Beam voltage dependence of Cu Kα k-ratio for various thicknesses of copper deposited on a nickel substrate. (a) logarithmic k-ratio scale. (b) linear k-ratio scale. The dotted vertical line is the copper K level energy below which no Kα emission is observed.

## 2.2 EPMA-EDS determination of alumina thickness on aluminium samples

Aluminium is a highly oxidizable metal: any aluminium sample is always covered by an alumina ($Al_2O_3$) layer. EPMA-EDS was used here to determine the alumina thickness on two aluminium samples. The first one was annealed at 500°C under air and a quite thick oxide layer is expected. The second one was stripped using nitric acid. EPMA-EDS analyses were carried out at 20 kV on 20×20 μm areas (rastering beam) throughout the aluminium sample. Fig. 5a shows the typical EDS spectra measured on both samples. The effect of stripping on the oxide layer and hence the oxygen Kα intensity is clearly demonstrated. The case depicted here is simple (film of known composition on a substrate of known composition) and a single voltage approach was used to determine the oxide layer. The quantification equation is the same as Eq. (2). Ten EDS spectra were acquired on each sample and an average oxygen k-ratio was determined using an $Al_2O_3$ standard material. Fig. 5b shows the oxygen Kα k-ratio measured at 20 kV for the two samples, as well as the corresponding oxide mass thickness $\mu$ determined from Eq. (2) using the Stratagem™ software, and the oxide thickness $e$ calculated from Eq. (3) assuming an alumina density $\rho = 3.96$ g cm$^{-3}$ (Bauccio, 1994). The oxide mass thicknesses found on the annealed sample and on the stripped one are 18.2 μg cm$^{-2}$ and 4.0 μg cm$^{-2}$ respectively, which corresponds to thicknesses of 46 nm and 10 nm.

It is clear from the spectra shown in Fig. 5a that even for an oxide layer as thin as 10 nm, the peak-to-background ratio of oxygen Kα line is still about one, which is undoubtedly above the limit of detection of EDS.

To ensure the reliability of the EPMA-EDS thickness measurements, XPS depth profiling was undertaken on the annealed aluminium using a ThermoVG Thetaprobe spectrometer equipped with a hemispherical analyser and a monochromatic AlKα primary beam (spot diameter of 400 μm). The sputtering sequences were carried out using 3 keV argon ions with an incidence angle of 45° onto a 4 mm$^2$ rastered area with a 9 A cm$^{-2}$ current density. Both oxide and metal components of the Al2p peak were acquired over sputtering time. Sputtering time was converted to depth using a sputtering rate previously measured on another aluminium sample. The depth profiles are shown in Fig. 6. No metal aluminium peak is detected within the first ~40 nm, which corresponds to the oxide thickness (the very smooth transition from pure oxide to pure metal, i.e. from ~40 to >~500 nm depth, observed in Fig. 6 is due to the sample roughness). It can be concluded that the oxide layer thickness determined on the annealed aluminium by the two techniques are in very good agreement: ~40 nm and 46 nm by XPS and EPMA-EDS respectively.

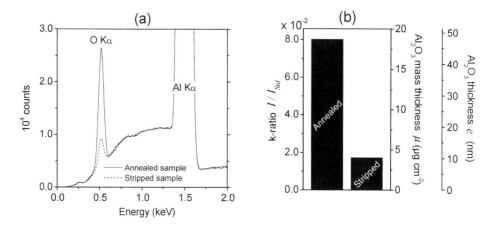

Fig. 5. (a) EDS spectra acquired on aluminium samples at 20 kV. The first sample was annealed at 500°C in air. The second one was stripped in nitric acid. (b): O Kα k-ratio ($I / I_{Std}$), oxide mass thickness ($\mu$) and oxide thickness ($e$) measured on the same two aluminium samples by EDS.

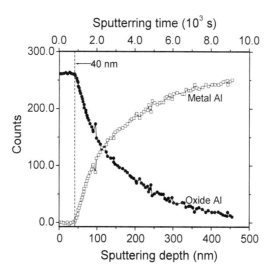

Fig. 6. Confirmation of the alumina thickness on the annealed aluminium by XPS depth profiling. A thickness of ~40 nm is found.

## 2.3 EPMA determination of gallium thickness on rough gallium-coated aluminium samples

The second example deals with gallium "thick" films deposited on aluminium substrates for brazing applications (Ferchaud, 2010). Solid gallium films were obtained by vapour deposition under vacuum. The typical film surface morphology is shown in Fig. 7a. It can be observed that the film thickness is not homogeneous at the micrometer scale. Thirteen aluminium samples (4x10x40 mm³) were coated with gallium using different deposition times. For each sample, the Ga film thickness was determined from the film mass measured using a high precision balance (0.01 mg) and from EPMA-EDS analysis. The EPMA-EDS analysis conditions were the same as those depicted in the previous section. Fig. 7b is a comparison of the gallium thicknesses determined by the two techniques. It can be observed that the EPMA-EDS measurements are slightly underestimated (within ~20%) with respect to the thicknesses determined from the film mass. That discrepancy could be due to the gallium film being partly oxidized, in which case the difference between the two techniques would correspond to the oxygen mass in the film. Anyway, it can be observed that the agreement between the two techniques is good although the conditions of EPMA-EDS analyses are far from ideal because of the high surface roughness.

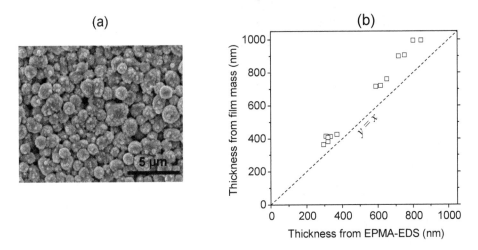

Fig. 7. (a) Typical SEM morphology of the gallium film vapour deposited on aluminium substrates. (b) Comparison of the gallium thickness determined for thirteen aluminium samples from film mass and from EPMA-EDS analysis.

## 2.4 EDS versus WDS

EPMA-WDS is known to achieve better quantitative results than EPMA-EDS. It is interesting to compare the two techniques with regards to the analysis of thin films. Four copper layers of various thicknesses were electro-deposited on nickel substrates. The copper thickness was determined using both EPMA-EDS and EPMA-WDS at 20 kV. The WDS data were acquired using a SX50 microprobe from Cameca. The results are shown in Fig. 8. EDS and WDS results are found to be consistent with one another within 5%, except for the smallest copper thickness.

Fig. 9 shows the EDS spectra acquired on the four copper thin films. It should be emphasized that EDS analysis of small amounts of copper in nickel is not an easy case because of the partial interference between Cu K$\alpha$ and Ni K$\beta$ (this interference is of course easily overcome using WDS whose spectral resolution is far better). Nevertheless, it can be observed from Fig. 8 that the copper thickness determination by EDS is very reliable with respect to the WDS one. The discrepancy between the two techniques is actually within 5% apart from the thinnest film, for which the Cu K$\alpha$ peak height is very small and undoubtedly affected by a high uncertainty (see red spectrum in Fig. 9). Anyway the generally good agreement between the two techniques demonstrates the good reliability of the routine peak deconvolution procedures implemented in most up-to-date EDS software packages.

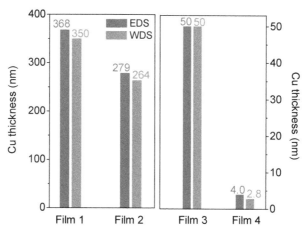

Fig. 8. Thickness determination of a Cu thin film deposited on a nickel substrate by EPMA at 20 kV. Comparison between EDS and WDS.

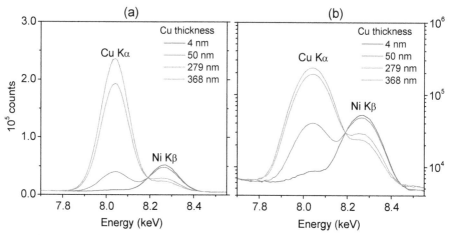

Fig. 9. EDS spectra acquired on four nickel samples coated with different thicknesses of copper. (a) linear counts scale. (b) logarithmic counts scale.

## 2.5 More complicated samples – Multiple voltage approach

The quantification of thin film thickness in the examples shown in the previous sections is quite easy as both the thin film and substrate compositions are known, and a single voltage approach is sufficient. EPMA-EDS can also be used for the analysis of thin films in more complicated situations including thin films and/or substrates of unknown composition, as well as multilayer structures. In those cases, a multiple voltage approach is usually required. This is out of the scope of this paper but nice examples of such complex samples can be found in (Pouchou & Pichoir, 1993).

## 3. EPMA-WDS: A tool for the characterization of surface thin films in the monolayer and sub-monolayer range

### 3.1 Principle

The principle of thin film analysis by EPMA-WDS is exactly the same as the one detailed for EPMA-EDS in the previous section. The difference is that WDS has a much better sensitivity than EDS and hence enables the quantification of much thinner films, down to the submonolayer range. The use of EPMA-WDS for the quantification of such thin layers has been described in details in (Christien & Le Gall, 2008; Nowakowski et al., 2011). For very thin surface layers, the quantification equation (Eq. (2)) can be simplified. As the layer mass thickness $\mu$ is very small, Eq. (2) reduces to:

$$\frac{I}{I_{Std}} = \frac{\mu \times \xi(0)}{\int\limits_{\rho z=0}^{\infty} \xi_{Std}(\rho z) d\rho z} \tag{4}$$

or, if the standard material is not a pure element:

$$\frac{I}{I_{Std}} = \frac{\mu \times \xi(0)}{X_{Std} \int\limits_{\rho z=0}^{\infty} \xi_{Std}(\rho z) d\rho z} \tag{5}$$

where $X_{Std}$ is the weight fraction of the analysed element in the standard material. So in the case of a very thin surface layer, the layer mass thickness $\mu$ is simply proportional to the measured k-ratio:

$$\mu = K \frac{I}{I_{Std}} \tag{6}$$

with:

$$K = \frac{X_{Std} \int\limits_{\rho z=0}^{\infty} \xi_{Std}(\rho z) d\rho z}{\xi(0)} \tag{7}$$

K can be easily determined for any system at a given beam voltage using the XPP model. Let us consider for example a substrate of nickel covered by a fractional monolayer of sulphur (the Ni-S system is chosen here as an example as the next sections are devoted to it). Fig. 10 shows the $\xi$ functions for the S K$\alpha$ line at 20 kV in nickel and in the FeS$_2$ compound which was used here as a standard material. The sulphur weight fraction in FeS$_2$ is $X_{Std}$ = 0.534. We finally obtain: $K$ = 256.1 $\mu$g cm$^{-2}$ at 20 kV. It should be emphasized that the $K$ constant is voltage-dependant: the higher the beam voltage, the higher the $K$ constant.

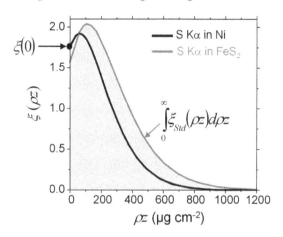

Fig. 10. $\xi$ functions calculated using the XPP model at 20 kV for the sulphur K$\alpha$ line in nickel and FeS$_2$ (used in this work as a standard material for sulphur element).

The use of EPMA-WDS to quantify surface monolayers and sub-monolayers is quite recent. It can be of great interest for the investigation of surface and grain boundary segregation phenomena in materials science (Christien & Le Gall, 2008; Nowakowski et al., 2011). The well-known case of sulphur segregation in nickel will be considered in the next sections. Nickel and nickel alloys always contain traces of sulphur at the level of some ppm or tens of ppm. When the material is annealed at an "intermediate" temperature (typically 400°C to 800°C), sulphur atoms tend to gather on the surfaces and at the grain boundaries as a monolayer or fraction of a monolayer. This is what we call "interface segregation" which is a very important metallurgical phenomenon as it can dramatically affect the metal properties (Saindrenan et al., 2002; Lejcek, 2010).

EPMA-WDS data presented in this section were acquired either on a Cameca SX50 microprobe equipped with four WDS spectrometers or on a Carl Zeiss Merlin high current SEM equipped with an Oxford Instrument INCA Wave spectrometer.

## 3.2 Surface segregation

A nickel sample containing 7.2 wt ppm of sulphur was annealed for 112 h at 800°C under high vacuum and then introduced in the microprobe. After annealing, the surface is expected to be covered by a sulphur fractionnal monolayer because of surface segregation. An example of sulphur K$\alpha$ peak acquired by EPMA-WDS on that sample at 10 kV is shown in Fig. 11. As demonstrated later, this peak is entirely due to the sulphur fractional monolayer covering the nickel substrate, and not to the sulphur located in the sample bulk. A good peak-to-background ratio is demonstrated, which illustrates the very good sensitivity of WDS.

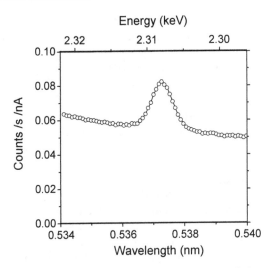

Fig. 11. WDS spectrum of the S Kα line acquired at 10 kV on a nickel sample covered with a fractional monolayer of sulphur.

Fig. 12 shows the voltage-dependence of the S Kα k-ratio measured at the sample surface. The strong decrease of the k-ratio over the beam voltage indicates that the detected sulphur signal is coming from the surface (and not from the bulk). The k-ratio corresponding to the sulphur bulk content (7.2 wt ppm) can be calculated and it is expected to be as low as ~10⁻⁵ (no matter what the beam voltage is) which is far below the experimental k-ratio values. So the bulk contribution to the sulphur WDS signal measured here can be considered negligible.

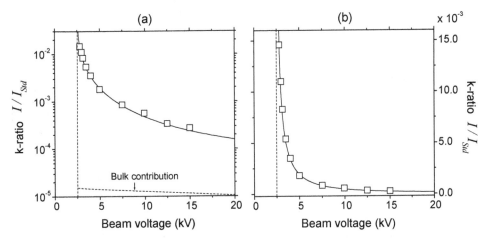

Fig. 12. Beam voltage dependence of the sulphur Kα k-ratio on a nickel substrate covered by a sulphur fractional monolayer. The points are WDS measurements and the curves were calculated using Eq. (5) for beam voltages ranging from 2.5 to 20 kV. (a) logarithmic k-ratio scale. (b) linear k-ratio scale. The dotted vertical line is the sulphur K level energy below which no Kα emission is observed.

The surface sulphur mass thickness $\mu$ can be quantified here from the experimental k-ratio measured at any voltage (again, a single-voltage approach would be sufficient here). The $\mu$ values determined at the ten different beam voltages investigated here were found to be consistent with each other within 5%. On average, we found:

$$\mu = 42.0 \ \text{ng cm}^{-2} \tag{8}$$

This mass thickness can be converted to a number of sulphur atoms. It corresponds to $7.9 \times 10^{14}$ atoms cm$^{-2}$, which is about half the density of a (100) nickel plane ($1.6 \times 10^{15}$ atoms cm$^{-2}$).

### 3.3 Grain boundary segregation

The conventional route for the analysis of grain boundaries using surface techniques such as Auger spectroscopy is to fracture the sample and to analyse the fractured surface. When elements such as sulphur or phosphorus are segregated, the material usually breaks along the grain boundaries which makes it possible to analyse them using surface analysis techniques. The same procedure was undertaken in this work for EPMA-WDS analysis. Nickel samples containing 5.4 wt ppm of sulphur in bulk were annealed at 550°C, 750°C and 900°C. Each of them was then fractured using tensile test in liquid nitrogen and one half of the broken specimen was introduced into the Merlin SEM chamber for WDS analysis. The WDS analyses were carried out using a beam voltage of 20 kV and a beam current of 400 nA. It should be emphasized that in this work the samples were fractured *ex-situ* with a conventional tensile machine, in contrast to the Auger spectroscopy procedure where the sample *must* be fractured *in-situ* (i.e. in the ultra-high vacuum of the spectrometer chamber) to avoid any surface contamination. Fig. 13a shows the principle of grain boundary segregation analysis by EPMA-WDS and Fig. 13b is one of the fracture surfaces observed in the Merlin SEM. For each of the three samples, about 15 to 20 grain boundary facets were analysed with a counting time of 6 minutes each.

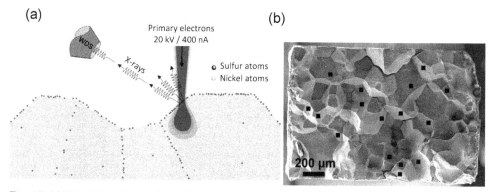

Fig. 13. (a) Principle of grain boundary segregation quantification by EPMA-WDS. (b) Example of grain boundary fracture surface observed in the SEM and analysed using WDS. The black squares indicate the analysed areas.

One difficulty of this kind of analysis is that the analysed facet is in general not perpendicular to the primary electron beam. In that case, Eq. (6) becomes:

$$\mu = K\frac{I}{I_{Std}}\cos\theta \tag{9}$$

where $\theta$ is the tilt angle (i.e. the angle between the facet normal and the primary beam). The tilt angle can be determined for each analysed facet from the measurement of the absorbed specimen current using Eq. (10) (a detailed description of that technique is given in (Nowakowski et al., 2011)):

$$\cos\theta = \frac{0.36}{1 - C_{ABS}/C_B} \tag{10}$$

where $C_B$ is the primary beam current and $C_{ABS}$ is the absorbed specimen current. It should be noticed that the 0.36 constant in Eq. (10) is correct only for a *nickel* sample and at a beam voltage of 20 kV. As the sample is fractured before analysis, the initial sulphur grain boundary concentration is randomly split on both fracture surfaces so that the sulphur concentration detected by WDS is only one half of the initial grain boundary concentration. This has to be taken into account in the quantification equation which becomes:

$$\mu = 2K\frac{I}{I_{Std}}\cos\theta \tag{11}$$

Fig. 14 shows the distributions of the sulphur mass thickness $\mu$ over the analysed grain boundary facets for the three nickel samples with different annealing temperatures. It is demonstrated that the average sulphur grain boundary concentration $\bar{\mu}$ significantly decreases with the annealing temperature, which is consistent with the Langmuir-McLean formalism of segregation (Saindrenan et al., 2002; Lejcek, 2010).

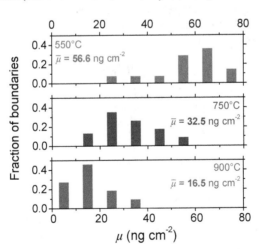

Fig. 14. Distributions of the sulphur concentration $\mu$ over the grain boundary facets analysed by EPMA-WDS. The colours correspond to three nickel samples with different annealing temperatures. $\bar{\mu}$ is the sulphur concentration averaged over all the analysed grain boundary facets.

The limit of detection of EPMA-WDS for the analysis of very thin surface layers has been determined in (Christien & Le Gall, 2008). For a primary beam current of 400 nA and a counting time of a few minutes, it is of the order of 1 ng cm⁻² which corresponds to about 1% of a monolayer.

The main advantage of EPMA-WDS with respect to Auger spectroscopy (which has been the most common technique for the analysis of interface segregation for decades) is that it is insensitive to surface contamination from the atmosphere. This is due to the "high" energy of Kα X-rays with respect to Auger electrons. For example, in the case of sulphur, the Kα line energy is 2.3 keV. This enables the X-rays to get through the thin contamination layer that forms when the sample is in contact with air (Nowakowski et al., 2011). On the contrary, the sulphur Auger electrons have an energy of 0.15 keV only and are totally absorbed in the atmospheric contamination layer. The consequence is that the sample preparation for Auger analysis has to be done *in-situ* (including cryogenic fracturing of the sample needed for grain boundary segregation analysis), whereas it can be done *ex-situ* for EPMA-WDS analysis, which is a very substantial advantage.

## 4. EPMA analysis of sandwich films

### 4.1 Principle

It has been demonstrated in the previous sections that EPMA is a powerful technique for the quantification of surface thin films. That technique can be extended to the case of "sandwich" films, which is a very common configuration in materials science (grain boundary segregation, multilayered structures in microelectronics...) as well as in geology or biology. Let us consider for example a gold sandwich film in a nickel matrix (Fig. 15). The technique consists in preparing a cross-section of material and acquiring an EPMA concentration profile across the film as shown in Fig. 15. The film is assumed thin enough (<~50 nm) to have no significant effect on the electron probe size and shape. In other words, the probe is supposed to be the same as it would be in pure nickel (or in nickel containing a very small concentration of gold).

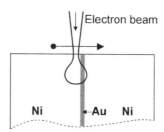

Fig. 15. Principle of EPMA quantification of a sandwich thin film.

As demonstrated in the following, the gold peak that is detected on the concentration profile can be used to quantify the gold mass thickness. The analysed volume $V$ is schematically represented on Fig. 16a. $l$ is the length of the concentration profile and $S$ is the surface of gold film in the analysed volume. $A_{Peak}$ is the peak area measured on the concentration profile (g cm⁻²) (Fig. 16b).

A general characteristic of microbeam techniques is that an object of real size $A$ analysed with a probe of size $B$ has an "observed" size $\sim A+B$. In our case, the size of the object (the

Au film) is negligible in comparison to the probe size (~ 1000 nm at 20 kV). The "observed" object size is then equal to the probe size. In other words, the peak width observed in the Au concentration profile is not the Au film width, but the electron probe size.

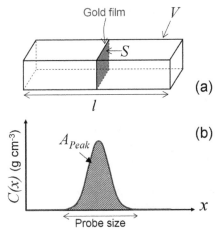

Fig. 16. a) Schematic shape of the volume analysed by EPMA during a profile along the $x$ direction. $V$ is the analysed volume, $l$ is the length of the concentration profile and $S$ is the surface of gold film in the analysed volume. b) Measured gold concentration profile across the film. $A_{Peak}$ is the profile peak area (g cm$^{-2}$).

The mean gold concentration $\overline{C}$ (g cm$^{-3}$) over the analysed volume is defined by (Fig. 16b):

$$\overline{C} = \frac{1}{l} \int_{x=0}^{l} C(x)\,dx \tag{12}$$

By definition, the peak area $A_{Peak}$ is:

$$A_{Peak} = \int_{x=0}^{l} C(x)\,dx \tag{13}$$

So that:

$$\overline{C} = \frac{A_{Peak}}{l} \tag{14}$$

Assuming that this mean concentration $\overline{C}$ is originating entirely from the gold film, we can determine the gold mass thickness $\mu$ expressed in g cm$^{-2}$:

$$\mu = \frac{\overline{C} \times V}{S} = \overline{C} \times l \tag{15}$$

From Eqs (14) and (15):

$$\mu = A_{Peak} \tag{16}$$

The Au mass thickness $\mu$ (g cm$^{-2}$) can be converted into an Au thickness (cm), as long as the layer density $\rho_{Au}$ is known:

$$e_{Au} = \frac{\mu_{Au}}{\rho_{Au}} = \frac{A_{Peak}}{\rho_{Au}} \qquad (17)$$

### 4.2 Example

A 23 nm thick Au sandwich film in a nickel matrix was analysed by EPMA-EDS. Fig. 17 shows the experimental EPMA-EDS gold concentration profiles obtained at 15, 20 and 25 kV. Data processing using Eq. (17) gives $e_{Au}$ = 25.6 nm at 15 kV, 26.2 nm at 20 kV and 24.8 nm at 25 kV. The results are in good agreement with the expected value of 23.0 nm. It is interesting to notice also that as expected the result does not depend on the beam voltage.

This technique has been for example used recently to measure gallium sandwich thin films in aluminium obtained by grain boundary penetration of liquid gallium. That work is to be published elsewhere.

The results presented in Fig. 17 are quite exploratory, but very promising. The development of the EPMA technique applied to very thin sandwich films is still in progress. In the near future, it should be possible (especially using WDS rather than EDS) to quantify much thinner sandwich films (down to the monolayer range), which is of great interest for many applications in materials science.

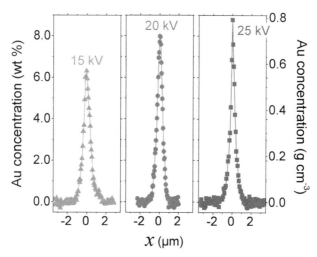

Fig. 17. EPMA-EDS Au concentration profiles acquired on a 23.0 nm thick Au sandwich film in a Ni matrix at 15, 20 and 25 kV.

## 5. Conclusion

It has been demonstrated in this paper that EPMA can be used to accurately determine the thickness of thin films. Examples have been given demonstrating that EDS is a very simple tool for the quantification of thin films in the range 10 nm to 1000 nm, including in complicated situation involving "light" elements (oxygen) and rough surfaces.

WDS offers much higher sensitivity than EDS and enables the determination of ultra-thin surface layers down to the sub-monolayer range, which makes it a competitive technique for surface analysis with respect to more usual techniques such as Auger spectroscopy. EPMA-WDS has even been very successfully used for the quantification of grain boundary segregation on fractured samples.

"Sandwich" thin films can also be quantified using EPMA. Some preliminary results on a 23 nm thick gold film in a nickel matrix were presented in this work. These promising results show that it should be possible in the near future to quantify sub-monolayer sandwich thin films using EPMA (especially with WDS rather than EDS), which is of great interest for various applications in materials science.

## 6. Acknowledgements

Vincent Barnier (SMS-MPI, Ecole Nationale Supérieure des Mines de Saint-Etienne) is sincerely acknowledged for careful XPS experiments.

## 7. References

Bastin G., Heijligers H. & Dijkstra J. (1990). Computer programs for the calculation of X-ray intensities emitted by elements in multiplayer structures. *Proceedings of the 25th Annual Conference of Microbeam Analysis Society*, Seattle, August 1990, pp. 159-160. ISSN: 02781727.

Bauccio M. (2nd Edition) (1994). *ASM Engineered Materials Reference Book*, Ed. ASM International, Materials Park, OH.

Castaing R. & Guinier A. (1950). Application des sondes électroniques à l'analyse métallographique. *Proceedings of First Europ. Conf. Electr. Microscopy*, Delft, July 1949.

Castaing R. (1951). *Application des sondes électroniques à une méthode d'analyse ponctuelle chimique et cristallographique*, Ph. D. Thesis, Université de Paris, France.

Christien F. & Le Gall R. (2008). Measuring surface and grain boundary segregation using wavelength dispersive X-ray spectroscopy. *Surface Science* Vol. 602, pp. 2463-2472.

Ferchaud E. (2010). *Brasage-diffusion de l'aluminium par le gallium*. PhD thesis, Université de Nantes, France

Grillon F. & Philibert J. (2002). The legacy of Raymond Castaing. *Mikrochim. Acta.* Vol. 138, No. 3-4, pp. 99-104. ISSN: 00263672.

Lejcek P. (2010). *Grain Boundary Segregation in Metals*, Springer, ISBN 978-3-642-12504-1.

Llovet X. & Merlet C. (2010). Electron probe microanalysis of thin films and multilayers using the computer program XFILM, *Microscopy and Microanalysis* Vol. 16, pp. 21-32. ISSN: 14319276.

Nowakowski P., Christien F., Allart M., Borjon-Piron Y. & Le Gall R. (2011). Measuring grain boundary segregation using Wavelength Dispersive X-ray Spectroscopy: Further developments. *Surface Science* Vol. 605, pp. 848-858.

Packwood R.H. & Brown J.D. (1981). A Gaussian expression to describe $\phi(\rho z)$ curves for quantitative electron probe microanalysis. *X-Ray Spectrom.*, Vol. 10, pp. 138-146.

Philibert J. & Tixier R. (1968). Electron penetration and the atomic number correction in electron probe microanalysis. *Brit. J. Appl. Phys.* Vol. 1, pp. 685-694. ISSN: 00223727.

Pouchou J.L. & Pichoir F. (1987). Basic expressions of PAP computation for quantitative EPMA, *Proceedings of ICXOM 11*, Ontario, August 1987, pp. 249-253.

Pouchou J.L., Pichoir F. & Boivin D. (1990). XPP procedure applied to quantitative EDS X-ray analysis in the SEM, *Proceedings of the 25th Annual Conference of Microbeam Analysis Society*, Seattle, August 1990, pp. 120-126. ISSN: 02781727.

Pouchou J.L. & Pichoir F. (1991). Quantitative analysis of homogeneous or stratified microvolumes applying the model "PAP", In: *Electron Probe Quantitation* Heinrich K., Newbury D., pp. 31-76, Plenum Press, ISBN 0306438240, 9780306438240, New-York.

Pouchou J.L. (1993). X-Ray microanalysis of stratified specimens, *Anal. Chim. Acta*, Vol. 283, pp. 81-97.

Pouchou J.L. & Pichoir F. (1993). Electron probe X-ray microanalysis applied to thin surface films and stratified specimens, *Scanning Microscopy Supplement*, Vol. 7, pp. 167-189. ISSN: 0892-953X.

Saindrenan G., Le Gall R. & Christien F. (2002) *Endommagement interfacial des métaux*, Editions Ellipses, ISBN 2-7298-0911-2, Paris.

Staub P., Jonnard P., Vergand F., Thirion J. & Bonnelle A.C. (1998). IntriX: A Numerical Model for Electron Probe Analysis at High Depth Resolution: Part II - Tests and Confrontation with Experiments, *X-Ray Spectrom.* Vol. 27, pp. 58-66. ISSN: 00498246

# X-Ray Analysis on Ceramic Materials Deposited by Sputtering and Reactive Sputtering for Sensing Applications

Rossana Gazia[1], Angelica Chiodoni[1], Edvige Celasco[2],
Stefano Bianco[1] and Pietro Mandracci[2]

[1]*Center for Space Human Robotics@PoliTo, Istituto Italiano di Tecnologia*
[2]*Materials Science and Chemical Engineering Department, Politecnico di Torino*
*Italy*

## 1. Introduction

Piezoelectric materials have been extensively studied and employed over the last decades with the aim of developing new sensors and actuators to be applied in a wide range of industrial fields. The continuous increase of miniaturization requirements led also to an increasing interest in the synthesis of piezoelectric materials in the form of thin films.

As an example, the growth of wireless mobile telecommunication systems favored the expansion of the design and fabrication of high-frequency oscillators and filters (Huang et al., 2005). Within this context, conventional surface acoustic wave (SAW) filters were gradually replaced by film bulk acoustic resonator (FBAR) devices. Their advantage in comparison with SAW devices resides in a higher quality factor (Lee et al., 2003) and lower fabrication costs (Huang et al., 2005). These kinds of devices are based on piezoelectric materials. Thus, an improvement in such material properties can have a strong impact in communication performances.

A similar situation exists in the field of sensing. The increase of the sensitivity in piezoelectric sensors is one of the main targets which could be achieved by improving material characteristics.

Therefore, the interest in this class of materials is not decreasing with time. On the contrary the enhancement of the piezoelectric material performances, together with the synthesis of new lead-free piezoelectric materials and the integration of ceramic materials with flexible substrates (Akiyama et al., 2006; Wright et al., 2011), is one of the latest research goals.

Most piezoelectric materials are metal oxide and few metal nitride crystalline solids and can be single crystals or polycrystalline materials. In both cases, piezoelectric materials are anisotropic and the determination of the response of the material to an external mechanical stress induced along a certain direction on its surface must be performed along different crystallographic axes. Quantitative information about how the material responds to external stresses is given by the piezoelectric constants.

Thus, the efficiency of the piezoelectric response can strongly vary with the crystal orientation of the material, and this occurs for bulk materials and thin films (Du et al., 1999; Yue et al., 2003).

The dependence of the piezoelectric response with the crystallographic orientation makes the control of the crystal properties during the growth of thin films one of the critical points in material synthesis.

Nevertheless, also material stoichiometry control is a relevant element for the enhancement of the piezoelectric efficiency because either the presence of impurities or the lack in some of the compound constituent elements can introduce a distortion in the crystal unit cell, which may be detrimental for the piezoelectric response (Hammer et al., 1998; Ramam & Chandramouli, 2011).

In many deposition techniques the control of film stoichiometry can be a difficult task, especially for high vacuum techniques, where the presence of residual gases in the vacuum chamber can affect the final quality of the thin films.

Among them, the sputtering technique is widely used in the synthesis of piezoelectric thin films. Two approaches can be used in the growth of such materials. In the simplest, the source of material (target) is a ceramic material with the same stoichiometry as that of the desired thin film. The other approach requires the presence in the vacuum chamber of a gas able to react with the material removed from the surface of the target and form a compound while the deposition is being carried out. The main difference between the two methods is that, with the latter, a control of stoichiometry can be achieved. On the other hand, the insertion of reactive gases leads to a decrease of the sputtering rate and to target contamination. In particular, target contamination can be avoided by changing the parameters during the deposition process but, as a drawback, an increase in process complexity is introduced.

The use of characterization techniques able to give detailed information about crystal structure and chemical composition is therefore necessary in order to assess the quality of the piezoelectric thin films. Among the numerous techniques available, X-Ray Diffraction (XRD), Energy-Dispersive X-ray spectroscopy (EDX or EDS) and X-ray Photoelectron Spectroscopy (XPS) analyses can be considered exhaustive enough for the determination of such information.

Each of these characterization techniques helps in determining those film properties which are at the basis of the correct operation of the materials as piezoelectrics.

Film orientation, as previously asserted, is probably the most important characteristic regarding materials with piezoelectric properties and XRD analysis is, therefore, one of the fundamental characterization techniques to be employed for providing such information. However, materials such as ceramics tend to grow in a polycrystalline form and the evaluation of the actual orientation of the crystals can be puzzling whenever impurities and contaminations are present in the films. As an example, some peaks present in XRD spectra related to films grown by sputtering, which is a technique that can likely produce films in a crystalline form without the need of post-annealing processes, can be attributed to some crystalline phase formed by the main compound with the residual oxygen still present into the vacuum chamber at the time that the deposition is carried out. Thus, the cross-check with information supplied by compositional analysis is of fundamental importance, in order to better interpret data from XRD analysis.

In particular, EDX can help in the first qualitative evaluation of the species present in the films, while XPS can give detailed information about the nature of the bonds occurring between atoms in the films and also semiquantitative information. Thus, if the orientation of the film does not correspond to what expected, compositional analysis can provide an explanation for that, indicating whether the problem resides in the fact that the preferred orientation of the film is different from the desired one, or in the presence of impurities in the film or in a deviation from the right stoichiometry introducing stresses in the thin film. Aim of this work is to compare the quality of films of aluminum nitride and zinc oxide, which are piezoelectric materials, grown by sputtering and reactive sputtering in order to understand how the different approaches adopted during the deposition processes affect the film quality.

## 2. Sputtering growth of ceramic materials

Sputtering technique is one of the most common PVD methods used for the growth of thin films. This technique is based on the removal of atoms from the surface of a target material by using inert gas ions such as Ar ions. The atoms so obtained are able to head towards a substrate, where they can nucleate and grow as thin film. A typical sputtering system consists of an ultra-high vacuum chamber containing a target material and a substrate holder, opportunely polarized. The system is equipped with mass flow controllers, valves and gas inlets in order to create a controlled atmosphere in the vacuum chamber, which is of fundamental importance for the creation of a stable plasma. Gas plasma is generated by the voltage applied between the cathode represented by the target material and the substrate. A bias voltage could be applied when required. The ionized gas atoms or molecules present in the plasma are accelerated towards the target and are able to sputter away from its surface some atoms that afterwards are deposited on the substrate surface. There, through nucleation and coalescence processes, the atoms create a first layer of material, which is followed by the formation of additional layers, as the deposition process is carried out. During the target bombardment, secondary electrons are also emitted, allowing the plasma to be maintained (Kelly & Arnell 2000).

The target is usually composed of a metallic material or other alloys and compounds with high conductivity as well as the substrates. In fact if an insulating material is used, the DC polarization applied between target (cathode) and substrate (anode) in a sputtering system induces a charge accumulation able to slowly reduce the ions acceleration towards the target, with the creation of an electric field in opposition to that generated by the DC polarization. For this reason it turns out to be necessary to modify the set up for the deposition of insulating materials. The application of a high frequency voltage can overcome the problem of charge accumulation, and this solution is found in RF sputtering systems. The typical voltage frequency for a standard RF sputtering system is 13.56 MHz. Such solution allows for the deposition of oxides, nitrides and other insulating materials on both metallic and insulating substrates.

The sputtering rate associated to each material depends on numerous parameters, including the sputtering efficiency of the gas which is strictly related to the mass of the gas atoms. For this reason gases with heavy atoms are preferred and argon is the most common gas used in sputtering processes. However, when compounds are to be obtained by sputtering, reactive gases are used, such as oxygen or nitrogen. Because of the role played by these gases during a thin film deposition, they are named "sputtering gases".

In a sputtering process the gas is inserted in the vacuum chamber only after a very low pressure (of the order of $10^{-5}$ Pa) has been reached. At these pressure levels, most of the contaminant species that could alter the film final composition are not present anymore, with some exceptions, and in principle the sole gas participating to the growth process is the one deliberately inserted in the vacuum chamber. In order to achieve low pressure values two kinds of vacuum pumps are required: one for the rough vacuum and one for the high vacuum. Typically, a rotary pump and a turbomolecular pump are employed for reaching high vacuum regime in a sputtering system. Rotary pumps usually allow to reach pressures of about $10^{-1}$ Pa while, with turbomolecular pumps, pressures down to $10^{-5}$ Pa can be achieved.

When the gas is inserted in the vacuum chamber at the desired flow, and the gas pressure is set to the value fixed for the deposition process, the plasma can be generated between the two electrodes present in the system. The target material is actually one of the two electrodes (the cathode). To provide a compensation for the local temperature increase caused by the energy transfer from the incident ions to the target surface, and during the high temperature processes, the cathode is usually equipped with a cooling system. The substrate is placed at the anode. This is the simplest sputtering system configuration, the so-called diode sputtering (Wasa et al., 2004a) and can be used with both DC and RF power supply. In order to increase the sputtering efficiency, a system of magnets with an appropriate configuration can be adopted. This system is intended to create a magnetic field parallel to the target surface that, affecting the trajectory of the secondary electrons, is able to confine them in proximity of the target surface, increasing the probability of collision between electrons and gas atoms and consequently of their ionization (Kelly & Arnell 2000). This region of denser plasma increases the sputtering rate of the target material. This modified system is called magnetron sputtering (Wasa et al., 2004b).

### 2.1 Sputtering growth of ceramic materials in reactive mode

The reactive sputtering technique is similar to the classic sputtering technique in all aspects, with one exception, that is the type of gas inserted in the vacuum system. In section 2, argon is indicated as the gas usually employed in a sputtering deposition process. However, for reasons that will be explained later in this section, it is necessary sometimes to be able to vary the stoichiometry of a compound to be grown in the form of thin film. In such cases, what is called a 'reactive gas' must be inserted in addition or in substitution of the sputtering inert gas. If the latter condition occurs, the gas is at the same time named sputtering and reactive gas.

At the base pressure reached with conventional vacuum systems, almost all those species that could alter the desired stoichiometry of a thin film are absent. However, in real systems there is always some residual unwanted gases that participate in the reaction between the target material and the reactive gas (Sproul et al., 2005). As previously asserted, the most commonly used reactive gases are oxygen and nitrogen, the former being used in oxide depositions, the latter in nitrides formation. In both cases, the target is usually a metallic material that reacts with the ionized gas atoms or molecules, generated by plasma, after its transition into vapor phase. The reason for using this approach is that it allows the control of the stoichiometry of the films formed on the substrate material. Although targets made of stoichiometric compounds are commercially available, sometimes it is required to adjust the stoichiometry of the films in order to gradually change their properties. By regulating the gas flow in the chamber or the concentration of the reactive gas in a mixture of sputtering and reactive gases, it is possible to regulate also the percentage of gas atoms or molecules bound to the metal ions.

The most important drawback in such a process is the phenomenon of the so-called target poisoning. It basically consists of the contamination of the target surface with the compound formed by the reaction between gas molecules and target atoms (Sproul et al., 2005), while in solid phase. When the target surface is totally covered by this compound, the target is poisoned (Sproul et al., 2005). Many works (Berg & Nyberg, 2005; Maniv & Westwood, 1980; Safi, 2000; Sproul et al., 2005; Waite & Shah, 2007) reported on the dependence on the reactive gas flow of both the target poisoning effect and the sputtering rate of the target material. Sputtering rates from compound targets are lower than that of pure metallic targets mainly because the sputtering yield of metal atoms from a compound on the target surface is lower than that from a pure metallic target. The ideal working point in terms of gas flow should be in between the elemental region (i.e., where the target is not contaminated at all), and the condition of target poisoning, but this region is not very stable and even weak variations of the growth conditions can lead to target poisoning (Sproul et al., 2005). Thus, it is very often necessary to work at low deposition rate and not having a full control of the stoichiometry of the thin films. The phenomenon of target poisoning described above occurs only when cathodes having uniform plasma density at their surface are used. In the case of magnetron cathodes, the formation of a compound at the target surface is position-dependent due to the low spatial homogenity of the magnetic field, and there is the possibility of the contemporary presence of three states across the surface: metallic, oxidized and partially oxidized (Safi, 2000). Many solutions for the process stabilization have been proposed, based on the variation of different parameters, such as pump speed or gas flow inserted into the vacuum chamber (Safi, 2000; Sproul et al., 2005).

## 3. Growth of piezoelectric ceramic materials on silicon substrates

The sputtering system used for depositing the films herein described and analyzed includes a stainless steel cylindrical reactor containing a magnetron sputtering source, capable of carrying a target of 10 cm in diameter. The source is fixed at the upper part of the reactor, facing downwards, whilst the silicon substrates are placed at the bottom part of the reactor, facing upwards, at a distance of 80 mm from the sputtering target. The system is pumped both by a turbomolecular pump and a mechanical pump in sequence (nominal pumping speeds 350 l/s and 15 m³/h respectively). The gases employed during the growth processes are injected into the reactor by means of mass flow controllers and the plasma discharge is lit by applying a radio frequency voltage at a frequency of 13.56 MHz between the target and the grounded substrate holder. The plasma impedance is then tuned by a matching network to reduce reflected power down to negligible values.

Before each deposition process the silicon substrates were cleaned in an ultrasonic bath with acetone (10 min) and ethanol (10 min) and dried under direct nitrogen flow. The substrates were then placed inside the vacuum chamber, which was evacuated down to a pressure of about $10^{-5}$ Pa.

### 3.1 Growth of ZnO thin films

Zinc oxide thin films were grown on [100]-oriented silicon substrates at room temperature, i.e., no intentional heating was applied to the substrates. Two different targets were used for the deposition processes: a Zn target (Goodfellow, purity 99.99+%) and a ZnO target (Goodfellow, purity 99.99%). The films were deposited according to three different approaches exploiting the two targets and different gas mixtures. In particular the films were grown by:

- sputtering the Zn target in a mixture of Ar and $O_2$
- sputtering the ZnO target in Ar atmosphere
- sputtering the ZnO target in a mixture of Ar and $O_2$

Two of the three approaches involve the reactive sputtering mode, while the other method avoids the use of a reactive gas, and the only source of oxygen should be the target itself. The complete sets of process parameters are reported in Table 1.

| Sample Name | Target | Base pressure (Pa) | Ar flow (sccm) | $O_2$ flow (sccm) | Pressure (Pa) | Power (W) | Deposition time (min) | Thickness (nm) |
|---|---|---|---|---|---|---|---|---|
| ZnO_MR | Zn | 5.3E-05 | 39 | 1 | 0.67 | 100 | 60 | 220 |
| ZnO_CN | ZnO | 1.6E-05 | 40 | - | 0.67 | 100 | 120 | 210 |
| ZnO_CR | ZnO | 2.9E-05 | 40 | 1 | 0.67 | 100 | 120 | 200 |

Table 1. Deposition conditions for the growth of ZnO thin films

The name of each sample reports the name of the material and two letters representing the nature of the target (Metallic, Ceramic) and the sputtering mode (Reactive, Non-reactive). The deposition time was chosen on the basis of a previous optimization work performed on the conditions of depositions of each class of films.

### 3.2 Growth of AlN thin films

Aluminum nitride thin films were grown on [100]-oriented silicon substrates at room temperature. Two different targets were used for the deposition processes: an Al target (Goodfellow, purity 99.999%) and an AlN target (Goodfellow, purity 99.5%). As for ZnO films, AlN films were deposited according to three different approaches involving the two targets and different gas mixtures. In particular the films were grown by:
- sputtering the Al target in $N_2$ atmosphere
- sputtering the AlN target in Ar atmosphere
- sputtering the AlN target in a mixture of Ar and $N_2$

The complete sets of process parameters are reported in Table 2.

| Sample Name | Target | Base pressure (Pa) | Ar flow (sccm) | $N_2$ flow (sccm) | Pressure (Pa) | Power (W) | Deposition time (min) | Thickness (nm) |
|---|---|---|---|---|---|---|---|---|
| AlN_MR | Al | 4.0E-05 | - | 40 | 0.35 | 100 | 180 | 210 |
| AlN_CN | AlN | 1.6E-05 | 40 | - | 0.67 | 100 | 120 | 60 |
| AlN_CR | AlN | 2.9E-05 | 40 | 1 | 0.67 | 100 | 120 | 55 |

Table 2. Deposition conditions for the growth of AlN thin films

For the names of the samples, the same criterion exposed in section 3.1 was applied. As well as in the case of ZnO thin films, the sputtering time and the gas pressure used for the growth of sample AlN_MR were chosen on the basis of the characterization results obtained from previous depositions of AlN films.

## 4. X-Ray analysis on ceramic thin films

PVD techniques usually allow the growth of thin films having the same stoichiometry of the source of material used during the deposition process. However, in the case of growth of compound thin films, the conditions in which the growth processes occur can cause either lacks of some of the elements making part of the compound, or contamination with other undesired elements. In other words, a key role is played by the deposition conditions in the growth process of a thin film with the correct stoichiometry.

The concentration of elements, and the way they are bound to each other, affects also the physical properties of materials. For this reason, chemical composition analysis is of fundamental importance since it helps to optimize those process parameters which directly influence the chemical composition of materials in order to obtain high-quality thin films, with optimum properties.

Energy Dispersive X-ray spectroscopy is a powerful technique for the investigation of the chemical species present in a material. It is based on the analysis of the energy and intensity distribution of the x-ray signal generated by the interaction of an electron beam with a specimen (Goldstein et al., 2003a). EDX is an optional tool installed on electron microscopes. In such a set up an electron beam is directed towards the sample to be analyzed. The electrons with a certain kinetic energy can interact with the nucleus or with the electrons of a specific atom. The Coulombic interaction with the positive nucleus causes a deceleration that is responsible for an energy loss, the so called bremsstrahlung radiation (Goldstein et al., 2003b). Since the electrons may lose any amount of energy between zero and the initial energy, the detection of this emitted radiation gives rise to a continuous electromagnetic spectrum that constitutes the background of the collected spectrum (Goldstein et al., 2003a). When the primary electrons interact with the electrons of the atom, ionization phenomena occur, usually involving K shell electrons. Since the atom in this excited state tends to reach the minimum of energy, the electrons from L or M shells can occupy the vacancy left from the K shell electron. The transition between L and K or M and K shells cause an emission of X-ray radiation designated as $K_\alpha$ and $K_\beta$, respectively (Goldstein et al., 2003c). The energy difference between L, M and K levels are well defined for each element, so the emission of $K_\alpha$ and $K_\beta$ X-rays identifies the elements present in the analyzed sample. The detection of this transition levels should in principle give a line in the continuous electromagnetic spectrum. However, the typical peak width of an EDX spectrum is about 70 times wider than the natural line width (Goldstein et al., 2003d). The spatial resolution of this technique is strictly related to the region of the sample interacting with the electron beam that is the X-ray generation volume. In fact, this volume depends on the material density and on the critical ionization energy, i.e., the energy that the primary electron beam should have in order to ionize atoms of the specimen (Goldstein et al., 2003e).

Both the depth and the lateral distribution of the X-ray production is relevant because in depth it determines the amount of photoelectric absorption of X-rays from the material atoms, and laterally it determines the spatial resolution of the X-ray microanalysis. However, the photoelectric absorption phenomenon is not significant in micron-sized samples so this side effect can be ignored (Goldstein et al., 2003e).

A fundamental aspect of this technique is the X-ray detection. It is usually performed using solid state detectors, especially lithium-drifted silicon detectors. The detection range is comprised between 0,2 and 30 KeV (Goldstein et al., 2003f). EDS analysis on ZnO and AlN thin films were performed with an Oxford INCA Energy 450 installed on a Field Emission Scanning Electron Microscope SUPRA 40 (ZEISS).

X-ray Photoelectron Spectroscopy is an advanced technique for the study of the surface composition of a wide range of materials, stated that they have to be vacuum-compatible. The sampling depth, defined as the depth from which 95% of all photoelectrons are scattered by the time they reach the surface, ranges between 5 and 10 nm below the surface. The phenomenon at the basis of this technique is the photoelectron emission that involves energy exchanges between a source of x-rays and a specimen, according to the following equation (Briggs, 1983):

$$E_K = h\nu - E_B - \phi \qquad (1)$$

where $E_K$ is the measured electron kinetic energy, $h\nu$ the energy of the exciting radiation, $E_B$ the binding energy of the electron in the specimen, and $\phi$ the work function, the latter having a specific value depending on the material and on the spectrometer (Briggs, 1983a). As stated above, the exciting radiation is derived from an X-ray source. The material responsible for the emission of X-rays must be chosen taking into account two aspects: the X-ray line width cannot exceed a certain value that otherwise would strongly limit the resolution of the measurement, and the energy associated to the X-ray must be high enough to allow the ejection of a sufficient range of core electrons for unambiguous analysis (Briggs, 1983b). Usually $K_\alpha$ radiation from aluminum or magnesium is employed to excite the material atoms. Photoelectron energy is measured by an appropriate analyzer. Many types of analyzers have been proposed over the last decades, although the simplest and most common ones are the cylindrical mirror analyzer (CMA) and the concentric hemispherical analyzer (CHA) (Briggs, 1983c). The energy associated to the photoemitted electrons precisely identifies a material. Since this energy varies with respect to the shell the electron is ejected from, each peak in the spectrum will correspond to the energy associated to the electrons ejected from all those levels that require an excitation energy lower than that of the incident X-rays. If the specimen is a compound material, each peak will represent the energy associated to the photoelectrons from all the elements present in that compound. However, peaks corresponding to the same element undergo what is called a chemical shift if the boundary conditions of the element change, i.e., the energy associated to photoelectrons ejected from an atom of an element changes with the molecular environment, with the oxidation state, or with its lattice site (Briggs, 1983d). Chemical shifts allow to understand how atoms are bound within compounds and alloys. Aluminum nitride and zinc oxide thin films were characterized with XPS technique in order to determine their exact chemical composition. The instrument used for XPS analysis was a PHI 5000 VersaProbe Scanning ESCA microprobe - Physical Electronics, equipped with an ion gun used in these experiments for cleaning the surface before each measurement, in order to remove any impurity from the analyzed area. A dual beam charge neutralization method was employed during the measurements, in order to reduce the charging effect on the samples. This charge neutralization method consists in a combination of low energy argon ions and electrons.

X-Ray Diffraction technique is employed for the analysis of the crystallographic properties of materials. The diffraction method is based on the following equation:

$$\lambda = 2d_{hkl} \sin\theta \qquad (2)$$

the well known Bragg law, where $\lambda$ is the x-ray wavelength which irradiates the sample to be analyzed, $d_{hkl}$ is the lattice spacing of a specific (hkl) family planes of the crystalline material, and $\theta$ is the diffraction angle of the X-ray beam with respect to the direction parallel to the (hkl) plane. For a given (hkl) family, characterized by a precise value of $d_{hkl}$, the conditions on $\lambda$ and $\theta$ are then very stringent. According to this equation, in a crystal it is possible to observe diffracted beams only along fixed directions, determined by the constructive interference between diffracted beams at the different crystal planes. Each single atom is actually able to scatter an incident beam of X-rays along every direction, but a periodic arrangement of atoms cancels, with a destructive interference phenomenon, all those beams scattered along the directions not satisfying equation (2). The analysis of materials with unknown values of lattice spacing is carried out by varying either the incidence angle or the wavelength of the X-ray beam. Depending on which of the two parameters is varied, the diffraction methods are named as follows (Cullity, 1956a):

- Laue method, where $\lambda$ is varied and $\theta$ fixed;
- Rotating-crystal method, used for monocrystals only, where $\lambda$ is fixed and the crystal rotates;
- Powder method, where $\lambda$ is fixed and $\theta$ is varied.

The latter has been used for the characterization of all the samples described in the next chapters, although thin films were not reduced in powder.

The instrument used to perform XRD measurements is called diffractometer, and it consists of a monochromatic X-ray source and an X-ray detector (commonly named counter), both placed on the circumference of a circle having its centre on the specimen. The specimen is held by an appropriate holder, placed on a table which can rotate about its axis, which is also the axis about which X-ray source and detector are able to rotate. X-rays are sent to the specimen where they are diffracted to form a convergent beam focused on a slit placed just before the X-ray detector.

A filter is also placed in the diffracted beam path in order to suppress $K_\beta$ radiation. The counter position can be read on a graduated scale of the goniometer and it corresponds to $2\theta$. Moreover the table where the specimen holder is fixed and the counter are mechanically coupled in a way that a rotation of the specimen of $x$ degrees corresponds to a rotation of the counter of $2x$ degrees (Cullity, 1956b). The values of the parameter $\theta$ correspond to the angular positions where peaks of diffracted X-ray intensity are registered. After the whole set of values for $\theta$ is determined, lattice spacings $d_{hkl}$ corresponding to different crystal planes can be calculated. Lattice spacing, however, is not the only parameter that can be obtained from an XRD measurement. The evaluation of the crystals size ($t$) is strictly related to the structural properties of materials, (Cullity, 1956c) and can be calculated as follows:

$$t = \frac{0.9\lambda}{B\cos\theta_B} \tag{3}$$

where $\lambda$ is the X-ray wavelength, $B$ is the peak width in radians, measured at the half the maximum intensity, and $\theta_B$ is the angle satisfying equation (2). Equation (3) is known as the Scherrer formula (Cullity, 1956d). Because of its formulation, equation (3) does not take into account the role of strain in the material.

The instrument used for the characterization of AlN and ZnO films was a Panalytical X'Pert MRD PRO diffractometer, equipped with a Cu $K_\alpha$ radiation source ($\lambda = 1.54056$ Å).

## 4.1 X-Ray analysis on ZnO thin films

ZnO was extensively studied over the last decades with the aim of producing high-quality thin films for the fabrication of SAW and FBAR devices (Ferblantier et al., 2005; Huang et al., 2005; Lee et al., 2003). These kinds of devices exploit the piezoelectric properties of their active layers. Materials exhibiting high piezoelectric coupling factor are usually highly oriented along one axis. For ZnO the $c$-axis orientation presents the highest piezoelectric coefficient and for this reason many works, more or less recent (Muthukumar et al., 2001; Sundaram et al., 1997; Yan et al., 2007), focused their attention on the XRD analysis and electrical characterization of the piezoelectric layers, often neglecting the analysis of the film composition. The lack of one of the elements in the film can be detrimental in the final response of the material when used in the fabrication of piezoelectric devices. For this reason it is crucial to grow highly oriented and highly stoichiometric materials for optimizing their piezoelectric properties. Moreover, techniques such as sputtering require very long optimization processes in order to satisfy the two requirements.

The combination of the results obtained from the X-ray analyses performed on the films described in this work will show how important is to combine the information collected by different techniques in order to precisely evaluate and improve the material properties.

### 4.1.1 EDS analysis on ZnO thin films

A first evaluation of the composition of the ZnO thin films grown by the three different approaches described in section 3.1 was assessed by EDS analysis. The system for the microanalysis was first calibrated with a Co sample, used as reference material to check the performance of the instrument. The spectra reported in Figure 1 refer to the whole area of the FESEM images shown below. All the spectra were collected by using the same conditions (electron energy, magnification, integration time).

The spectra were analyzed in standardless mode and the related semiquantitative information about the elements and their concentration is reported in Table 3. Four elements were detected. The Si peak is related to the substrate contribution, the C peak is related to organic impurities on the surface of the sample. The Zn peak is obviously related to zinc oxide, while the oxygen peak is partially related to this oxide.

| Element | C K Atomic concentration (%) | O K Atomic concentration (%) | Si K Atomic concentration (%) | Zn K Atomic concentration (%) |
|---|---|---|---|---|
| ZnO_MR | 6.79 | 31.28 | 45.97 | 15.97 |
| ZnO_CN | 5.90 | 31.35 | 46.72 | 16.03 |
| ZnO_CR | 5.31 | 30.00 | 49.96 | 14.73 |

Table 3. Atomic percentages of the elements detected by EDS analysis on the ZnO thin films

The excess of oxygen with respect to the desired stoichiometry of the films can be attributed to the fact that silicon is bound to oxygen, creating an oxide phase at the interface between the substrate and the film. However, since the oxygen peak area is the result of the contributions of the oxygen in the zinc oxide film and the oxygen present in the native silicon oxide layer present at the substrate surface, it is not possible to precisely determine the amount of oxygen actually bound to zinc in the compound. With the exception of carbon, usually present on the surface of materials exposed to air, no element different from the expected ones was detected with EDS analysis. A more precise evaluation of the film composition must be supplied with the help of other techniques.

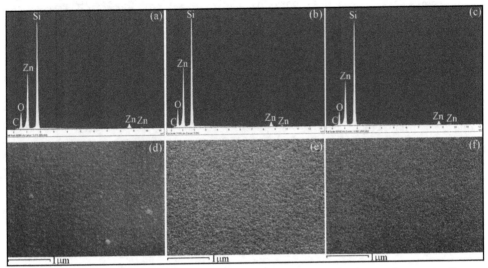

Fig. 1. EDX spectra and FESEM images of ZnO thin films grown from metal target (a, d), ceramic target in Ar atmosphere (b, e) and ceramic target in a mixture of Ar and $O_2$ (c, f).

### 4.1.2 XPS analysis on ZnO thin films

The XPS technique can give precise information about the composition of the first atomic layers of a material. This capability is of fundamental importance when dealing with applications where the study and control of surface properties is crucial. In the present case, instead, the desired information is not related to the actual surface composition: the piezoelectric constants of the materials are not related to the surface composition but to the crystallographic orientation of grains with the right stoichiometry. The exposure of the samples to air promotes the contamination of the surfaces (as shown from the EDX spectra reported above). As a result, the elements present on the sample surface are not the same that constitute the overall material layer. For this reason, in order to obtain information about the composition of the whole film, the removal of surface contamination before the acquisition of each XPS spectrum is mandatory and is usually performed by sputtering the surface with argon ions. The contamination of surfaces (adventitious carbon) due to exposure to air mainly generates C-C and C-H bonds. Carbon peak C1s is commonly used as a reference for the compensation of the XPS spectrum shift caused by the charge accumulation effect due to the charged particles reaching the sample surface during the analysis. However after the removal of all the surface contaminations the carbon peak is not detectable anymore since no source of this element is present during the film growth process. On the basis of these considerations, XPS spectra were acquired before and after the surface cleaning. With a high resolution acquisition before the surface cleaning the C1s peak position was precisely determined. The second high resolution acquisition determined the position of the peaks related to each constituent element in the film. The difference between the energy the C1s peak is centered at and the energy the carbon peak from database is centered at (fixed at 284.6 eV) gives the shift to be applied to all the acquired peaks in the XPS spectrum. This procedure was applied to all the samples herein described. For the determination of the exact peak positions all the peaks were fitted with a Gaussian-Lorentzian function. The concentration of the elements in the film was calculated by the MultiPak v.9.0 software according to the following equations:

$$\text{Oxygen concentration (\%)} = \frac{A_O \times SF_O}{\left(A_O \times SF_O\right) + \left(A_{Zn} \times SF_{Zn}\right)} \times 100 \qquad (4)$$

$$\text{Zinc concentration (\%)} = \frac{A_{Zn} \times SF_{Zn}}{\left(A_O \times SF_O\right) + \left(A_{Zn} \times SF_{Zn}\right)} \times 100 \qquad (5)$$

where $A_O$ and $A_{Zn}$ are the areas of the O1s and Zn2p$^{3/2}$ peaks, respectively, and $SF_O$ and $SF_{Zn}$ are the sensitivity factors for oxygen and zinc, respectively. Figure 2 shows the O1s and Zn2p$^{3/2}$ peaks acquired with the post-cleaning high-resolution analysis on the ZnO_MR sample.

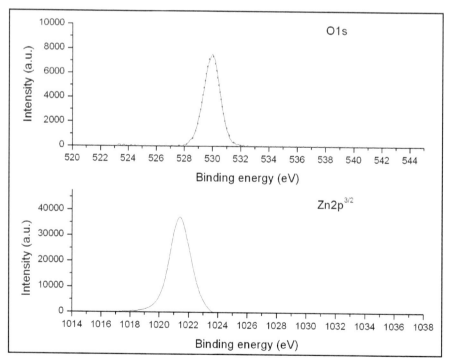

Fig. 2. O1s and Zn2p$^{3/2}$ XPS peaks acquired after surface cleaning on ZnO_MR sample. The dotted red line represents the peak fitting the O1s peak.

The O1s peak is centered at 529.9 eV, the typical position of the oxygen peak in Zn-O bonds (Haber et al., 1976). The calculated concentrations are reported in Table 4. The excess of Zn with respect to oxygen reveals that the film contains also Zn-Zn bonds. In order to evaluate the percentage of such bonds, the zinc peak should be fitted with two appropriate curves. The areas of the fitting curves centered at the typical binding energies of Zn in the Zn-Zn and Zn-O bonds represent the percentages of the metallic phase and the oxide phase. Unfortunately these binding energies are 1021.5 eV and 1021.7 eV, respectively (Klein & Hercules, 1983), and are too close to univocally fit the Zn2p$^{3/2}$ peak. However, for the absence of any other contaminant in the film, it is reasonable to assert that the overall Zn excess constitutes a metallic zinc phase in the film.

Figure 3 reports the high resolution peaks of ZnO_CN sample. Again it is not possible to precisely determine the metallic phase percentage by fitting the peaks revealed during the analysis. However, again in this case, the amount of impurities can be neglected. The carbon concentration is very low indeed (Table 4) and there is no evidence that a consistent portion of it could be bound to oxygen atoms. The peaks fitting the O1s peak and centered at 529.7 eV, 531.6 eV and 532.3 eV can be in fact identified with the binding energies of O-Zn bonds (Chen et al., 2000), O-H bonds and oxygen bound in the $H_2O$ molecule, respectively (Avalle et al., 1992). The latter is present in particular because of the possible presence of humidity inside the vacuum chamber which is not equipped with a load-lock chamber and requires to be opened every time a substrate has to be loaded for a new process.

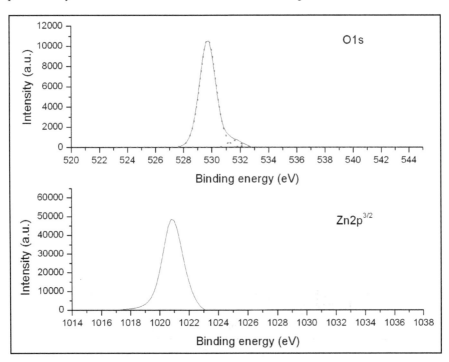

Fig. 3. O1s and Zn2p$^{3/2}$ XPS peaks acquired after surface cleaning on ZnO_CN sample. The dotted red lines represent the peaks fitting the O1s peak.

However the total area of the two peaks related to the binding energy of oxygen bound to hydrogen atoms is less than the 7% of the total O1s peak area. This means that the 93% of oxygen atoms are bound to zinc. The Zn excess produces a metallic phase in the film as for ZnO_MR sample. Although the target used for the growth of sample ZnO_CN is stoichiometric, the film presents a lack of oxygen which needs to be compensated.

In the case of ZnO_CR sample, grown in a mixture of Ar and $O_2$, there was still an excess of zinc (see Table 4). However the concentration of oxygen in the gas mixture was very low (1 sccm over a total flow of 41 sccm) and probably the amount of gas required to obtain the right stoichiometry should be higher. High resolution peaks for ZnO_CR sample are reported in Figure 4. Zinc peak was not fitted because of the impossibility to exactly determine the position of the two peaks representing the binding energies associated to Zn-O and Zn-Zn bonds. The O1s peak was fitted with two curves, one centered at 529.7 eV (O-Zn bonds), and the other centered at 531.7 eV (O-H bonds), very close to those chosen for fitting the O1s peak in the XPS spectrum of ZnO_CN sample. The amount of contamination due to the humidity present in the chamber is basically the same as that of ZnO_CN. It is identified by the small peak on the right-bottom part of the O1s peak shown in Figure 4, with a peak area of about the 4% of the total O1s peak area.

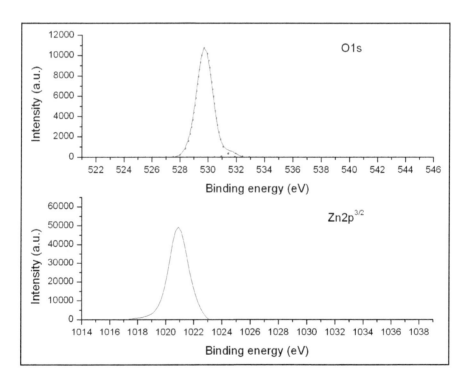

Fig. 4. O1s and Zn2p[3/2] XPS peaks acquired after surface cleaning on ZnO_CR sample. The dotted red lines represent the peaks fitting the O1s peak.

| Sample name | C atomic concentration | O atomic concentration | Zn atomic concentration |
|---|---|---|---|
| | (%) | (%) | (%) |
| ZnO_MR | - | 40.9 | 59.1 |
| ZnO_CN | 0.7 | 44.0 | 55.3 |
| ZnO_CR | - | 43.9 | 56.1 |

Table 4. Atomic concentrations obtained by fitting the O1s and Zn2p$^{3/2}$ peaks from the XPS spectra of ZnO thin films

### 4.1.3 XRD analysis on ZnO thin films

XRD analysis on ZnO thin films was carried out in order to assess the crystallographic orientation of the material since from this information it is possible to predict how efficient the material will be in terms of piezoelectric response. XRD spectra acquired on ZnO are shown in Figure 5 and the peak positions are reported in Table 5. The peak positions were determined by fitting the spectra with Pseudovoigt curves.

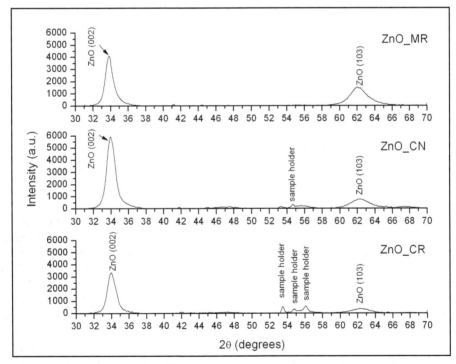

Fig. 5. XRD patterns of ZnO thin films

Two main peaks were identified in the XRD patterns of all the three samples, one related to the [002] orientation and one to the [103] orientation of ZnO. Peaks detected in the ZnO_MR sample are the most shifted with respect to the database peak positions (JCPDS, 1998a). These data are in agreement with the fact that the metallic phase in this film was the largest in comparison with the other samples. The presence of a high amount of metallic phase might have induced a tensile stress in the zinc oxide unit cells with the effect of producing a shift towards lower angle values. To this end it is worth to notice that no peak related to any zinc crystallographic orientation was detected. Thus we can consider the metallic zinc as an impurity instead of a single phase.

For what concerns the peak relative intensities, the [002] peak intensity of sample ZnO_CR is about 10 times higher than the [103] peak intensity, meaning that during the growth from ceramic target the presence of oxygen favors the growth of crystals along the [002] orientation.

The ratio between the intensities of the two peaks in sample ZnO_CN was in fact lower (about 8). The sample grown from the metal target (ZnO_MR) showed the lowest intensity ratio (about 3).

It can be concluded that, on the basis of the whole set of results, ZnO_CR sample could in principle perform better than the other samples in terms of piezoelectric properties.

| Sample name | Peak position (2θ) | |
| --- | --- | --- |
| | ZnO [002] | ZnO [103] |
| | (degrees) | (degrees) |
| Database (JCPDS, 1998a) | 34.379±0.001 | 62.777±0.001 |
| ZnO_MR | 33.86±0.02 | 62.17±0.02 |
| ZnO_CN | 34.03±0.02 | 62.33±0.02 |
| ZnO_CR | 34.03±0.02 | 62.22±0.02 |

Table 5. Position of the peaks detected during XRD analysis performed on ZnO thin film

### 4.2 X-Ray analysis on AlN thin films

Aluminum nitride is a semiconductor of the III-V semiconductor group, with an hexagonal wurtzite crystalline structure (Penza et al., 1995; Xu et al., 2001), lattice constants of a = 0.3110 nm e c =0.4980 nm (Xu et al., 2001), and is characterized by a broad direct energy gap (6.2 eV) (Cheng et al., 2003; Hirata et al., 2007; Penza et al., 1995), chemical stability (Cheng et al., 2003; Penza et al., 1995), high thermal conductivity (3.2 W/mK) (Cheng et al., 2003), low thermal expansion coefficient (4.5 ppm/°C) (Ruffner et al., 1999), high breakdown voltage (Hirata et al., 2007; Xu et al., 2001), high acoustic speed (Cheng et al., 2003; Xu et al., 2001), high refractive index (n = 2.1) (Penza et al., 1995), high electrical resistivity ($10^{11}$-$10^{14}$ Ωcm) (Ruffner et al., 1999) and, above all, good piezoelectric response (Cheng et al., 2003; Hirata et al., 2007; Penza et al., 1995; Ruffner et al., 1999; Xu et al., 2001) (piezoelectric coefficient of 5.4 pm/V) (Ruffner et al., 1999). Piezoelectric behaviour of AlN strongly depends on its crystallographic orientation (Cheng et al., 2003) and, in particular, films grown in the [002] orientation are preferred as the highest piezoelectric stress constant is associated to this orientation (Cheng et al., 2003). Piezoelectricity in AlN has been demonstrated only when in the form of a thin film (Gautschi, 2002). Analysis performed by many research groups on reactive-sputtered AlN films (Cheng et al., 2003; Penza et al., 1995; Xu et al., 2001) revealed the strong dependence of film crystal orientation on growth parameters, such as target-substrate distance, power applied to the target, substrate temperature, process pressure, and gas relative concentration when a mixture of Ar and $N_2$ was employed for the depositions of AlN. An interesting advantage of growing AlN by reactive sputtering is the possibility to obtain polycrystalline films at low substrate temperatures (from room temperature up to ~200 °C (Penza et al., 1995; Xu et al., 2001).

As for ZnO, AlN thin films were characterized by EDS, XPS and XRD techniques, in order to assess the film quality and to find out relationships between film properties and growth conditions.

### 4.2.1 EDS analysis on AlN thin films

EDX spectra and FESEM images obtained from the analysis of AlN thin films, deposited with the three different approaches described in section 3.2, are shown in Figure 6. The results from microanalysis are reported in Table 6.

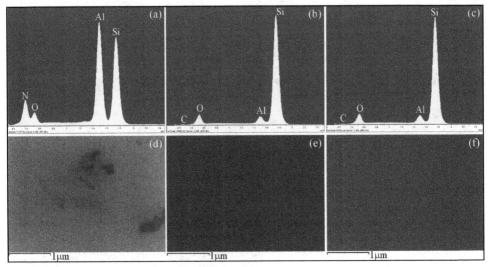

Fig. 6. EDX spectra and FESEM images of AlN thin films grown from metal target (a, d), ceramic target in Ar atmosphere (b, e) and ceramic target in a mixture of Ar and $N_2$ (c, f).

The information acquired with this kind of analysis is of fundamental importance for the interpretation of the results obtained from the other analysis performed on the AlN samples. It can be observed indeed that the nitrogen peak was not detected in any of the films deposited from ceramic target, while it is present in the spectrum of the sample grown from a metallic aluminum target in nitrogen atmosphere. Also in this case the data acquired with the EDS technique are not enough to assert that AlN_CN and AlN_CR samples do not contain nitrogen. A relevant issue for the energy-dispersive X-Ray spectroscopy technique is that release of volatile elements (such as nitrogen) can occur during the analysis (Bohne et al., 2004). Thus, in samples characterized by low nitrogen concentration the detection of the nitrogen peak can be difficult.

In the case of the two samples grown from ceramic target only aluminum and oxygen were detected (as well as the silicon peak from the substrate and carbon traces as impurities).

For what concerns the film grown from a metal target, nitrogen and oxygen peaks are strong enough to be clearly distinguished. Oxygen peak detected on the sample comprises the contribution of the signals generated by the oxygen contained in the film and by the oxygen contained in the layer of native oxide present on the silicon substrate, and it is not possible to estimate which is the percentage of each contribution for the two layers (Bohne et al., 2004). In consideration of the limit of this characterization further analysis are required in order to better evaluate the composition of each film.

| Element | N K | C K | O K | Al K | Si K |
|---|---|---|---|---|---|
| | Atomic concentration | Atomic concentration | Atomic concentration | Atomic concentration | Atomic concentration |
| | (%) | (%) | (%) | (%) | (%) |
| AlN_MR | 38.76 | - | 9.72 | 24.11 | 27.40 |
| AlN_CN | - | 2.17 | 16.32 | 3.91 | 77.60 |
| AlN_CR | - | 2.50 | 16.83 | 4.20 | 76.47 |

Table 6. Atomic percentages of the elements detected by EDS analysis on the AlN thin films

### 4.2.2 XPS analysis on AlN thin films

AlN samples were treated with the same surface cleaning process as ZnO samples (see section 4.1.2). XPS spectra acquired on sample AlN_MR confirmed the presence of nitrogen found by means of the EDS analysis. The peaks of the elements present in the film are shown in Figure 7, while the values of elemental concentration obtained for all the AlN samples are reported in Table 7. If qualitative information is basically the same for the two techniques, with the exception of the presence of the silicon peak that is not found in the XPS spectrum for the intrinsic properties of such a technique, quantitative information are different. By comparing the values reported in Table 6 and Table 7 it is evident that the concentrations of each element are strongly different. For what concerns the oxygen concentration, which results to be much higher from the XPS analysis than from EDS analysis, it has to be noticed that the surface of the material can contain a higher amount of this element if compared with bulk because of the capability of the surface to be oxidized. In principle, the oxygen concentration calculated from EDX spectra could be more accurate when considering the whole film volume composition. On the other hand the surface cleaning process performed just before the XPS analysis should have prevented from the analysis of the very first layers of the sample which could have undergone a higher oxidation with respect to deeper layers. For this reason information from the XPS analysis are expected to be more reliable.

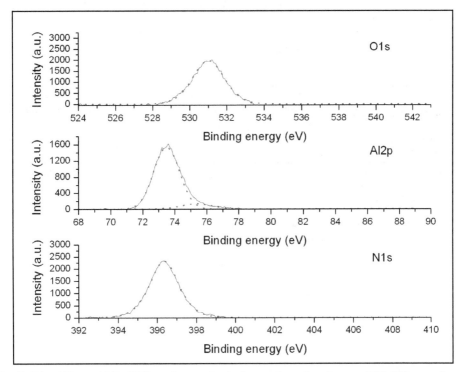

Fig. 7. O1s, Al2p and N1s XPS peaks acquired after surface cleaning on AlN_MR sample. The dotted red lines represent the peaks fitting the O1s peak.

The curves used for fitting the peaks, reported in Figure 7, represent a possible way to fit the XPS peaks found in the sample in order to understand how the atoms are bound in the film. The peaks used to fit the spectrum are those reported in literature, corresponding to Al-O and Al-N bonds. However, by using this method, the resulting percentage of Al-O bonds is too low (about 8%) if considering the oxygen concentration in the film (26.7 %). For this reason it is possible to assume that the film contains an $AlO_xN_y$ phase which cannot be exactly estimated because a small variation of the position of each peak used for the fit of the main Al peak determines a high variation in the relative percentages of the different bonds.

The results obtained from the XPS analysis performed on AlN thin films deposited from a ceramic target confirm what already shown by EDS analysis: nitrogen was not detected in the films. XPS spectra for AlN_CN and AlN_CR are shown in Figure 8 and Figure 9. In spite of the absence of nitrogen in the films, a deeper investigation on the spectra can lead to interesting information. There is, in fact, a slight difference in the two spectra.

Although an accurate estimation of the concentration is not allowed because of the low and noisy signal obtained in the spectral region corresponding to the nitrogen peak position, a faint peak can be distinguished in Figure 9 at binding energies around 397.5 eV, the typical binding energy at which the N1s peak is centered when nitrogen is bound to aluminum (Manova et al., 1998).

The collection of the nitrogen peak in sample AlN_CR allows to assert that the presence of nitrogen gas during the sputtering process is necessary in order to include nitrogen in the film. Nonetheless the amount of gas used during the growth process was too low, and that is the reason why only noise was acquired in the spectral region of the N1s peak.

| Sample name | C atomic concentration (%) | N atomic concentration (%) | O atomic concentration (%) | Al atomic concentration (%) |
|---|---|---|---|---|
| AlN_MR | - | 34.2 | 26.7 | 39 |
| AlN_CN | - | 0.03 | 66.91 | 33.06 |
| AlN_CR | 0.28 | 0.35 | 65.73 | 33.63 |

Table 7. Atomic concentrations obtained by fitting the C1s, N1s, O1s and Al2p peaks from the XPS spectra of AlN thin films

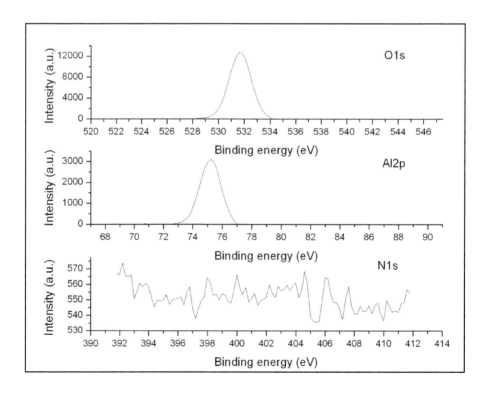

Fig. 8. O1s, Al2p and N1s XPS peaks acquired after surface cleaning on AlN_CN sample

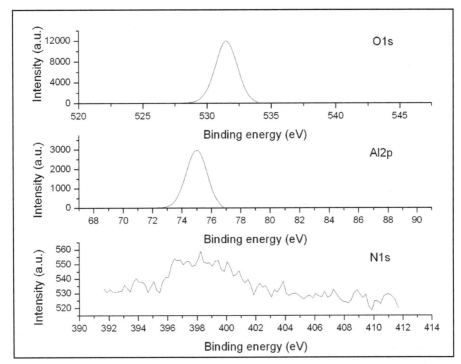

Fig. 9. O1s, Al2p and N1s XPS peaks acquired after surface cleaning on AIN_CR sample

### 4.2.3 XRD analysis on AIN thin films

Without the support of the EDS and XPS analyses on the AIN thin films grown from metal and ceramic targets, the interpretation of the XRD patterns would be more complicated. It is not usual in sputtering processes to obtain materials not maintaining a stoichiometry similar to that of the material source. However, for some classes of compounds, such as nitrides, this phenomenon is more frequent because one of the constituting elements is volatile and is less reactive than oxygen. In vacuum chambers used for the growth of both oxides and nitrides the chances of introducing contaminations in the films are numerous because of the residual oxygen trapped at the chamber walls. Moreover the residual humidity in the chamber containing oxygen is also responsible for the presence of oxide phases in the thin films. With such a low amount of nitrogen in the chamber, the compensation for the loss of nitrogen during the film growth cannot be achieved.

XRD patterns for all the analyzed samples are shown in Figure 10, while the peak positions are reported in Table 8. On the basis of the results discussed above, it was obvious that any peak acquired by XRD analysis of AIN_CN and AIN_CR samples should be identified as a peak corresponding to crystalline alumina phase and not to AIN crystalline phases. The chances that the 0.3 % of nitrogen found by fitting the XPS peak might form a crystalline phase with aluminum that could be detected by XRD are in fact very low. For this reason, all the diffraction peaks of the samples deposited from ceramic target were compared with the database peaks of $Al_2O_3$. The only detectable peak (with the exception of the peaks

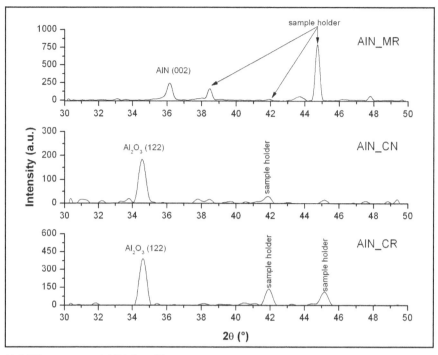

Fig. 10. XRD patterns of AlN thin films

| Sample name | Peak position (2θ) | |
| --- | --- | --- |
| | AlN [002] | Al₂O₃ [122] |
| | (degrees) | (degrees) |
| Database (JCPDS, 1998b) | 35.981±0.001 | 34.734±0.001 |
| AlN_MR | 36.15±0.02 | - |
| AlN_CN | - | 34.57±0.02 |
| AlN_CR | - | 34.62±0.02 |

Table 8. Position of the peaks detected during XRD analysis performed on AlN thin film

produced by the sample holder) and present in the spectra of both films, was attributed to the [122] orientation of crystalline alumina.

For sample Al_MR instead, a peak corresponding to the [002] orientation of crystalline AlN was detected. In the view of obtaining films with good piezoelectric properties this is a good result since the highest piezoelectric coefficient for AlN is found to be along the $c$-axis orientation. There is certainly a problem of oxygen contamination that has to be solved in order to minimize the amorphous $AlO_xN_y$ phase and increase the quality of the material but it is strictly related to technological issues.

The correct interpretation of XRD patterns was possible only on the basis of the results obtained from EDS and XPS analysis. If considering only the growth conditions of the films, the attribution of the sole detected peak to an alumina phase could not be so obvious. The assignment of the detected peak to an aluminium oxide nitride compound, for example, would have been straightforward. From the database, in fact, it is possible to find a peak corresponding to the [103] orientation of such a compound centered at 34.357° (JCPDS, 1998c), very close to the position of the peaks detected in the AlN_CN and AlN_CR samples. On the basis of the sole experimental set up the material could be logically thought as a nitride. The misinterpretation of the XRD results was avoided just because of the availability of additional data related to the film composition.

## 5. Conclusions

Aluminum nitride and zinc oxide thin films were grown by the RF magnetron sputtering technique. Three methods were used for depositing each of the two materials, involving the use of different material sources and gas mixtures. Both materials are piezoelectric and this property is guaranteed and maximized when the material has certain characteristics. The analysis based on X-ray spectroscopy of all the samples was aimed at the comparison of the quality of films grown with the different methods and to select the most effective. In addition, the characterization results were used to underline the differences in the technological processes used for the growth of different classes of materials (oxides and nitrides). In particular, the growth of nitrides is characterized by issues due to the nature of the material itself, especially for what concerns nitrogen which is a volatile element and is characterized by a higher ionization energy than oxygen, that make it less reactive, lowering the possibilities to include such element in the films.

With reference to the growth of zinc oxide thin films, the approach that guaranteed a good stoichiometry and at the same time favored the growth along the c-axis was that characterized by the use of a ceramic target in the presence of a gas mixture containing Ar and $O_2$. Although a slight excess of zinc with respect to the right stoichiometry was detected, the film was mainly [002]-oriented. The future optimization of the parameters will be aimed at the growth of fully [002]-oriented ZnO thin films in order to maximize the piezoelectric coefficient of the material.

An aluminum nitride thin film was successfully grown only by reactive sputtering in nitrogen atmosphere. The growth from ceramic target led instead to the growth of aluminum oxide as confirmed by the X-ray spectroscopy analysis performed on the samples. The amount of nitrogen required for its inclusion in the film is higher than that used for the growth of the oxide samples. In spite of the not fully positive results related to the presence of nitrogen in the films, these experiments were useful for assessing the great importance of the combination of results coming from different analysis. The correct interpretation of XRD patterns of samples grown from ceramic target was possible only because of the presence of EDS and XPS data.

## 6. References

Akiyama, M., Morofuji, Y., Kamohara, T., Nishikubo, K., Tsubai, M., Fukuda, O. & Ueno, N. (2006). Flexible piezoelectric pressure sensors using oriented aluminum nitride thin films prepared on polyethylene terephthalate films. *Journal of Applied Physics*, Vol. 100, No. 11, pp. 114318

Avalle, L., Santos, E., Leiva, E. & Macagno, V. A. (1992). Characterization of TiO$_2$ films modified by platinum doping. *Thin Solid Films*, Vol. 219, No. 1-2, pp. 7-17, ISSN 0040-6090

Berg, S. & Nyberg, T. (2005). Fundamental understanding and modeling of reactive sputtering processes. *Thin Solid Films*, Vol. 476, No. 2, pp. 215-230, ISSN 0040-6090

Bohne, W., Röhrich, J., Schöpke, A., Selle, B., Sieber, I., Fuhs, W., del Prado, Á., San Andrés, E., Mártil, I. & González-Díaz, G. (2004). Compositional analysis of thin $SiO_xN_y$:H films by heavy-ion ERDA, standard RBS, EDX and AES: a comparison. *Nuclear Instruments and Methods in Physics Research Section B: Beam Interactions with Materials and Atoms*, Vol. 217, No. 2, pp. 237-245, ISSN 0168-583X

Briggs, D. & Seah, M. P. (1983a). Practical Surface Analysis: Auger and X-ray photoelectron spectroscopy., Briggs, D. & Seah, M. P., pp. 112, John Wiley & Sons, Chichester

Briggs, D. & Seah, M. P. (1983b). Practical Surface Analysis: Auger and X-ray photoelectron spectroscopy., Briggs, D. & Seah, M. P., pp. 51, John Wiley & Sons, Chichester

Briggs, D. & Seah, M. P. (1983c). Practical Surface Analysis: Auger and X-ray photoelectron spectroscopy., Briggs, D. & Seah, M. P., pp. 69, John Wiley & Sons, Chichester

Briggs, D. & Seah, M. P. (1983d). Practical Surface Analysis: Auger and X-ray photoelectron spectroscopy., Briggs, D. & Seah, M. P., pp. 120, John Wiley & Sons, Chichester

Chen, M., Wang, X., Yu, Y. H., Pei, Z. L., Bai, X. D., Sun, C., Huang, R. F. & Wen, L. S. (2000). X-ray photoelectron spectroscopy and auger electron spectroscopy studies of Al-doped ZnO films. *Applied Surface Science*, Vol. 158, No. 1-2, pp. 134-140, ISSN 0169-4332

Cheng, H., Sun, Y., Zhang, J. X., Zhang, Y. B., Yuan, S. & Hing, P. (2003). AlN films deposited under various nitrogen concentrations by RF reactive sputtering. *Journal of Crystal Growth*, Vol. 254, No. 1-2, pp. 46-54, ISSN 0022-0248

Cullity, B. D. (1956a). Elements of x-ray diffraction, Cullity, B. D., pp. 92, Addison-Wesley Publishing Company, Reading, Massachusetts, USA

Cullity, B. D. (1956b). Elements of x-ray diffraction, Cullity, B. D., pp. 189-190, Addison-Wesley Publishing Company, Reading, Massachusetts, USA

Cullity, B. D. (1956c). Elements of x-ray diffraction, Cullity, B. D., pp. 281, Addison-Wesley Publishing Company, Reading, Massachusetts, USA

Cullity, B. D. (1956d). Elements of x-ray diffraction, Cullity, B. D., pp. 102, Addison-Wesley Publishing Company, Reading, Massachusetts, USA

Du, X.-h., Wang, Q.-M., Belegundu, U., Bhalla, A. & Uchino, K. (1999). Crystal orientation dependence of piezoelectric properties of single crystal barium titanate. *Materials Letters*, Vol. 40, No. 3, pp. 109-113, ISSN 0167-577X

Ferblantier, G., Mailly, F., Al Asmar, R., Foucaran, A. & Pascal-Delannoy, F. (2005). Deposition of zinc oxide thin films for application in bulk acoustic wave resonator. *Sensors and Actuators A: Physical*, Vol. 122, No. 2, pp. 184-188, ISSN 0924-4247

Gautschi, G. (2002). *Piezoelectric Sensorics* (1st edition), Springer-Verlag Berlin Heidelberg New York, ISBN 978-3-540-42259-4, Germany.

Goldstein, J., Newbury, D.E., Joy, D.C., Lyman, C.E., Echlin, P., Lifshin, E., Sawyer, L.C. & Michael, J.R. (2003a). Scanning Electron Microscopy and X-ray Microanalysis, Goldstein, J., pp. 297, Springer Science + Business Media, ISBN 0306472929, New York, USA

Goldstein, J., Newbury, D.E., Joy, D.C., Lyman, C.E., Echlin, P., Lifshin, E., Sawyer, L.C. & Michael, J.R. (2003b). Scanning Electron Microscopy and X-ray Microanalysis, Goldstein, J., pp. 271, Springer Science + Business Media, ISBN 0306472929, New York, USA

Goldstein, J., Newbury, D.E., Joy, D.C., Lyman, C.E., Echlin, P., Lifshin, E., Sawyer, L.C. & Michael, J.R. (2003c). Scanning Electron Microscopy and X-ray Microanalysis, Goldstein, J., pp. 275, Springer Science + Business Media, ISBN 0306472929, New York, USA

Goldstein, J., Newbury, D.E., Joy, D.C., Lyman, C.E., Echlin, P., Lifshin, E., Sawyer, L.C. & Michael, J.R. (2003d). Scanning Electron Microscopy and X-ray Microanalysis, Goldstein, J., pp. 282, Springer Science + Business Media, ISBN 0306472929, New York, USA

Goldstein, J., Newbury, D.E., Joy, D.C., Lyman, C.E., Echlin, P., Lifshin, E., Sawyer, L.C. & Michael, J.R. (2003e). Scanning Electron Microscopy and X-ray Microanalysis, Goldstein, J., pp. 286-288, Springer Science + Business Media, ISBN 0306472929, New York, USA

Goldstein, J., Newbury, D.E., Joy, D.C., Lyman, C.E., Echlin, P., Lifshin, E., Sawyer, L.C. & Michael, J.R. (2003f). Scanning Electron Microscopy and X-ray Microanalysis, Goldstein, J., pp. 349, Springer Science + Business Media, ISBN 0306472929, New York, USA

Haber, J., Stoch, J. & Ungier, L. (1976). X-ray photoelectron spectra of oxygen in oxides of Co, Ni, Fe and Zn. *Journal of Electron Spectroscopy and Related Phenomena*, Vol. 9, No. 5, pp. 459-467, ISSN 0368-2048

Hammer, M., Monty, C., Endriss, A. & Hoffmann, M. J. (1998). Correlation between Surface Texture and Chemical Composition in Undoped, Hard, and Soft Piezoelectric PZT Ceramics. *Journal of the American Ceramic Society*, Vol. 81, No. 3, pp. 721-724, ISSN 1551-2916

Hirata, S., Okamoto, K., Inoue, S., Kim, T. W., Ohta, J., Fujioka, H. & Oshima, M. (2007). Epitaxial growth of AlN films on single-crystalline Ta substrates. *Journal of Solid State Chemistry*, Vol. 180, No. 8, pp. 2335-2339, ISSN 0022-4596

Huang, C.-L., Tay, K.-W. & Wu, L. (2005). Fabrication and performance analysis of film bulk acoustic wave resonators. *Materials Letters*, Vol. 59, No. 8-9, pp. 1012-1016, ISSN 0167-577X

Joint Committee on Powder Diffraction Standards (JCPDS) database—International Center for Diffraction Data (1998a) PCPDFWIN v.2.01. The JCPDS card number of ZnO is 89-1397

Joint Committee on Powder Diffraction Standards (JCPDS) database—International Center for Diffraction Data (1998b) PCPDFWIN v.2.01. The JCPDS card number of AlN is 89-3446 and JCPDS card number of Al2O3 is 88-0107

Joint Committee on Powder Diffraction Standards (JCPDS) database—International Center for Diffraction Data (1998c) PCPDFWIN v.2.01. The JCPDS card number of aluminum oxide nitride is 48-1581

Kelly, P. J. & Arnell, R. D. (2000). Magnetron sputtering: a review of recent developments and applications. *Vacuum*, Vol. 56, No. 3, pp. 159-172, ISSN 0042-207X

Klein, J. C. & Hercules, D. M. (1983). Surface characterization of model Urushibara catalysts. *Journal of Catalysis*, Vol. 82, No. 2, pp. 424-441, ISSN 0021-9517

Lee, J. B., Kim, H. J., Kim, S. G., Hwang, C. S., Hong, S.-H., Shin, Y. H. & Lee, N. H. (2003). Deposition of ZnO thin films by magnetron sputtering for a film bulk acoustic resonator. *Thin Solid Films*, Vol. 435, No. 1-2, pp. 179-185, ISSN 0040-6090

Maniv, S. & Westwood, W. D. (1980). Oxidation of an aluminum magnetron sputtering target in Ar/O₂ mixtures. *Journal of Applied Physics*, Vol. 51, No. 1, pp. 718-725, ISSN 0021-8979

Manova, D., Dimitrova, V., Fukarek, W. & Karpuzov, D. (1998). Investigation of d.c.-reactive magnetron-sputtered AlN thin films by electron microprobe analysis, X-ray photoelectron spectroscopy and polarised infra-red reflection. *Surface and Coatings Technology*, Vol. 106, No. 2-3, pp. 205-208, ISSN 0257-8972

Muthukumar, S., Gorla, C. R., Emanetoglu, N. W., Liang, S. & Lu, Y. (2001). Control of morphology and orientation of ZnO thin films grown on SiO₂/Si substrates. *Journal of Crystal Growth*, Vol. 225, No. 2-4, pp. 197-201, ISSN 0022-0248

Penza, M., De Riccardis, M. F., Mirenghi, L., Tagliente, M. A. & Verona, E. (1995). Low temperature growth of r.f. reactively planar magnetron-sputtered AlN films. *Thin Solid Films*, Vol. 259, No. 2, pp. 154-162, ISSN 0040-6090

Ramam, K. & Chandramouli, K. (2011). Dielectric and piezoelectric properties of rare-earth gadolinium modified lead lanthanum zirconium niobium titanate ceramics. Ceramics International, Vol. 37, No. 3, pp. 979-984, ISSN 0272-8842

Ruffner, J. A., Clem, P. G., Tuttle, B. A., Dimos, D. & Gonzales, D. M. (1999). Effect of substrate composition on the piezoelectric response of reactively sputtered AlN thin films. *Thin Solid Films*, Vol. 354, No. 1-2, pp. 256-261, ISSN 0040-6090

Safi, I. (2000). Recent aspects concerning DC reactive magnetron sputtering of thin films: a review. *Surface and Coatings Technology*, Vol. 127, No. 2-3, pp. 203-218, ISSN 0257-8972

Sproul, W. D., Christie, D. J. & Carter, D. C. (2005). Control of reactive sputtering processes. *Thin Solid Films*, Vol. 491, No. 1-2, pp. 1-17, ISSN 0040-6090

Sundaram, K. B. & Khan, A. (1997). Characterization and optimization of zinc oxide films by r.f. magnetron sputtering. *Thin Solid Films*, Vol. 295, No. 1-2, pp. 87-91, ISSN 0040-6090

Waite, M. M. & Shah, S. I. (2007). Target poisoning during reactive sputtering of silicon with oxygen and nitrogen. *Materials Science and Engineering: B*, Vol. 140, No. 1-2, pp. 64-68, ISSN 0921-5107

Wasa, K., Kitabatake, M. & Adachi, H. (2004a). Sputtering systems, in: *Thin film materials technology-sputtering of compound materials*. K. Wasa, M. Kitabatake, H. Adachi, pp. 135, William Andrew Publishing, Norwick, NY/Springer-Verlag, ISBN 3-540-21118-7, Heidelberg, Germany

Wasa, K., Kitabatake, M. & Adachi, H. (2004b). Sputtering systems, in: *Thin film materials technology-sputtering of compound materials*. K. Wasa, M. Kitabatake, H. Adachi, pp. 139, William Andrew Publishing, Norwick, NY/Springer-Verlag, ISBN 3-540-21118-7, Heidelberg, Germany

Wright, R. V., Hakemi, G. & Kirby, P. B. (2011). Integration of thin film bulk acoustic resonators onto flexible liquid crystal polymer substrates. *Microelectronic Engineering*, Vol. 88, No. 6, pp. 1006-1009, ISSN 0167-9317

Xu, X.-H., Wu, H.-S., Zhang, C.-J. & Jin, Z.-H. (2001). Morphological properties of AlN piezoelectric thin films deposited by DC reactive magnetron sputtering. *Thin Solid Films*, Vol. 388, No. 1-2, pp. 62-67, ISSN 0040-6090

Yan, Z., Zhou, X. Y., Pang, G. K. H., Zhang, T., Liu, W. L., Cheng, J. G., Song, Z. T., Feng, S. L., Lai, L. H., Chen, J. Z. & Wang, Y. (2007). ZnO-based film bulk acoustic resonator for high sensitivity biosensor applications. *Applied Physics Letters*, Vol. 90, No. 14, pp. 143503

Yue, W. & Yi-jian, J. (2003). Crystal orientation dependence of piezoelectric properties in LiNbO₃ and LiTaO₃. *Optical Materials*, Vol. 23, No. 1-2, pp. 403-408, ISSN 0925-3467

# 11

# Employing Soft X-Rays in Experimental Astrochemistry

Sergio Pilling and Diana P. P. Andrade

*Inst. de Pesquisa & Desenvolvimento, Universidade do Vale do Paraíba (UNIVAP),*
*São Jose dos Campos*
*Brazil*

## 1. Introduction

Young stars and protostars are hot enough to produce a strong ionizing field with the maximum at ultraviolet and soft X-rays (Koyama et al. 1996). These objects are usually associated with regions in which several stars are borning together, called star-forming regions, inside of molecular clouds. Examples of such regions are Sgr B2, Orion KL, and W51. In these star-forming regions, the presence of widespread UV and X-ray fields could trigger the formation of photodissociation regions (PDRs). X-ray photons are capable of traversing large column densities of gas before being absorbed, which produce X-ray-dominated regions (XDRs) where the interface between the ionized gas and the self-shielded neutral layers could influence the selective heating of the molecular gas (Goicoechea et al. 2004). The typical energy flux in XDRs inside molecular clouds is about $\sim 10$ erg cm$^{-2}$s$^{-1}$, which decreases as the visual extinction increases within the clouds. The complexity of these regions may allow a combination of different scenarios and excitation/ionization mechanisms to coexist.

Figure 1 shows a typical photodissociation region (Gaseous Pillars from M16 nebulae) sculpted by radiation pressure of young stars inside molecular clouds obtained by the hubble space telescope.

The photochemistry induced by soft X-rays on molecular gas produces excited states (both valence and core levels), radicals, ions (cations and anions), and excited ions, as well as promotes the release of fast electrons. Such electrons can also trigger further excitation/ionization processes. In solid phase (such as interstellar ices), the main effects of soft X-rays are local heating, molecular excitation/ionization, molecular bonds rupture (photodissociation), and photodessorption (releasing of ion or neutral species from the frozen phase to the gas phase). In addition, interstellar grain charging can occur due to the photoelectron effect on the grain surfaces induced by soft X-rays.

In both gas and solid phases, the presence of molecules in excited states, radicals, and ions (cations and anions) is highly enhanced due to soft X-rays in comparison with UV photons. Those species react with each other and with neutral gas, enhancing the molecular complexity of the region.

Due to the high absorbance of soft X-rays by the Earth's atmosphere, most of the information of extraterrestrial X-rays sources comes from space telescopes, such as, ROSAT, Chandra, XMM-Newton, Beppo-SAX, Swift, ASCA, and others.

Fig. 1. Photodissociation region known as Gaseous Pillars associated with recent star forming region inside M16 nebulea (Figure inset). The fingerlike structures are the remnants of the molecular clouds which were sculpted by radiation pressure of young stars. (From Hubble Space Telescope 2003)

## 2. Soft X-rays in experimental astrochemistry

Soft X-rays have been used to simulate the interaction between stellar photons and matter in different astrophysical scenarios, including both gas-phase and solid-phase (ices). For these experiments, synchrotron light sources have been employed because of their high intensity and wide wavelength range (coverage). Most of the molecular species investigated were detected in different astrophysical environments (e.g. interstellar medium, comets). In this chapter, we report several experiments performed at the Brazilian Synchrotron Light Laboratory (LNLS) in Campinas, São Paulo, Brazil.

### 2.1 Gas-phase experiments

Soft X-ray photons ($\sim 10^{12}$ photons/s) from a toroidal grating monochromator (TGM) beamline (100–310 eV) perpendicularly intersected the gas sample inside a high vacuum chamber. The gas needle was kept at ground potential. The photon flux was recorded by a light sensitive diode. Several species were investigated in the gas-phase mode, including formic acetic, formic acid, methanol, ethanol, methyl formate, methylamine, acetone, and benzene, as well as some amino acids and nucleobases. These species have been detected or predicted to be present in interstellar medium.

Conventional time-of-flight mass spectra (TOF-MS) were obtained using the correlation between one Photoelectron and a Photoion Coincidence (PEPICO). Details of Time of flight mass spectrometry are given elsewhere (e.g. Boechat-Roberty et al. 2005, Pilling et al. 2006, Pilling 2006). A schematic diagram of the experimental setup employed in gas-phase experiments is shown in Fig. 2a. Figure 2b is a photograph of the equipment coupled to the soft X-ray beamline of LNLS.

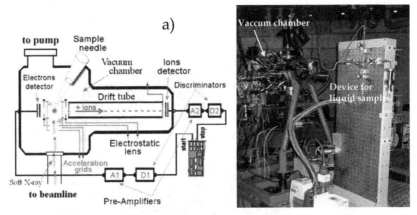

Fig. 2. a) Schematic diagram of the experimental setup employed in gas-phase experiments. b) Photography of equipment employed at the soft X-ray beamline of LNLS.

Briefly, the ionized recoil fragments produced by the interaction with the photon beam were accelerated by a two-stage electric field and detected by two microchannel plate detectors in a chevron configuration, after mass to-charge (m/q) analysis by a time-of-flight mass spectrometer. They produced up to three stop signals for a time-to-digital converter (TDC) started by the signal from one of the electrons accelerated in the opposite direction and recorded without energy analysis by two micro-channel plate detectors. The integrated area of the given peak (ion) in mass spectra divided by the total area (total collected ions) times 100 percent is called the branching ratio (BR). The measurements were done at room temperature. Figure 3 illustrates the particle tracks inside the TOF spectrometers and the potentials employed in these experiments. The simulations were performed using the particle track software SIMION®.

Considering that the kinetic energy of the ionic fragments inside the interaction region is negligible compared with the extraction electron field, the equations for the time of flight of ions inside the spectrometer can be given by

$$t_{voo_i} = t_d + t_L + t_D$$

$$t_{voo_i} = \frac{\sqrt{\frac{q}{m}E_d d}}{\frac{q}{m}E_d} + \frac{\sqrt{\frac{q}{m}(E_d d + E_L L)} - \sqrt{\frac{q}{m}E_d d}}{\frac{q}{m}E_L} + \frac{D}{\sqrt{\frac{q}{m}(E_d d + E_L L)}} \quad (1)$$

and, for the electrons, by the equation

$$t_{voo_e} = t'_d + t_{L_1} + t_{L_2}$$

$$t_{voo_e} = \frac{\sqrt{\frac{|q_e|}{m_e}E_d d}}{\frac{|q_e|}{m_e}E_d} + \frac{\sqrt{\frac{|q_e|}{m_e}(E_d d + E_{L_1}L_1)} - \sqrt{\frac{|q_e|}{m_e}E_d d}}{\frac{|q_e|}{m_e}E_{L_1}} + \frac{\sqrt{\frac{|q_e|}{m_e}(E_d d + E_{L_1}L_1 + E_{L_2}L_2)} - \sqrt{\frac{|q_e|}{m_e}(E_d d + E_{L_1}L_1)}}{\frac{|q_e|}{m_e}E_{L_2}} \quad (2)$$

where L2, L1, d (extraction region), L, and D (drift tube) represent the length of specific regions inside the TOF spectrometer (see Fig. 3).

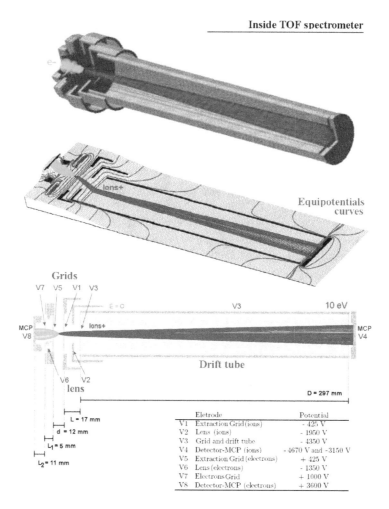

| Eletrode | | Potential |
|---|---|---|
| V1 | Extraction Grid (ions) | - 425 V |
| V2 | Lens (ions) | - 1950 V |
| V3 | Grid and drift tube | - 4350 V |
| V4 | Detector-MCP (ions) | -4670 V and -3150 V |
| V5 | Extraction Grid (electrons) | + 425 V |
| V6 | Lens (electrons) | - 1350 V |
| V7 | Electrons Grid | + 1000 V |
| V8 | Detector-MCP (electrons) | + 3600 V |

Fig. 3. Particle tracks inside the TOF spectrometers and the potentials employed in the experiments. The simulations were performed using the particle track software SIMION®.

However, due to the employed coincidence technique, the ion's time of flight is not necessarily the time of their detections, $t_{det}$, because one photoelectron, produced during the photoionization process, must reach the detector to start counting the time of the ion´s flight ($t_i \gg t_e \neq 0$). By knowing the electric field inside the TOF spectrometer (Ed = 850/0.012 = 70833 V/m, EL = 3950/0.0017 = 230882 V/m, EL1 = 575/0.005 = 115000 V/m and EL2= 2600/0.011 = 236364 V/m), we can estimate the time of flight of protons ~ 415 ns and of electrons ~1 ns to reach their respective detectors.

Strictly, the time of detection, $t_{det}$, of a given ion is the difference between the ion's time of flight , $t_{vooi}$, and the time of flight of its respective photoelectron, $t_{vooe}$,

$$t_{det} = t_{voo_i} - t_{voo_e} \tag{3}$$

Therefore, employing this condition in the Eq. [1] and [2] we obtain

$$t_{det} = \frac{a'}{\sqrt{\frac{q}{m}}} + b' \tag{4}$$

where $a' = \sqrt{E_d d}/E_d + (\sqrt{E_d d + E_L L} - \sqrt{E_d d})/E_L + D/(\sqrt{E_d d + E_L L})$ and $b' = -t_{voo_e}$

And finally, we obtain the quadratic relation between the m/q ratio (mass/charge) and the time of flight of ions:

$$\frac{m}{q} = \left(\frac{t_{det} - b'}{a'}\right)^2 = a t_{det}^2 + b t_{det} + c \tag{5}$$

where $a = 1/a'^2$, $b = -2b'/a'^2$ and $c = b'^2$. are constants to be determined during the calibration of TOF spectrometer by employing time of flight of fragments with known m/q (eg. Ar+ at 40 u, Ar++ at 20u and, Ar+++ at 13.3u). From Eq. [5] we observe that the spectrometer's resolution is related to the m/q ratio, decreasing for higher values of m/q.

### 2.1.1 Single coincidence mass spectrum (PEPICO)

The simple configuration of the TOF spectrometer is to produce a single coincidence mass spectrum, or in other words, the coincidence between one photo-electron and one photo-ion (PEPICO). The typical reaction scene that occurs inside the ionization region is illustrated by

$$M + h\nu \longrightarrow M^+ + e^-$$
$$M^+ \longrightarrow M_1^+ (+ \text{ neutrals}).$$

After the absorption of a soft X-ray photon, the molecule releases one electron and becomes positively charged. Following this stage, the excited ion can also dissociate to produce ionic and neutral fragments. Each molecule has its specific pattern of fragment production which depends also on the incoming photon and the molecular orbital configuration.

Depending on the incoming soft X-ray photon, the ejected electron is from inner valence orbitals (less energetic photons) or core orbitals (more energetic photons). The absorption cross sections decrease as a function of photon energy, however, at some specific energies, the absorption cross sections increase abruptly. These energies are called resonances and are associated with the exact energies of electrons to be released from the molecular orbitals. Figure 4 shows a schematic view of the absorption probability, as a function of photon energy, from visible to soft X-rays. The absorption probability is highest at UV wavelengths and decreases to the higher energies. At resonance energies, the absorption probability also increases. These resonances represent the exact energies to promote electrons from valence or inner-shell orbital to unoccupied orbitals (photoexcitation) or to the continuum (photoionization).

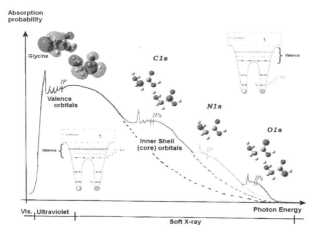

Fig. 4. Schematic view of absorption probability as a function of photon energy for polyatomic molecule (e.g. glycine). IP indicates the ionization potential of a given orbital.

Two examples of PEPICO spectra of water fragments produced by soft X-rays are given in Fig 5. The left spectrum is the original time-of-flight spectra (counts vs. time in ns). The right spectrum was obtained after calibration employing known fragments. This can be performed employing Eq. 5 together with data from two different situations: i) By employing previous experiment with Ar, or ii) by employing the SIMION® program to simulate time-of-flight of different ions. A comparison between these two spectra shows that the peak resolution changes with mass/charge ratio, being worst for higher mass/charge values.

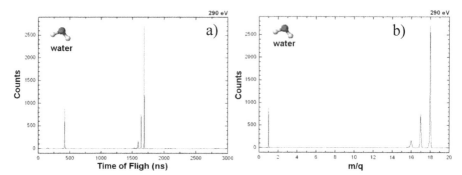

Fig. 5. Sample of time of flight spectra of water fragments from the irradiation with 290 eV before m/q calibration (a) and after m/q calibration (b).

Figure 6 presents some samples of PEPICO mass spectra of studied molecules of astrophysical interest (Pilling et al. 2006; Pilling et al 2011; Ferreira Rodrigues et al. 2008). All spectra were obtained with soft x-rays. The employed soft X-ray energies are shown in the top of each spectrum. Labels close to the peaks indicate possible fragment attributions.

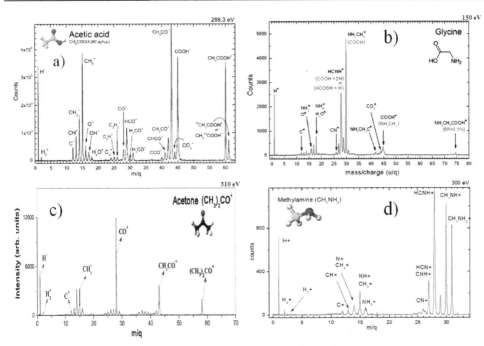

Fig. 6. Time of flight mass spectra (PEPICO) of the fragments produced from the impact of soft X-rays with different gaseous molecules of astrophysical interest. a) acetic acid (Pilling et al 2007) , b) glycine (Pilling et al. 2011), c) Acetone (Ferreira Rodrigues et al. 2008),  and  d) methylamine.

From the analysis of the peaks area, we derived the partial ion yield (PIY) or relative intensities, which correspond to the relative contribution of each cationic fragment in the PEPICO mass spectra. The PIY for each fragment $i$ were obtained by

$$PIY_i = \left( \frac{A_i}{A_t^+} \pm \frac{\sqrt{A_i} + A_i \times ER/100}{A_t^+} \right) \times 100\% \qquad (6)$$

where $A_i$ is the area of a fragment peak, $A_t^+$ is the total area of the PEPICO spectrum. The error factor $ER$ (2%) is the estimated error factor due to the data treatment.

Depending on the energy of the incoming photon, a given excitation channel and/or ionization channel will be opened. Figure 7 shows five mass spectra of the fragments produced by the irradiation of the simplest interstellar alcohol (methanol) at different photon energies between 100 and 310 eV (Pilling et al 2007a). The analysis of relative area of individual peaks shows that the area of the molecular ion (parent molecule that lost a electron - $M^+$) peak decreases with the enhancement of photon energy. This suggests that although the absorption probability in soft X-ray range decreases for higher photon energy, the amount of fragmentation produced increases. In other words, the more energy the incoming photon has (in the soft X-ray range), the more the molecule is dissociated in small fragments (atomization).

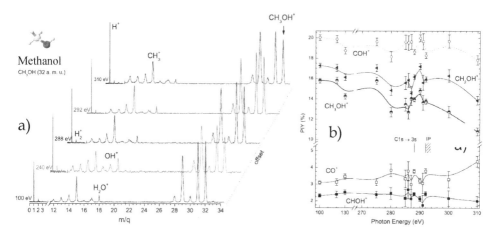

Fig. 7. a) Mass spectra of methanol fragments obtained at different photon energies between 100 and 310 eV. b) Partial ion yield (PIY) of the fragments release by the methanol as a function of photon energy (From Pilling et al 2007a).

Figure 8 shows the mass spectrum (PEPICO) of the fragments of the two simplest interstellar alcohols, methanol and ethanol, irradiated with 292 eV (~42 Angstroms) soft X-rays photons. The arrows indicate the molecular ion in each mass spectrum. From the analysis of the relative area of each peak, we derived the photoionization and photodissociation cross section and also the postproduction cross section of a given fragment. These procedures will be discussed further.

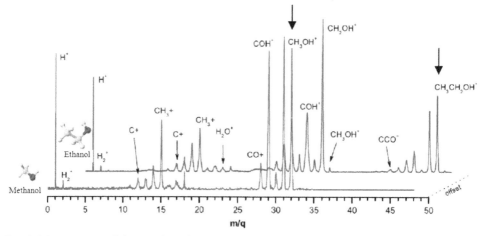

Fig. 8. Mass spectra of the methanol and ethanol fragments produced during irradiation with 292 eV soft X-ray photons.

The mass spectrum of the two simplest carboxylic acids observed in space is given in Figure 9, which presents the fragments produced during the irradiation with 290 eV (~43 Angstroms) soft X-rays photons. The arrows indicate the molecular ion in each mass spectrum. By comparing the relative area of both molecular ions, we observe that formic acid is much more sensitive to this photon energy than acetic acid. Therefore, if the same amount of these molecules is exposed to soft X-rays, after a certain amount of time, the region will be richer in acetic acid than formic acid.

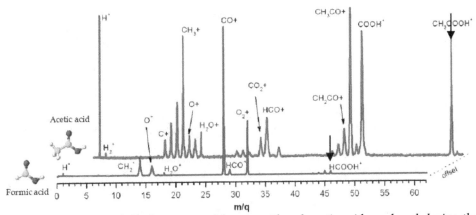

Fig. 9. Mass spectra of the fragments of formic acid and acetic acid produced during the irradiation with 293 eV soft X-ray photons.

## Cross sections

The absolute cross section values for both photoionization ($\sigma_{ph-i}$) and photodissociation ($\sigma_{ph-d}$) of organic molecules are extremely important as input for molecular abundance models. Sorrell (2001) presented a theoretical model in which biomolecules are formed inside the bulk of icy grain mantles photoprocessed by starlight (ultraviolet and soft X-rays photons). However, the main uncertainty of this equilibrium abundance model comes from the uncertainty of the $\sigma_{ph-d}$ value. Therefore, precisely determining the $\sigma_{ph-d}$ of biomolecules is very important for properly estimating the molecular abundance of those molecules in the interstellar medium that were produced by this mechanism. Moreover, knowing the photon dose $I_0$ and $\sigma_{ph-d}$ values makes it possible to determine the half-life of a given molecule, as discussed by Bernstein et al. (2004).

In order to put the data on an absolute scale, after subtraction of a linear background and false coincidences coming from aborted double and triple ionization (see Simon et al. 1991), we have summed up the contributions of all cationic fragments detected and normalized them to the photoabsorption cross sections available in the literature (e.g. Ishii & Hitchcook 1988). Assuming a negligible fluorescence yield (due to the low carbon atomic number, Chen et al. 1981) and anionic fragment production in the present photon energy range, we assumed that all absorbed photons lead to cation formation. The absolute cross section determination is described elsewhere (Boechat Roberty 2005, Pilling et al. 2006).

Briefly, the non-dissociative single ionization (photoionization) cross-section $\sigma_{ph-i}$ and the dissociative single ionization (photodissociation) cross section $\sigma_{ph-d}$ of studied species can be determined by

$$\sigma_{ph-i} = \sigma^+ \frac{PIY_{M^+}}{100} \qquad (7)$$

and

$$\sigma_{ph-d} = \sigma^+ \left( 1 - \frac{PIY_{M^+}}{100} \right) \tag{8}$$

where $\sigma^+$ is the cross section for single ionized fragments.

From Eq. [7], we can also determine the production cross section of a given cationic fragment, $j$, from

$$\sigma_{ph-i} = \sigma^+ \frac{PIY_j}{100} \tag{9}$$

Figure 10 presents the ionization and dissociation cross section of simplest interstellar alcohol and carboxylic acids in the soft X-ray energy range around carbon inner shell (C1s).

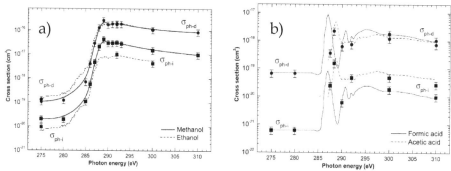

Fig. 10. Ionization and dissociation cross section of a) simples interstellar alcohol and b) carboxylic acids in the soft X-ray energy range around carbon inner shell (C1s).

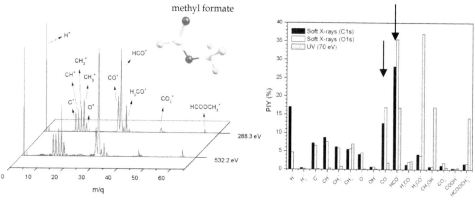

Fig. 11. a) Mass spectra of the methyl formate fragments during irradiation with soft X-ray photons with energies of 288.3 eV (near C1s edge) and 532.2 eV (near O1s edge). b) PIY of methyl formate fragments due to dissociation by soft X-rays (288.3 eV and 532.2 eV) and by UV analogous photon field.

A comparisons between the fragments produced by the dissociation of methyl formate (an isomer of acetic acid) due to soft X-rays with energy around carbon inner shell (C1s) and oxygen inner shell (O1s) is illustrated in Fig. 11a (from Fantuzi et al. 2011). Figure 11b shows a comparison between partial ion yields (PIY) of methyl formate fragments due to dissociation by soft X-rays, at 288.3 eV and 532.2 eV and by 70 eV electrons from the NIST database. The dissociation induced by 70 eV electrons is very similar to the dissociation induced by UV photons

From the PIY analysis at both energies, we observe that for this molecule, the production of CO+ and HCO+ is enhanced when photons at O1s are employed (see arrows). This production is also very enhanced when compared with UV photons. The molecular ion did not have significant changes in this photon energy range.

Another important interstellar molecule studied was benzene ($C_6H_6$). This species may be taken as the basic unit for the polycyclic aromatic hydrocarbons (PAHs), which are believed to play an important role in the chemistry of the interstellar medium (Woods et al. 2003). It may also serve as a precursor molecule to more complex organic compounds, such as amino acids like phenylalanine and tyrosine. Figure 12a shows a comparison of two PEPICO Mass spectra of benzene molecule recorded at (a) UV photons (21.21 eV) and (b) soft X-ray photons (289 eV) (Boechat-Roberty et al. 2009). Figure 12b presents the photoionization and photodissociation cross sections of benzene around carbon C1s edge. The photoabsorption cross-section, $\sigma_{ph-abs}$ (solid line), taken from Hitchcock et al. (1987), is also shown.

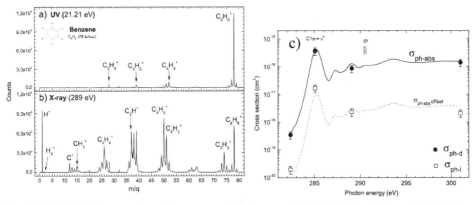

Fig. 12. PEPICO Mass spectra of benzene molecule recorded at (a) UV photons (21.21 eV) and (b) soft X-ray photons (289 eV). c) Photoionization cross section and photodissociation cross section of benzene around C1s edge as a function of photon energy. The dashed line is an off-set of photoabsorption cross-section and is only to guide the eyes (from Boechat-Roberty et al. 2009).

### Kinetic Energy of ionic fragments

The present time-of flight spectrometer was designed to fulfill the Wiley-McLaren conditions for space focusing (Wiley & McLaren 1955). Within the space focusing conditions, the observed broadening of peaks in spectra is mainly due to kinetic energy release of fragments. A Schematic view of the peak broadening promoted by different orientation of produced fragments with the same initial velocity ($v_0$) at the interact region inside the time-of-flight spectrometer is shown in Figure 13.

Considering that the electric field in the interaction region is uniform, we can determine the released energy in the fragmentation process ($U_0$) from each peak width used by Simon et al. (1991), Hansen et al. (1998), and Santos et al. (2001)

$$U_0 = \left(\frac{qE\Delta t}{2}\right)^2 \frac{1}{2m} \tag{10}$$

where $q$ is the ion fragment charge, $E$ is the electric field in interaction region, $m$ is the mass of fragment, and $\Delta t$ is the time peak width (FWHM) taken from PEPICO spectra.

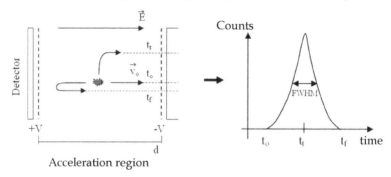

Fig. 13. Schematic view of the peak broadening promoted by different orientation of produced fragments with the same initial velocity ($v_0$) at the interact region inside the time-of-flight spectrometer.

| Fragments | | PIY (%) / $U_0$ (eV) | | | | | | |
| m/q | Attribution | 200 eV | 230 eV | 275 eV | 280 eV | 290 eV | 300 eV | 310 eV |
|---|---|---|---|---|---|---|---|---|
| 1 | $H^+$ | – | 0.71 / 0.24 | 0.19 / 0.24 | 1.33 / 2.17 | 1.32 / 2.17 | 2.36 / 2.87 | 2.83 / 4.80 |
| 7 | $CH_3^+$; $CO^{++++}$? | – | 1.57 / 30.5 | 0.86 / 11.0 | 0.72 / 13.8 | 0.47 / 8.84 | 0.52 / 9.97 | 0.64 / 10.0 |
| 8 | $O^{++}$ | – | – | 0.65 / 2.45 | 0.57 / 12.0 | 0.30 / 10.9 | 0.38 / 8.71 | 0.49 / 2.45 |
| 9.3 | $CO^{+++}$ ? | – | – | 1.21 / 52.0 | 0.91 / 87.8 | 0.83 / 67.5 | 0.91 / 33.5 | 0.95 / 45.8 |
| 12 | $C^+$ | – | – | 0.26 / 1.28 | 0.56 / 0.50 | 0.49 / 0.40 | 0.93 / 0.98 | 1.19 / 0.98 |
| 13 | $CH^+$ | – | – | – | 0.30 / 1.18 | 0.30 / 1.04 | 0.49 / 1.04 | 0.47 / 0.16 |
| 14 | $CH_2^+$; $CO^{++}$ | 26.7 / 0.07 | 24.6 / 0.11 | 22.6 / 0.72 | 20.6 / 0.72 | 17.7 / 0.61 | 17.1 / 0.43 | 18.2 / 0.96 |
| 16 | $O^+$ | 20.9 / 0.30 | 20.0 / 1.99 | 16.8 / 1.99 | 15.9 / 1.65 | 13.1 / 0.96 | 13.4 / 1.22 | 14.9 / 1.50 |
| 17 | $OH^+$ | – | – | – | 0.45 / 0.06 | 0.49 / 0.17 | 0.59 / 0.13 | 0.75 / 0.17 |
| 18 | $H_2O^+$ | – | 0.66 / 0.12 | 0.25 / 0.05 | 0.55 / 0.05 | 0.67 / 0.03 | 0.64 / 0.03 | 0.60 / 0.05 |
| 28 | $CO^+$ | 37.9 / 0.02 | 37.4 / 0.02 | 39.9 / 0.02 | 37.7 / 0.02 | 41.1 / 0.02 | 39.2 / 0.02 | 37.3 / 0.03 |
| 29 | $HCO^+$ | – | 0.64 / 0.10 | 0.31 / 0.03 | 2.89 / 0.13 | 3.08 / 0.13 | 3.24 / 0.17 | 3.19 / 0.16 |
| 32 | $O_2^+$ | 14.4 / 0.03 | 13.4 / 0.03 | 14.1 / 0.03 | 13.2 / 0.02 | 15.5 / 0.02 | 14.4 / 0.02 | 13.3 / 0.01 |
| 44 | $CO_2^+$ | – | – | 0.16 / 0.03 | 0.63 / 0.02 | 0.56 / 0.02 | 0.72 / 0.02 | 0.75 / 0.03 |
| 45 | $COOH^+$ | – | 0.76 / 0.01 | 0.14 / 0.16 | 1.18 / 0.03 | 1.26 / 0.03 | 1.29 / 0.02 | 1.19 / 0.03 |
| 46 | $HCOOH^+$ | – | 0.14 / 0.11 | 0.11 / 0.06 | 0.91 / 0.01 | 0.97 / 0.01 | 1.02 / 0.02 | 0.94 / 0.01 |

Table 1. Relative intensities (partial ion yield – PIY) and kinetic energy $U\_0$ *release by* fragments in the formic acid mass spectra, as a function of photon energy. Only fragments with intensity >0.1% were tabulated. The estimated experimental error was 10% (from Boechat-Roberty et al. 2005)

In order to test the above equation, we measured the argon mass spectrum under the same conditions. The kinetic energy value achieved for the Ar+ ions is in agreement with the mean kinetic energy $(3/2)KT$ obtained assuming Maxwell's distribution law. For example, the calculated values for kinetic energy release ($U_0$) of formic acid fragments produced by the interaction with soft X-ray photons in the range from 200 eV to 310 eV is shown in Table 1 (from Boechat-Roberty et al. 2005). We observe that, in the case of formic acid, the highest kinetic energy release was associated with the lightest fragment H+ ($m/q = 1$), as expected. Extremely fast ionic fragments ($U_0 > 10$ eV), usually associated with dissociation of doubly or multiply-charged ions, were also observed at high photon energies. These observations point to the important role of Auger process ion fragmentation of core-ionized polyatomic molecules. The Coulomb explosion associated with the Auger process should explain the increase in kinetic energy of the ionic fragments, reflected in the increasing broadening of several fragments. This broadening observed in simple coincidence spectra (PEPICO) and its consequence on the shape of peaks in mass spectra was discussed by Simon et al. (1993).

### 2.1.2 Multicoincidence spectra (PE2PICO, PE3PICO)

In addition to the PEPICO spectra, two kinds of other coincidence mass spectra can be obtained simultaneously, PE2PICO spectra (PhotoElectron Photoion Photoion Coincidence) and PE3PICO spectra (PhotoElectron Photoion Photoion Photoion Coincidence). These spectra have ions from double and triple ionization processes, respectively, which arrive coincidentally with photoelectrons. Of all signals received by the detectors, only about 10% come from PE2PICO and 1% from PE3PICO spectra, indicating that the majority of the contribution is due to single event coincidence. Typical PE2PICO and PE3PICO spectra are given in Figure 14a and 14b, respectively. Inset figures show the *coincidence figures* (parallelogram or elliptical shape-like set of points) of a given set of fragments. The slopes of the coincidence figures in PE2PICO and PE3PICO spectra provide the information about the dissociation mechanism of the fragments that reach the detector in coincidence with the photoelectron (Pilling et al. 2007b; Pilling et al 2007c).

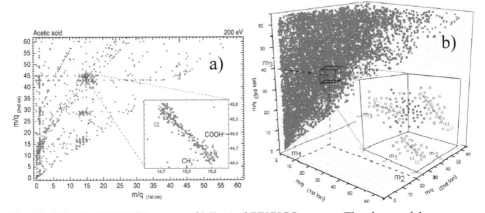

Fig. 14. a) Typical PE2PICO spectra. b) Typical PE3PICO spectra. The slopes of the coincidence figures in PE2PICO and PE3PICO spectra give the information about the dissociation mechanism of selected fragments.

The multi-coincidence detection technique was introduced by Fransiski et al. (1986) and Eland et al. (1986, 1987) in experiments involving energies at valence level near the ionization of inner shell electrons. In such regions, the probability of multiple ionizations increases due to the ionization processes such as Auger. In PE2PICO, the photoexitation/photoionizaton mechanism results a double ionization of parent molecule ($M^{++}$) which can dissociate or not. A typical reaction is illustrated by

$$M + h\nu \longrightarrow M^+ + e^-$$
$$M^+ \xrightarrow{Auger} M^{++} + e^-$$
$$M^{++} \longrightarrow M_1^+ + M_2^+ \text{ (+ neutrals)}.$$

The production of negative ions as the result of multi-ionized species is also observed but this probability is negligible compared with positively charged ions.

Following Eland (1989) and Simon et al. (1993) the dissociation dynamics of double charge species can be studied by analyzing the coincidence figure (see inset Fig. 14a) of the ions in double coincidence mode in PE2PICO spectra. These authors suggested several mechanisms that may occur from the ionization with soft X-rays photons. Some examples are given below:

i.   Two Body dissociation: This is the simplest dissociation case: The slope of coincidence figure (see inset figure 14a) is exactly - 45 degrees ($\alpha=-1$).

$$AB^{++} \longrightarrow A^+ + B^+$$

ii.  Tree body dissociation: In this situation a third neutral fragment is produced from the photodissociation. Some examples are given below:

-    Concerted dissociation: In this case, the kinetic energy distribution (or momentum) between the three bodies in not necessarily unique. This gives one coincidence figure with a ovoid (or circular) shape, in particular if the momentum of the fragments were not aligned (Eland et al. 1986).

$$ABC^{++} \longrightarrow A + B^+ + C^+$$

-    Deferred charge separation: In this case, the dissociation occurs in two steps. In the first the double charge molecule is broke into two species, but the charge is retained in one of the instable fragment. After that, this double charged fragment dissociates into two small single charge fragments.

$$ABC^{++} \longrightarrow A + BC^{++}$$
$$BC^{++} \longrightarrow B^+ + C^+$$

The kinetic energy released in the last reaction (columbic explosion) is much higher than involved in the first step. Therefore, the momenta of the fragments can be treated as $p_A \approx 0$, and $p_B = -p_C$. As in the two body dissociation, this results in a coincidence figure with exactly - 45 degrees ($\alpha=-1$).

-    Secondary decay: In this situation the molecular ion first brakes in two charged fragments and then one of these fragments brakes producing a neutral fragment and another small charged species.

$$ABC^{\cdot++} \longrightarrow B^+ + AC^+$$
$$AC^+ \longrightarrow A + C^+$$

Following Simon et al. (1993), the slope of coincidence figure in PE2PICO spectra in this case is given by:

$$\alpha = -\frac{m_c}{m_a + m_c} \quad \text{for} \quad m_b > m_c \quad \text{or} \quad \alpha = -\frac{m_a + m_c}{m_c} \quad \text{for} \quad m_b < m_c$$

In PE3PICO, the photoexitation/photoionizaton mechanism results a triple ionization of parent molecule ($M^{+++}$) which can dissociate or not. A typical reaction is illustrated by

$$M + h\nu \longrightarrow M^+ + e^-$$
$$M^+ \xrightarrow{Auger} M^{+++} + 2e^-$$
$$M^{+++} \longrightarrow M_1^+ + M_2^+ + M_3^+ \text{ (+ neutrals)}.$$

In this case, the coincidence figure involving a given set of three fragments has three dimensions (See inset Fig. 14b). Some information about the dissociation mechanism involving these fragments can be given by analyzing the projection of its coincidence figure projected in each plane, however, the dissociation mechanism in this case is difficult to determine.

From the analysis of the fragment's produced mass spectra, we determined the branching ratio of the fragments. Employing a similar methodology, as previously discussed, the photodissociation and photoionization cross sections in double and triple ionization can also be determined.

The relative intensities of each ion pair coincidences in each spectrum can be determined directly from spectra analysis. In the case of PE2PICO spectra, we obtained the partial double coincidence yield (PDCY) by:

$$\text{PDCY}_{i,j} = \left( \frac{A_{i,j}}{A_t^{2+}} \pm \frac{\sqrt{A_{i,j}} + A_{i,j} \times \text{ER}/100}{A_t^{2+}} \right) \times 100\% \tag{11}$$

where $A_{i,j}$ is the number of events in double coincidence of a given $i$ and $j$ ion pair, the $A_t^{2+}$ is the total number of counts of PE2PICO spectra, and ER = 2-4% is the estimated uncertainty due to the data acquisition and the data treatment (Pilling et al 2007b).

In a similar manner, for the PE3PICO spectra, we determined the Partial Triple Coincidence Yield (PTCY) (Pilling et al 2007b, Pilling et al 2007c).

From the cross section, we can derive the photo-dissociation rate and photo-production rate (of a given fragment), if we know the soft X-ray flux in a give astrophysical region. For example, the H$^{3+}$ photoproduction rate due to the dissociation of methyl compound molecules by soft X-rays (200–310 eV) is given by the simple expression

$$k_{\text{ph}} = \int \sigma_{\text{H}_3^+}(\varepsilon) F(\varepsilon) \, d\varepsilon \sim \sigma_{\text{H}_3^+} F_{\text{softX}} \quad (\text{s}^{-1}) \tag{12}$$

where $\sigma_{H_3^+} = \sigma_{H_3^+}^+ + \sigma_{H_3^+}^{++}$ and $F_{\text{softX}}$ is the averaged $H^{+3}$ photoproduction cross-section and photon flux over the soft X-ray energy (200–310 eV), respectively (Pilling et al 2007d).

The photodissociation rate can be obtained employing the photodissociation cross section $\sigma_{\text{ph-d}}$ and soft X-ray photon flux in the given astrophysical scenario.

$$k_{ph-d} = \int\limits_{softX} \sigma_{ph-d}(\varepsilon)F(\varepsilon)d\varepsilon \approx \sigma_{ph-d} \times F_{softX} \ (s) \tag{13}$$

The half-life $(t_{1/2})$ of the studied compounds in the presence of interstellar soft X-rays can be obtained directly from

$$t_{1/2} = \frac{\ln 2}{k_{ph-d}} \approx \frac{0.69}{\sigma_{ph-d} \times F_{softX}} \ (s) \tag{14}$$

## 2.2 Condensed-phase (ice) and solid-phase experiments

In the cold and dense interstellar regions in which stars are formed, CO, $CO_2$, $H_2O$, and other molecules collide with and stick to cold (sub)micron-sized silicate/carbon particles, resulting in icy mantles. Inside these regions, called molecular clouds, the dominant energy source is cosmic rays and soft X-rays from embedded sources or the interstellar radiation field. These kinds of radiation can also produce an expressive amount of fast electrons, which can also promote chemical differentiation on ices in a different way when compared with photons.

Several experiments employing the interaction between soft X-rays and frozen (or solid) compounds were performed to investigate the effects of stellar ionizing radiation in astrophysical ices. Two different techniques were employed, one to investigate the species, which desorbs from the surface due to the irradiation, and another to analyze the bulk (of the ice) during the irradiation.

## 2.2.1 Photo stimulation ion desorption

In an attempt to investigate the fragments that were released into gas phase due to the impact of soft X-rays on astrophysical ices, we employed the photo stimulation ion desorption (PSID) technique by using the spherical grating monochromator (SGM) beam line in single-bunch mode (pulse period of 311 ns, with a width of 60 ps) at the Brazilian synchrotron light source (LNLS). By using soft X-rays photons, one can excite specific atoms inside a molecule by tuning the incoming radiation as a consequence of the different chemical shifts. Therefore, this technique is element- and site-specific. Several species were investigated by this technique including formic acid and methanol.

Formic acid (HCOOH) is the simplest carboxylic acid, while methanol ($CH_3OH$) is the simplest alcohol. These molecules have been found abundantly in icy mantles on interstellar and protostellar dust grains. These small molecules together with glycine (the smallest amino acid) serve as model systems to understand the properties of larger and more complex amino acids and proteins. Formic acid, for example, has been observed in several astronomical sources such as comets (Kuan et al. 2003, Snyder et al. 2005, Jones et al. 2007), protostellar ices NGC 7538:IRS9 (Cunningham et al. 2007), condritic meteorites (Cronin and Pizzarrelo, 1997), dark molecular clouds (Turner, 1991), and regions associated with stellar formation (Nummelin et al. 2000; Ehrenfreund et al. 2001).

In some massive star-forming regions such as Sgr B2, Orion KL, and W51, formic acid has been observed. Kuan et al (2003) suggest that glycine may be detected in these regions, but more observation are necessary to confirm this possibility Methanol has been detected through infrared spectroscopy in some low- and high-mass protostars such as W33A and RAFGL 7009. In these regions, $CH_3OH$ is the most abundant solid-state molecule after $H_2O$ (Dartois et al. 1999; Pontoppidan et al. 2003; Boogert et al. 2008).

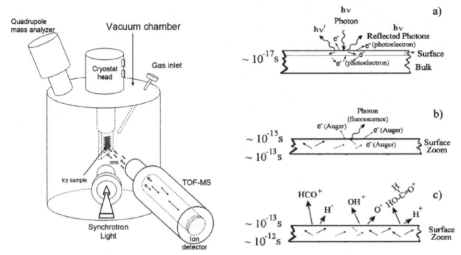

Fig. 15. Schematic diagram of the experimental set-up that was employed in the photo stimulation ion desorption of astrophysical ice analogs (left). Scheme of desorption process due to X-ray impact on the ice surface, simulating the interstellar ice (right).

When X-ray photons interact with the grain surfaces, the icy molecules can dissociate, producing small neutral or ionized species and atoms. If the surface temperatures are around 50 K, the radicals can diffuse and associate to form larger molecules. Afterwards, these neutral and ionic species can sublimate from the surface by thermal or non-thermal desorption mechanisms. The non-thermal desorption mechanisms are those stimulated by photons or energetic charged particles. Each ionizing agent will promote a different fragmentation in the molecule, favoring the formation of ionic species rather than neutral species, depending on the impact energies. The inner-shell photoexcitation process, for example, may produce instabilities on the molecular structure leading to peculiar dissociation pathways. The molecules and ions formed and desorbed into the gas phase, as well as the complex molecules formed from these species, will depend on the interaction of radiation field in the considered region. To quantify the complex organic molecules incorporated in grains or desorbed into the gas phase and to predict the chemical evolution, it is necessary to establish the main formation pathways, which can be tested in laboratories. Therefore, studying the effects of different ionization agents on the ices is crucial in the knowledge of evolution of interstellar chemistry.

In this chapter, the main dissociate channel in the molecules are produced by hole excitation/ionization. When grains are processed by soft X-ray photons, the photons can be transmitted, absorbed, reflected, or scattered by the surface. Figure 15 shows a schematic diagram of desorption process by X-ray photons on the ice surface, simulating the effects of the radiation on interstellar/cometary ices. In Figure (15a), an X-ray photon is absorbed ($t \approx$ $10^{-17}$ s), promoting an electron into an excited state or to the continuum, creating a core hole (a hole in an internal orbital of the atom) on the molecule. As shown in Figure 15b, after relaxation, X-ray fluorescence or Auger electron emission can occur, depending on the atomic number of the atom involved. If the atomic number of the involved atom is below ~16, the Auger decay has probability ~1. On the other hand, if the atomic number is above 16 the X-ray fluorescence process can occur with high probability (Jenkins 1999). For molecules such as formic acid and methanol, which will be shown here, the atoms have a low atomic number. In this case, the probability of fluorescence vanishes and the Auger decay is dominant (Andrade et al, 2010). Mainly because of Coulomb repulsion of positive holes in valence orbitals, molecular dissociation and desorption occur (Figure 15c). The soft X-ray photons interact directly with surface molecules. In this way, the fragments are desorbed mainly from the interaction region. The energy deposited into the system is enough to produce resonant core level transition, which means localized excitation. The final step in this process is bond scission at the site of excitation and consequently desorption.

The experimental setup, presented here, includes a sample manipulator (where the molecules are condensed onto an Au thin film) connected at a helium cryostat and a time-of-flight mass spectrometer (TOF-MS) housed in an Ultra High Vacuum (UHV) chamber with a base pressure around $10^{-9}$ Torr, which increased by two orders of magnitude (1.5x $10^{-7}$ mbar) when the gas was introduced for condensation for around six min. The TOF-MS consists basically of an electrostatic ion extraction system, a drift tube and a pair of microchannel plate (MCP) detectors, disposed in the Chevron configuration. After extraction, the ions (positive and negative) travel through three metallic grids before reaching the MCP. The grids have a nominal transmission of ~90 per cent.

The time-of-flight (TOF) of a specific desorbed ion is given by its position in the spectrum plus an integer multiple of the synchrotron radiation (SR) pulse interval (311 ns) corresponding to the number of cycles that have passed from desorption of the ion to its detection. The mass resolution was limited by the resolution of the TOF spectrometer and by the time resolution of the TDC (250 ps ch$^{-1}$). The typical measured full width at half-maximum FWHM found for ionic species was 1.5 < FWHM < 20 ns, being 1.5 for H$^+$. In the photon case, the FWHM was ~250 ps.

The attribution of PSID spectra obtained in single-bunch mode is not an easy task. The time window of 311 ns is small enough even for the hydrogen ion. Thus H$^+$ and heavier fragments are detected at different cycles (or starts) of the experiment, causing the overlapping of many different species. However, because of the very good reproducibility of the SR pulses and the good resolution achieved in these experiments, it was possible to identify different structures and suggest the assignment of the TOF spectra.

In order to determine the photodesorption yield (Y) for the ions from condensed molecules, we obtained the area of each peak. $Y_i$ is the number of ions desorbed per incident photon for each ion $i$ and it was determined by dividing the peak area ($A$) by the number of bunches ($N_b$) and by the number of photons per bunch ($n_{ph}$):

$$Y_i = \frac{A}{N_b \, n_{ph}}. \tag{15}$$

The number of photons incident on a surface of 2 mm$^2$ during 60 ps (duration of the pulse), at the single-bunch mode, is about 1550 photons bunch$^{-1}$, which corresponds to a photon flux of $5 \times 10^{11}$ photons s$^{-1}$ cm$^{-2}$. The errors associated with ion yield measurements depend on peak position. For single peaks, the error was around 10 per cent. For blended peaks, the estimated errors were as high as 35 %.

The icy sample temperature was around 55-56 K during all measurements. The ice thickness was in the micrometer range. Due to the temperature and pressure conditions in the chamber, the formed ice was expected to be in its amorphous phase, similar to the icy organic found in comets and in the interstellar medium.

Figure 16 shows the total ion yield (TIY) of Near-edge X-ray Absorption Fine Structure (NEXAFS) spectrum at the oxygen 1s-edge for condensed formic acid (HCOOH), covering the photon energy range from 528 to 553 eV (Andrade et al. 2009). The NEXAFS spectrum has high-pitched peaks (B and D) and one broad band (F), which are related to electronic transitions from the oxygen 1s electron to unoccupied molecular orbitals. In comparison with gas phase NEXAFS data on formic acid (Prince at al. 2002; Hergenhahn et al. 2003), the peaks B can be assigned in the figure. The broad band F is located above the ionization threshold.

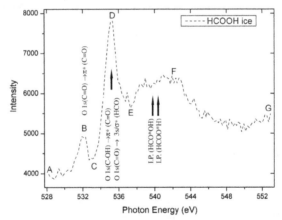

Fig. 16. The total ion yield (TIY) of Near-edge X-ray Absorption Fine Structure (NEXAFS) spectrum at the oxygen 1s-edge for condensed formic acid, covering the photon energy range from 528 to 553 eV.

Figure 17 shows typical positive (a) and negative (b) time-of-flight mass spectra of condensed formic acid obtained at the excitation energy labeled D (535.1 eV) (Andrade et al. 2009) and (c) positive ions and (d) negative ions TOF spectra from methanol ice obtained at 537 eV (Andrade et al. 2010). In both cases, $CH_2^+$ and/or $CO_2^+$ represent the main ionic molecular signal desorbed after photon excitation at the O 1s-edge followed by intense H$^+$. In the negative spectra, only H$^-$ and O$^-$ appear. Heavier fragments with lower intensities, such as C$^+$, CH$^+$, O$^+$, CO$^+$, and HCO$^+$ were found in the HCOOH spectra, and the (HCOOH)H$^+$ cluster is probably present. In the methanol case, $CH_3^+$ species was also formed with low intensity, and unlike in the case of formic acid, CH$^+$ was not seen.

Measurements of TOF spectra from HCOOH at different photon energies (A–G) around the O 1s-edge were also performed in an attempt to understand the ionic desorption process from condensed formic acid. Table 2 shows the edge-jump (the $I_{above}/I_{below}$ ratio) values for the total ion yield photoabsorption spectrum for several ions, where $I_{above}$ and $I_{below}$ represent the intensity signals above and below the absorption edge, respectively. Moreover, this table shows the positive and negative desorption yield (desorbed ion for photon incident) to the formic acid and methanol.

In the HCOOH as well as the $CH_3OH$ spectra, $OH^+$ ion peak overlaps with $H_3^+$ and $C_2^+$ peaks, thus it is difficult to separate their contributions. In the case of HCOOH, these species, as well as the $CH_2^+$ (and/or $CO_2^+$), $CO^+$, and $C_2^+$ do not show a significant increase at the D resonance, suggesting that the X-ray induced electron stimulated desorption (XESD) process, which competes with the ASID (Auger stimulated ion desorption) process, is responsible for the desorption signal. In the XESD process, desorption of surface species occurs due to outgoing energetic Auger and photoelectrons, most of which originate in the bulk (Ramaker et al. 1988; Purdie et al. 1991). The XESD process is proportional to the total ion yield and the desorbed ions appear below and above the absorption edge. Its ratios follow those of the photoabsorption spectrum.

Taking into account the positive and negative desorbed ions from formic acid, some possibilities for the $O^-$ and $H^-$ formation from singly charged desorbed ions are suggested below (from the ions seen in the positive ion spectra) (Andrade et al. 2009):

$$
\begin{aligned}
HCOOH^+ &\rightarrow O^- + HCO^+ + H^+ & [1]\\
&\rightarrow H^- + C^+ + H^+ + O_2 & [1]\\
&\rightarrow O^- + CH^+ + H^+ + O & [3]\\
&\rightarrow H^- + CH^+ + O^+ + H & [4]\\
&\rightarrow O^- + C^+ + H_2 + O^- & [5]\\
&\rightarrow H^- + HCO^+ + O^- & [6]\\
&\rightarrow O^- + C^+ + H^+ + O & [7]
\end{aligned}
$$

It is believed that the suggested reactions contribute most to formic acid dissociation, because we are working around the O 1s-edge. Thus, mechanisms like electron capture are not considered in our discussion and non-detected positive ions do not participate in the above reactions. OH species is probably released in the neutral form or HCOOH breaks giving H and O ions, rather than OH cations and anions, because this species showed low intensity in the formic acid positive spectra and was absence in the negative spectra. Because $O^+$, $O^-$, and $H^-$ yields are very intense at D resonance, this second option appears favorable. In the gaseous phase, the formation of the anionic OH fragment, via resonant excitation at the carbon 1s-edge and its complete absence near the oxygen 1s-edge was reported by (Stolte et al. 2002). $C_2^+$ was detected with a weak intensity in the formic acid spectra and large amounts of $CH_2^+$, $HCO^+$, $H^+$, and $O^+$ were present, but all of these fragments can be formed due to many different pathways, not only as counterparts of $O^-$ and $H^-$.

By comparing Table 2 and the cation yields for $O^-$ and $H^-$, we observe that $O^+$, $C^+$, and $CH^+$ have similar behavior as these anions at the D resonance, suggesting that Reactions [1]-[7] are strong candidates for $H^-$ and $O^-$ formation at O 1s-edge excitation. In addition, taking into account that $C^+$, $CH^+$, and $O^+$ ions show enrichment at the D resonance, the $O^-$ and $H^-$ formation from the Reactions [4] and [5] seems to be more favorable (Andrade et al. 2009).

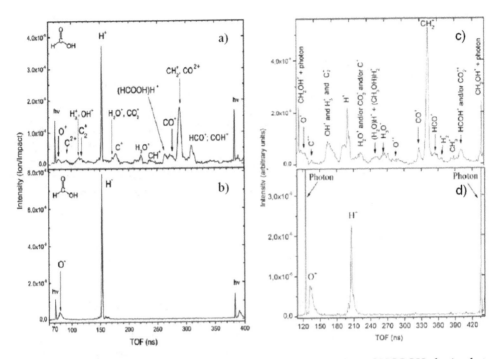

Fig. 17. a) positive ions and b) negative ions TOF spectra of condensed HCOOH obtained at 535.1 eV energy and c) positive ions and d) negative ions TOF spectra from methanol ice obtained at 537 eV.

| Ion assignment | ID/IB (HCOOH) | ID/IF (HCOOH) | $Y_i$ (x10$^{-9}$) HCOOH | $Y_i$ (x10$^{-9}$) CH$_3$OH |
|---|---|---|---|---|
| NEXAFS (TIY) | 1.60 | 1.22 | | |
| H$^+$ | 1.31 | 0.76 | 2.88 | 10.26 |
| O$^+$ | 2.33 | 1.88 | 1.14 | 5.006 |
| C$_2^+$/OH$^+$/H$_3^+$ | 1.82 | 1.22 | 8.45 | - |
| C$^+$ | 2.20 | 1.79 | 1.83 | - |
| CH$^+$ | 1.90 | 1.95 | 0.65 | - |
| CH$_2^+$/CO$^{2+}$ | 1.36 | 1.29 | 11.5 | 19.10 |
| (HCOOH)H$^+$ | 1.21 | 0.72 | 0.82 | - |
| HCO$^+$ | 1.47 | 0.91 | 1.39 | 1.619 |
| CO$^+$ | 1.47 | 1.11 | 0.47 | 1.342 |
| H- | 3.28 | 1.51 | 11.0 | 5.550 |
| O- | 2.25 | 1.22 | 1.70 | 4.832 |

Table 2. Edge-jump (the $I_{above}/I_{below}$ ratio) values for the total ion yield photoabsorption spectrum for several ions and $Y_i$ desorption ion yield (desorbed ion for photon incident) to HCOOH and CH$_3$OH ice.

After photon impact, desorption from the surface of multiple charged ions is not the most probable event because fast neutralization compete efficiently with ion desorption. In addition, there is the possibility of reactions involving doubly charged ions. $S_2^+$ was measured in the photofragmentation study of poly(3-methylthiophene) (Rocco et al. 2004; Rocco et al. 2006) and condensed thiophene (Rocco et al. 2007) following sulphur K-shell excitation and $Cl_2^+$ was observed from solid $CCl_4$ at the Cl 1s → $\sigma^*(7a_1)$ excitation (Babba et al. 1997). To explain the primary H- and O- production mechanism from doubly-charged ions, the most produced positive ion yields are that of $CO^{2+}$ and $C^{2+}$, which are possibly present in the positive PSID spectrum (Reactions [8]–[10]). However, enrichment of neither $CO^{2+}$ nor $C^{2+}$ associated with H- and O- increase at the D resonance is seen, suggesting that H- and O- formation from formic acid occur mainly from Reactions [4] and [5]. In the case of methanol, $CO^{2+}$ and $O^{2+}$ are detected and $C^{2+}$ cannot be excluded because its peak overlaps with $CH_2^+$, which is predicted to be a more abundant product.

$$HCOOH^* \rightarrow H + CO^{2+} + OH \qquad [8]$$
$$\rightarrow O^- + C^{2+} + OH + H \qquad [9]$$
$$\rightarrow H^- + C^{2+} + O_2 + H \qquad [10]$$

Neither $COOH^\pm$ nor $HCOOH^\pm$ is formed from HCOOH fragmentation. Therefore, if an O 1s electron is excited, HCOOH dissociate into smaller or atomic fragments. Pelc et al. (2002) showed, from theoretical calculations, that it is not possible by the thermodynamical point of view to bind an extra electron to HCOOH, because it has a negative adiabatic electron affinity.

In contrast to comparable gas-phase studies, $CH_3OH^+$ was not detected in the methanol positive mass spectra. This implies that either $CH_3OH^+$ is less stable in the ice phase compared to in the gas phase or that $CH_3OH^+$ is produced in the ice but lacks sufficient energy to desorb and thus escapes detection. Ionized formaldehyde, $H_2CO^+$, could also be identified and is intense in our methanol positive ion spectrum. This species has also been identified by other authors, in neutral form, using Fourier transform infrared spectroscopy (FTIR; Gerakines et al. 1996; Hudson & Moore 2000; Bennett et al. 2007). Palumbo et al. (1999) also observed a band at 1720 cm$^{-1}$ after ion irradiation with 3 keV helium ions in pure methanol, but admitted that there could be other species, such as acetone, contributing to this feature. Acetone was not found in our spectra.

$$
\begin{aligned}
CH_3OH^+ &\longrightarrow & CH_2^+ + H^- + OH^+ & \qquad [11] \\
&\longrightarrow & O^- + H^+ + CH_2^+ + H & \qquad [12] \\
&\longrightarrow & H^- + HCOH^+ + H^+ & \qquad [13] \\
&\longrightarrow & H^+ + H_2 + H^- + CO^+ & \qquad [14] \\
&\longrightarrow & H_3^+ + H^- + CO^+ & \qquad [15] \\
&\longrightarrow & HCO^+ + H^- + H_2^+ & \qquad [16] \\
&\longrightarrow & C^+ + OH + H^- + H^+ & \qquad [17] \\
&\longrightarrow & C^+ + O^- + H^+ + H_2 & \qquad [18] \\
&\longrightarrow & C^+ + O^- + H + H_2^+ & \qquad [19] \\
&\longrightarrow & C^+ + O^- + H + H_2^+ & \qquad [20] \\
&\longrightarrow & O^+ + H^- + CH_3^+ & \qquad [21] \\
&\longrightarrow & O^- + H^+ + CH_3^+ & \qquad [22]
\end{aligned}
$$

Some possible pathways for the O⁻ and H⁻ formation from singly charged desorbed ions, which were observed in the methanol positive ion spectra, are suggested above. For the reaction pathways, only the detected positive ions were taken into account and other mechanisms, such as electron capture, were not considered in our discussion. Examining the cation yields in Table 2, we suggest that the Reactions [11]-[13] are strong candidates for H⁻ and O⁻ formation from methanol at the O1s excitation energy. Moreover, taking into account that C⁺ and CH₃⁺ ions show low abundance, the O⁻ and H⁻ formation from the Reactions [18]-[22] seems to be less favorable.

In the cases of HCOOH and CH₃OH, the fragmentation pattern of the positive ions is clearly more pronounced than that of the negative ions, showing that positive ions are more easily formed than the negative one.

The astrophysical implication of the x-ray photon flux and the production of ions in the protoplanetary discs will be discussed. A protoplanetary disc is a rotating circumstellar disk of matter, including dust and gas, that surrounds a very young star and from which planets may eventually form or be in the process of forming, representing the early stages of planetary system formation. Measurements of high and intense X-ray fluxes from young stars have been obtained by Space telescopes such as *ROSAT, Chandra,* and *XMM–Newton*. These fluxes can be up to three orders of magnitude higher than for main-sequence stars (Gorti et al. 2009).

The closest known young T Tauri star is TW Hya, which has a massive, face-on optically thick disc. Using the infrared spectrograph onboard the *Spitzer* space telescope, Najita et al. (2010) suggested that a planet was probably formed in the inner region of the disc, at approximately 5 au. and also showed that TW Hya has a rich spectrum of emission of atomic ions (Ne II and Ne III) and molecules resulting from K-shell (Ne 1s) photoionization of neutral neon by X-rays.

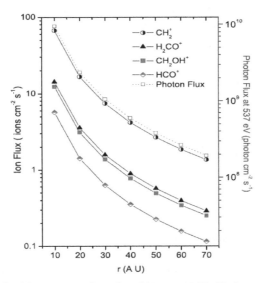

Fig. 18. Ion flux desorbed from icy methanol and X-ray at 537 eV photon flux in TW Hya as a function of the distance from a point in the disc to central star (from Eq. 16 for $\tau = 1$) (Andrade et al. 2010).

TW Hya has a X-ray luminosity ($L_X$) of $2 \times 10^{30}$ erg s$^{-1}$ , integrated from 0.2 to 2 KeV (Kastner et al. 2002). At a specific energy the photon flux $F_X$ can be obtained by:

$$F_X = \frac{L_X}{h_v . 4\pi r^2} e^{-\tau} \tag{16}$$

where $\tau$ is the X-ray optical depth given by $\tau = \sigma_{ab}(E)N_H$ and $r = (R^2 + z^2)^{0.5}$ is the distance from a point in the disc to the star in centimeter . $R$ is the distance from the star to disc and $z$ is the height from the mid-plane (Gorti & Hollenback 2004). $\sigma_{ab}(E)$ is the absorption cross-section as a function of the photon energy $E$, and $N_H$ is the hydrogen column density obtained by the product of the number density ($n$H) and the distance from the central star (Andrade et al. 2010). It was assumed that $\tau = 1$ using the $\sigma_{ab}(E)$ per H atom given by Gorti & Hollenback (2004) model and $N_H \sim 10^{21}$ cm$^{-2}$. $n$H varies from $6.6 \times 10^6$ to $0.95 \times 10^6$ cm$^{-3}$ in the distances range from 10 to 70 au. The ion flux (ions cm$^{-2}$s$^{-1}$) can be estimated by $f_{ion} = F_X Y_i$ ($Y_i$ is the photodesorption yield, ions/photon) from equation (18). Figure 18 shows the photon flux at 537 eV in TW Hya and the flux of the most important ions which are desorbed from the methanol ice as a function of the distance from the central star.

The most abundant ion desorbed by methanol is the $CH_2^+$ followed by ionized formaldehyde and $CH_2OH+$. For example, at 30 au, a flux of $\sim 10^9$ photons cm$^{-2}$ s$^{-1}$ may produce $\sim 9$ $CH_2^+$ ions cm$^{-2}$ s$^{-1}$ and desorb $\sim 0.6$ $HCO^+$ ion cm$^{-2}$ s$^{-1}$. Taking into account the half-life of the region ($1 \times 10^6$ yr), the amount of desorbed $HCO^+$ (formyl ion) and $CH_2^+$ will be around $2 \times 10^{14}$ , and $3 \times 10^{15}$ ions cm$^{-2}$, respectively.

The ion production rate $R_{ion}$ (cm$^{-3}$ s$^{-1}$) from grain surface, is dependent on the ion flux $f_{ion}$, the amount of grains $n_{gr}$ (cm$^{-3}$), their surface cross section $\sigma_{gr}$ (in cm$^2$ which is equal to $\pi a^2$, where $a$ is the radius of the dust particles), and the fraction $X$ of the surface covered by methanol (which can be archived at a range from zero to unity) by the expression:

$$R_{ion} = n_{gr}\sigma_{gr}f_{ion}X \tag{17}$$

Assuming that $n_{gr} \sim 10^4$ cm$^{-3}$ (Gorti et al. 2009), $a = 0.1$ µm, and $X = 1$ (the surface is totally covered by methanol), the $R_{HCO+}$ will be approximately $1 \times 10^{-6}$ cm$^{-3}$ s$^{-1}$. If $X = 0.3$ (70 per cent of the surface is covered by water ice), the $R_{HCO+}$ will be approximately $3 \times 10^{-7}$ cm$^{-3}$ s$^{-1}$.

The half-life of the methanol can be obtained by Eq. 14, which does not depend on the molecular number density. Considering that the photodissociation cross-section at 537 eV is of the same order of magnitude as the photoabsorption cross-section for the gaseous methanol ($\sim 1,5 \times 10^{-18}$ cm$^2$, Ishii & Hitchcook, 1988), the methanol molecules survives $\sim 15$ years at 30 au.

## 2.2.2 Photolysis of astrophysical ices analogs

To complete the investigation about the effects of soft X-rays on the surface of interstellar ices, we also study the chemical evolution inside the ices by employing Fourier transform infrared (FTIR) spectrometry. In these experiments, *in-situ* analysis of the bulk of the astrophysical ice analogs are performed by a FTIR spectrometer, coupled to the experimental chamber, at different X-ray photon doses.

In these experiments, a gaseous sample (mixture) was deposited on a cold substrate (NaCl or CaF₂) coupled to closed-cycle cryostat, inside a high vacuum chamber and exposed to soft X-rays. Most of the samples were investigated at 13 K. Pure samples and several mixtures were investigated including $N_2+CH_4$, $CO_2+H_2O$, and $CO+H_2O$.

Figures 19a-b show a schematic diagram of the experimental setup employing soft X-rays in solid samples of astrophysical interest. In the irradiation, the substrate with the ice sample is in front of the photon beam. After each irradiation dose, the target is rotated by 90° for FTIR analysis.

a)                                                          b)

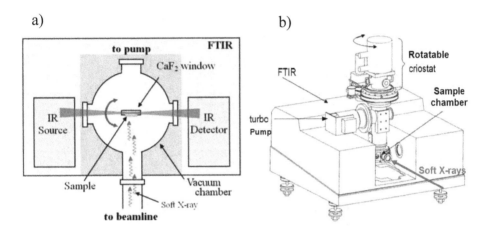

Fig. 19. Schematic diagram of the experimental setup employing soft X-rays on solid samples of astrophysical interest. a) top view. b) perspective view.

In the first set of experiments, we attempted to simulate the photochemistry induced by soft X-rays (from 0.1- 5keV) at aerosol analogs in the upper atmosphere of Titan, the largest moon of Saturn. The experiments were performed inside a high vacuum chamber coupled to the soft X-ray spectroscopy (SXS) beamline, at Brazilian Synchrotron Light Laboratory (LNLS), employing a continuum wavelength beam from visible to soft X-rays with a maximum flux in the 0.5-3 keV range. The frozen sample (Titan´s aerosol analog) was made by a frozen gas mixture containing 95% $N_2$ + 5% $CH_4$ (and traces of $H_2O$ and $CO_2$). The in-situ analysis was done using FTIR spectrometry. Among the different chemicals species produced, we observe several nitrogen-rich compounds including one of the DNA nucleobases, the adenine molecule (Pilling et al. 2009). This detection was confirmed also by ex-situ chromatographic and RMN analysis of the organic residue produced. Unlike UV photons, the interaction of soft X-rays with interstellar ices produces secondary electrons in and on the surface, which allows different chemical pathways.

Figure 20a shows the evolution of simulated Titan aerosol at different exposure times up to 73h (1 h ~ 3 × 10¹⁰ erg/cm²). A comparison between FTIR spectra of 73 h irradiated sample at 15, 200, and 300 K is given in Figure 20b. The vertical dashed lines indicate the frequency of some vibration modes of crystalline adenine. Figure 20c shows a comparison between LNLS´s SXS beamline photon flux and the solar flux at the Titan orbit.

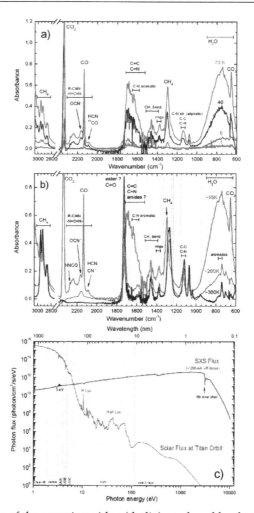

Fig. 20. a) FTIR spectra of the organic residue (tholin) produced by the irradiation of condensed Titan atmosphere analog at 15 K NaCl surface at different exposure times. b) Evolution of FTIR spectra of 73 h irradiated sample at three different temperatures: 15, 200, and 300 K (organic residue). c) Comparison between SXS beamline photon flux and the solar flux at the Titan orbit (adapted from Pilling et al 2009).

In a second experiment, we investigate the photochemistry effects of soft X-rays in CO and $CO_2$ ices covered by water ice (cap) within a soft X-ray field analogous to the ones founds in astrophysical environments. The measurements were taken at the Brazilian Synchrotron Light Laboratory at the soft X-ray spectroscopy (SXS) beamline, employing a continuum wavelength beam from visible to soft X-rays with a maximum flux between 0.5-3 keV range ($10^{12}$ photons cm$^{-2}$ s$^{-1}$ ~ $10^4$ ergs cm$^{-2}$ s$^{-1}$). Briefly, the samples were deposited onto a NaCl substrate cooled at 12 K under a high vacuum chamber ($10^{-7}$ mbar) and exposed to different radiation doses up to 3 h. In-situ sample analysis was performed by a Fourier transform infrared spectrometer (FTIR) coupled to the experimental chamber. Figure 21 shows a) the evolution of the abundance of selected species in the irradiation of CO + Water cap 12 K ice by soft X-rays; b) Evolution of the abundance of selected species in the irradiation of $CO_2$ + Water cap 12 K ice by soft X-rays and c) infrared spectra between 1600-600 cm-1 of the two samples after and before irradiation.

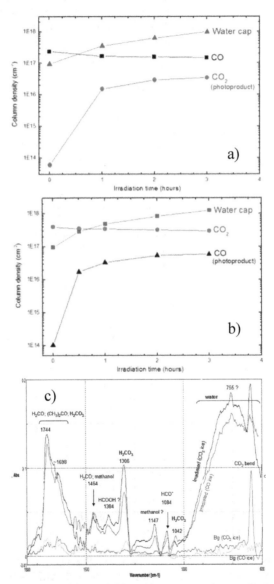

Fig. 21. a) Evolution of the abundance of selected species in the irradiation of CO + Water cap 12 K ice by soft X-rays. b) Evolution of the abundance of selected species in the irradiation of $CO_2$ + Water cap 12 K ice by soft X-rays. c) Infrared spectra between 1600-600 cm-1 of the two samples after and before irradiation.

The results showed that deep inside a typical dense cloud (e.g. AFGL 2591) the half-life of the studied species due to the photodissociation by penetrating X-rays photons was found to be about $10^3$ -$10^4$ years, which is in agreement with the chemical age of the cloud. Moreover, from the photodissociation of condensed CO, $CO_2$, and water molecules several organic species were also produced including $H_2CO$, $H_2CO_3$ and possibly formic acid and methanol. These results suggest that inside dense molecular clouds or other dense regions, where UV photons can not penetrate, the photochemistry promoted by soft X-rays photons (and cosmic rays) could be very active and plays an important role.

In these experiments, the analyzed samples were thin enough (i) to avoid saturation of the FTIR signal in transmission mode and (ii) to be fully crossed by the soft X-ray photon beam. The spectra were obtained in the 600 – 4000 $cm^{-1}$ wavenumber range with a resolution of 1 $cm^{-1}$. From the analysis of the IR spectra, we determined the amount of each molecular species present and its evolution as a function of soft X-ray dose. In addition, we determined the dissociation cross section of the sample and estimated the half-life of the studied species in solid-phase in astrophysical environments. For the low temperature experiments, we identified and quantified newly formed compounds.

In another experiment, we performed a photochemistry study of solid phase amino acids (glycine, DL-valine, DL-proline) and nucleobases (adenine and uracil) under a soft X-ray field in an attempt to test their stabilities against high ionizing photon field analogous to the ones found in dense molecular clouds and protostellar disks. Due to the large hydrogen column density and the amount of dust inside these environments, the main energy sources are the cosmic rays and soft X-rays. In this experiment, 150 eV photons (~ $4 \times 10^{11}$ photons $cm^{-2}$ $s^{-1}$ ~ $10^3$ erg $cm^{-2}$ $s^{-1}$) were employed from the toroidal grating monocromator (TGM) beamline of LNLS (Pilling et al. 2011).

The search for amino acids, nucleobases, and related compounds in the interstellar medium/comets has been performed for the last 30 years. Recently, some traces (upper limits) of these molecules (e.g. glycine and pyrimidine) have been detected in molecular clouds, protostars, and comets. The search for these biomolecules in meteorites, on the contrary, has revealed an amazing number, up to several parts per million! This chemical dichotomy between meteorites and interstellar medium/comets remains a big puzzle in the field of astrochemistry and in the investigation about the origin of life.

The diluted samples were deposited onto a $CaF_2$ substrate by drop casting following solvent evaporation. The sample thicknesses were measured with a Dektak perfilometer and were of the order of 1-3 microns. The samples were placed into a vacuum chamber ($10^{-5}$ mbar) and exposed to different radiation doses up to 20 h. The experiments were performed at room temperature and took several weeks. *In-situ* sample analysis were performed by a Fourier transform infrared spectrometer (FTIR) coupled to the experimental vacuum chamber.

Figure 22a shows the FTIR spectra of amino acid DL-proline after different doses of 150-eV soft X-rays up to 7h. The FTIR spectra of RNA nucleobase uracil after different doses of 150-eV soft X-rays up to 17 h is showed in Figure 22b. Figure 22c shows the integrated absorbance spectra of the solid-phase samples as a function of irradiation time (adapted from Pilling et al. 2011).

From the variation observed in the integrated area of IR spectrum (3000 to 900 cm-1) along the different radiation doses, photodissociation rates and half-life were determined by employing Eq. 16 and Eq. 17, respectively. The results showed that amino acids can survive at least ~ $7x10^5$ and ~$7x10^8$ years in dense molecular clouds and protoplanetary disks, respectively. For the nucleobases, the photostability is even higher, being about 2-3 orders of magnitude higher than found for the most radiation sensitive amino acids (Pilling et al.

2011). This high degree of survivability could be attributed to the low photodissociation cross section of these molecules at soft X-ray, combined with the protection promoted by the dust which reduces the X-ray field inside denser regions. During planetary formation (and after), these molecules, trapped in and on dust grains, meteoroids, and comets, could be delivered to the planets/moons possibly allowing pre-biotic chemistry in environments where water was also found in liquid state.

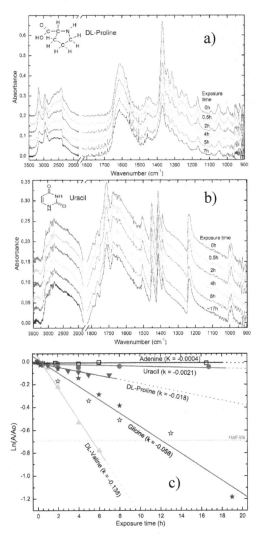

Fig. 22. a) FTIR spectra of amino acid DL-proline after different doses of 150-eV soft X-rays up to 7h. b) FTIR spectra of RNA nucleobase uracil after different doses of 150-eV soft X-rays up to 17 h. c) Integrated absorbance spectra of the solid-phase samples as a function of irradiation time. For each compound, the photodissociation rate (k) is also indicated (adapted from Pilling et al. 2011).

## 3. Conclusion

The presence of soft x-rays is very important for the chemical evolution of interstellar medium and other astrophysical environments close to young and bright stars. Soft X-rays can penetrate deep in molecular clouds and protostellar disks and trigger chemistry in regions in which UV stellar photons do not reach. The effects of soft X-rays in astrophysical ices are also remarkable because they release secondary electrons in and on the surface of the ices, which trigger a new set or chemical reactions. In this chapter, we presented several techniques employing soft X-rays in experimental simulation of astrophysical environments, which help us to understand the chemical evolution of these regions.

## 4. Acknowledgment

The authors would like to the thank the Brazilian agency FAPESP (#2011/14590-9) for the financial support. The authors also thank Ms. A. Rangel for the English revision of this book chapter.

## 5. References

Andrade D.P.P., Boechat-Roberty H. M., Pilling, S., da Silveira E. F., Rocco M. L. M. (2009) Positive and negative ionic desorption from condensed formic acid photoexcited around the O 1s-edge: Relevance to cometary and planetary surfaces. *Surface Science* vol. 603, pp. 3301–3306.

Andrade D. P. P., Rocco M.L.M., Boechat-Roberty H.M. (2010) X-ray photodesorption from methanol ice. *Monthly Not. Royal Astron. Soc.* vol. 409, pp. 1289–1296.

Baba Y., Yoshii K., Sasaki T.A. (1997) Desorption of molecular and atomic fragment-ions from solid $CCl_4$ and $SiCl_4$ by resonant photoexcitation at chlorine K-edge. *Surf. Sci.* vol. 376, pp. 330-338.

Bennett C. J., Chen S. H., Sun B. J., Chang A. H. H., Kaiser R. I. (2007) Mechanistical studies on the irradiation of methanol in extraterrestrial ices. *Astrophysical Journal*, vol. 660, pp. 1588-1608.

Bernstein M.P., Ashbourn S.F.M., Sandford S.A., Allamandola L.J., (2004) The Lifetimes of Nitriles (CN) and Acids (COOH) during Ultraviolet Photolysis and Their Survival in Space. *Astrophysical Journal*, vol. 601, pp. 365–370.

Boechat-roberty H.M., Pilling S. & Santos A.C.F. (2005). Destruction of formic acid by soft X-rays in star-forming regions. *Astronomy and Astrophysics*, vol. 438, pp. 915–922.

Boechat-Roberty H.M., Neves R., Pilling S., Lago A.F. & de Souza G.G.B. et al. (2009). Dissociation of the benzene molecule by ultraviolet and soft X-rays in circumstellar environment. *Mon. Not. Royal Astron. Soc.*, vol. 394, pp. 810–817.

Boogert A.C.A., Pontoppidan K.M., Knez C., Lahuis F., Kessler-Silacci J., van Dishoeck E.F., Blake G.A., Augereau J.-C., Bisschop S.E., Bottinelli S., Brooke, T.Y., Brown J., Crapsi A., Evans N.J., Fraser H.J., Geers V., Huard T.L., Jørgensen J.K., Öberg K.I., Allen L.E., Harvey P.M., Koerner D.W., Mundy L.G., Padgett D.L., Sargent A.I., Stapelfeldt K.R. (2008) The c2d Spitzer Spectroscopic Survey of Ices around Low-Mass Young Stellar Objects. I. $H_2O$ and the 5-8 μm Bands. *Astrophysical Journal*, vol. 678, pp. 985-1004.

Chen M. H., Crasemann B. & Mark H. (1981) Widths and fluorescence yields of atomic L-shell vacancy states. *Phys. Rev. A*, vol. 24, pp. 177-182.

Cunningham, M.R., Jones, P.A., Godfrey, P.D., Cragg, D.M., Bains, I., Burton, M. G. et al. (2007) A search for propylene oxide and glycine in Sagittarius B2 (LMH) and Orion. *Monthly Not. Royal Astron. Soc.*, vol. 376, pp. 1201-1210.

Cronin, J.R. , Pizzarello, S. (1997) Enantiomeric excesses in meteoritic amino acids. *Science*, vol. 275, pp. 951-955.

Dartois E., Schutte W., Geballe T. R., Demyk K., Ehrenfreund P., D'Hendecourt L. (1999) Methanol: The second most abundant ice species towards the high-mass protostars RAFGL7009S and W 33A , *Astronomy and Astrophysics*, vol. 342, pp. L32-L35.

Eland J. H. D., Wort F. S. & Royds R. N. (1986). A photoelectron-ion-ion triple coincidence technique for the study of double photoionization and its consequences. *J. Elect. Spectrosc. Relat. Phenom.*, vol. 41,pp. 297-309.

Eland J. H. D. (1987) The dynamics of three-body dissociations of dications studied by the triple coincidence technique PEPIPICO. *Mol. Phys.*, vol. 61, pp. 725-745.

Eland J. H. D., (1989) A new two-parameter mass spectrometry. *Acc. Chem. Res.*, vol. 22, pp. 381-387.

Ehrenfreund P., d'Hendecourt L., Charnley S.B., Ruiterkamp, R. (2001) energetic and thermal processing of interstellar ices. *J. Geophys. Res.*, vol. 106, pp. 33291-33302.

Frasinski L. J., Stankiewitz M., Randal K. J, Hatherley P. A. & Codling K (1986). Dissociative photoionization of molecules probed by triple coincidence-double time-of-flight techniques. *J. phys. B At. Mol. Opt.*, vol 19, pp. L819-L824.

Fantuzzi F., Pilling S., Santos A.C.F., Baptista L., Rocha A. B. & Boechat-Roberty H. M. (2011). Methyl formate under X-ray interaction in circumstellar environments. *Monthly Not. Royal Astron. Society*, In press.

Ferreira-Rodrigues A.M., Pilling S., Santos A.C.F., Souza G.G.B., Boechat-Roberty H.M. (2008). Production of $H_3^+$ and $D_3^+$ from $(CH_3)_2CO$ and $(CD_3)_2CO$ in SFRs., *Proceedings of the Precision Spectroscopy in Astrophysics, ESO/Lisbon/Aveiro*. Edited by N. C. Santos, L. Pasquini, A.C.M. Correia, and M. Romanielleo. Garching, Germany, pp. 279-282.

Gerakines P. A., Schutte W. A., Ehrenfreund P. (1996) Ultraviolet processing of interstellar ice analogs. I. Pure ices. *Astronomy and Astrophysics*, vol. 312, pp. 289-305.

Goicoechea J. R., Rodríguez-Fernández N. J. & Cernicharo J., 2004). The Far-Infrared Spectrum of the Sagittarius B2 Region: Extended Molecular Absorption, Photodissociation, and Photoionization. *Astrophysical Journal*, vol. 600, pp. 214–233.

Gorti U. & Hollenback D. (2004) Models of Chemistry, Thermal Balance, and Infrared Spectra from Intermediate-Aged Disks around G and K Stars. *Astrophysical Journal*, vol. 613, pp. 424-447.

Gorti U., Dullemond C. P., Hollenbach D. (2009) Time Evolution of Viscous Circumstellar Disks due to Photoevaporation by Far-Ultraviolet, Extreme-Ultraviolet, and X-ray Radiation from the Central Star. Astrophysical Journal, vol. 705, pp. 1237-1251.

Hergenhahn U., Rüdel A., Maier K., Bradshaw A. M., Fink R. F., Wen A. T. (2003) The resonant Auger spectra of formic acid, acetaldehyde, acetic acid and methyl formate. *Chem. Phys.*, vol. 289, pp. 57-65.

Hitchcock A. P., Fisher P., Gedanken A., RobinM. B. (1987). Antibonding .sigma.* valence MOs in the inner-shell and outer-shell spectra of the fluorobenzenes. *J. Phys. Chem.*, vol. 91, pp. 531–540.

Hudson R. L., Moore M. H. (2000) Note: IR spectra of irradiated cometary ice analogues containing methanol: A new assignment, a reassignment, and a nonassignment *Icarus*, vol. 145, pp. 661-663.

Ishii I. & Hitchcook A. P. (1988) The oscillator strengths for C1s and O1s Excitation of some saturated and unsaturated organic alcohols, acids and esters. *J. Electron Spectrosc. Rel. Phenomen.*, vol. 64, pp. 55-84.

Jenkins R., 1999, X-Ray Fluorescence Spectrometry. *Wiley*, New York.

Jones P.A., Cunningham M.R., Godfrey P.D., Cragg D.M. (2007) A Search for biomolecules in Sagittarius B2 (LMH) with the Australia Telescope Compact Array. *Monthly Not. Royal Astron. Soc.*, vol. 374, pp. 579-589.

Kastner J.H., Zuckerman B., Weintraub D.A., Forveille T. (1997) X-ray and molecular emission from the nearest region of recent star formation. *Science*, vol. 277, pp. 67-71.

Koyama K., Hamaguchi K., Ueno S., Kobayashi N. & Feigelson E. D. (1996), Discovery of Hard X-Rays from a Cluster of Protostars, *Publ. Astron. Soc. Japan.* vol 48, pp L87-L92, plate 22-23.

Kuan Y.-J., Charnley S.B., Huang H.-C., Tseng W.-L., Kisiel Z. (2003) Interstellar Glycine. *Astrophysical Journal.*, vol. 593, pp. 848-867.

Leach S., Schwell M., Dulieu F., Chotin J. L., Jochims  H. W., Baumgärtel H. (2002) Photophysical Studies of Formic Acid in the VUV: Absorption spectrum in the 6-22 eV region, *Phys. Chem. Chem. Phys.*, vol. 4, pp. 5025-5039.

Nummelin A., Bergman P., Hjalmarson A., Friberg P., Irvine W.M., Millar T.J., Ohishi M., Saito S. (2000) A Three-Position Spectral Line Survey of Sagittarius B2 between 218 and 263 GHZ. II. Data Analysis. *Astrophysical Journal Supplement Series, vol.* 128, pp. 213-243.

Pelc A., Sailer W., Scheier P., Probst M., Mason N. J., Illenberger E., Märk T.D. (2002) Dissociative electron attachment to formic acid (HCOOH), *Chem. Phys. Lett.* , vol. 361, pp. 277-284.

Pilling S.. Santos A.C.F.. Boechat-Roberty H.M. (2006). Photodissociation of organic molecules in star-forming regions II. Acetic acid. *Astronomy and Astrophysics*, vol. 449, pp. 1289–1296.

Pilling S. (2006), Fotoionização e Fotodissociação de ácidos e álcoois em regiões de formação estelar, *PhD thesis*, Chemical Institute/UFRJ, Rio de Janeiro, Brazil.

Pilling S., Neves R., Santos A. C. F. & Boechat-Roberty H. M. (2007a) Photo-dissociation of organic molecules in star-forming regions III: Methanol. *Astronomy & Astrophysics*, vol. 464, pp. 393 - 398.

Pilling S., Boechat-Roberty H.M., Santos A.C.F. & de Souza G.G.B. (2007b) Ionic yield and dissociation pathways from soft X-ray double-ionization of alcohols. *J. Electron Spectrosc. Relat. Phenom.* vol. 155, pp. 70–76.

Pilling S., Boechat-Roberty H.M., Santos A.C.F., de Souza G.G.B. & Naves de Brito A. (2007c). Ionic yield and dissociation pathways from soft X-ray multi-ionization of acetic acid. *J. Electron Spectrosc. Relat. Phenom.* vol. 156–158. pp. 139–144.

Pilling S., Andrade D. P. P., Neves R., Ferreira-Rodrigues A. M., Santos A. C. F. & Boechat-Roberty H. M. (2007d). Production of $H_3^+$ via photodissociation of organic molecules in interstellar clouds. *Monthly Not. Royal Astron. Society,* vol. 375, pp. 1488-1494.

Pilling S., Andrade D.P.P., Neto A.C., Rittner R. & Naves de Brito (2009). DNA nucleobase synthesis at titan atmosphere analog by soft X-rays. *J. Phys. Chem. A,* vol. 113, pp. 11161 -11166.

Pilling S., Andrade D.P.P., do Nascimento E.M., Marinho R.T., Boechat-Roberty H.M., Coutinho L.H., de Souza G.G.B., de Castilho R.B., Cavasso-filho R.L., Lago A.F. & de Brito A.N. (2011). Photostability of gas- and solid-phase biomolecules within dense molecular clouds due to soft X-rays. *Monthly Not. Royal Astron. Society,* vol. 411, pp. 2214 -2222.

Prince K. C., Richter R., Simone M., Alagia M. & Coreno M. (2003). Near Edge X-ray Absorption Spectra of Some Small Polyatomic Molecules. *J. Phys. Chem. A,* vol. 107, pp. 1955-1963.

Pontoppidan K.M., Dartois E., van Dishoeck E. F., ThiW.-F., d'Hendecourt, L. (2003) detection of abundant solid methanol toward young low mass stars. *Astronomy and Astrophysics,* vol. 404, pp. L17-L20.

Purdie D., Muryn C. A., Prakash N. S., Wincott P. L., Thornton G., Law D.S.L. (1991)Origin of the photon induced $Cl^+$ yield from Si(111)7×7-Cl at the Cl and Si K-edges. *Surface Science,* vol. 251, pp. 546-550.

Ramaker D.E., Madey T.E., Kurtz R.L., Sambe H. (1988) Secondary-electron effects in photon-stimulated desorption. *Phys. Rev. B,* vol. 38, pp. 2099-2111.

Rocco M.L.M., Weibel, D.E., Roman, L. S., Micaroni, L. (2004) Photon stimulated ion desorption from poly(3-methylthiophene) following sulphur K-shell excitation. *Surface Science,* vol. 560, pp. 45-52.

Rocco M.L.M., Sekiguchi T., Baba Y. (2006) Photon stimulated ion desorption from condensed thiophene photoexcited around the S 1s-edge. *Journal of Vacuum Science and Technology A,* vol. 24, pp. 2117-2221.

Rocco M.L.M., Sekiguchi T., Baba Y. (2007) Photon stimulated ion desorption from condensed thiolane photoexcited around the S 1s-edge. *J. Electron Spectrosc. Relat. Phenom.,* vol. 156, pp. 115-118.

Sorrell W.H. (2001) Origin of Amino Acids and Organic Sugars in Interstellar Clouds *Astrophysical Journal,* vol. 555, pp. L129- L132.

Simon M., LeBrun T., Morin P., Lavolée M., Maréchal J.L. (1991) A photoelectron-ion multiple coincidence technique applied to core ionization of molecules. *Nucl. Instrum. Methods B,* vol. 62, pp. 167-174

Simon M., LeBrun T., Martins R., de Souza G. G. B., Nenner I., Lavollee M., Morin P. (1993) Multicoincidence mass spectrometry applied to hexamethyldisilane excited around the silicon 2p edge. *J. Phys. Chem.,* vol. 97, pp. 5228–5237.

Snyder L.E., Lovas F.J., Hollis J.M., Friedel D.N., Jewell P.R., Remijan A., Ilyushin V.V., Alekseev E.A., Dyubko S.F. (2005) A Rigorous Attempt to Verify Interstellar Glycine. *Astrophysical Journal,* vol. 619, pp. 914-930.

Stolte W.C., Öhrwall G., Sant'Anna M. M., Dominguez Lopez I., Dang L.T.N., Piancastelli M. N., Lindle D.W. (2002) 100% site-selective fragmentation in core-hole-photoexcited methanol by anion-yield spectroscopy. *J. Phys. B: At. Mol. Opt. Phys.*, vol. 35, pp. L253-L259.

Turner B. E. (1991) A molecular line survey of Sagittarius B2 and Orion-KL from 70 to 115 GHz. II - Analysis of the data. *Astrophysical Journal Supplement Series*, vol. 76, pp. 617-686.

Woods P.M., Millar T. J., Herbst E., Zijlstra A. A. (2003) The chemistry of protoplanetary nebulae, *Astronomy & Astrophysics*, vol. *Astronomy and Astrophysics*, vol. 402, pp. 189–199.

Wiley W.E., McLaren I.W. (1955) Time of flight mass spectrometer with improved resolution, *Rev. Sci. Instrum.* vol. 26, pp. 1150-1157.

# 12

# Nanoscale Chemical Analysis in Various Interfaces with Energy Dispersive X-Ray Spectroscopy and Transmission Electron Microscopy

Seiichiro Ii
*National Institute for Materials Science*
*Japan*

## 1. Introduction

The properties of materials strongly depend on their microstructure such factors as lattice defect type and density, size, distribution of phases present. During last half century, the development of transmission electron microscopy (TEM) has provided an advance in our understanding of the details of not only the microstructure but also the crystallographic feature of materials on a sub-micrometer scale (Edington, 1974; Fultz & Howe, 2002; Wiliams and Carter, 1996). On the other hand, many of materials generally consist of more than two elements and include impurities and those are applied to the industrial field. Abovementioned lattice defects in those alloys are one of preferential site of segregation of alloying elements and impurities. In order to clarify the mechanical and functional behaviors of the alloys, we should investigate not only the microstructure but also the elemental distribution of alloying elements and impurities. For latter purpose, TEM equipped with a spectroscope is considered to be one of powerful tool. Typical spectroscopes installed to TEM are energy dispersive X-ray spectroscopy (EDS), wavelength dispersive X-ray spectroscopy (WDS) and electron energy loss spectroscopy (EELS). These spectroscopes are utilized some signals resulting from an interaction between electron and specimen as shown in Fig. 1. For EDS, characteristic X-ray shown as red colored characters in Fig. 1 is analyzed. In this chapter, the EDS is focused. EELS is also one of important spectroscopy in TEM as well as EDS. The EELS is detected and analyzed an inelastic electron interacted with the specimen. EELS has an advantage for analysis of light elements such as B, C, N and O. Additionally, obtained EEL spectrum provides the chemical bonding information of the specimen. Details can be referred to some references (Brydson, 2001; Egerton, 1996). EDS combined with TEM is one of useful spectroscopy in the materials science and widely used for chemical analysis such as identification and composition of the elements in desirable region on a submicrometer scale. Especially, the field emission (FE) type electron source in the electron gun of TEM has been developed recently, and it has been enabled to analyze the chemical composition on a nanoscale, since the electron probe size with the FE type electron gun can be easily converged to less than 1nm compared with conventional thermal filament type electron gun such as tungsten (W) and lanthanum hexaboride (LaB$_6$). So far, we had investigated the chemical analysis around the interface and surface in some materials and composite by using TEM with FE type gun. In this chapter, principles of EDS will be overviewed briefly and the validity of the analytical technique with the results for metal/ceramic clad materials and zirconia ceramics focused on their interfaces such as bonding interface, surface of the particles and interphase boundary are shown.

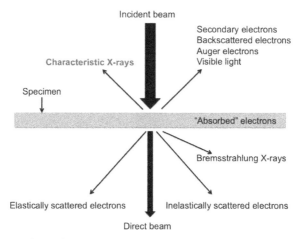

Fig. 1. An interaction of a high voltage electrons and specimen. For a qualitative and quantitative analysis with EDS, characteristic X-rays shown as bold are used.

## 2. Principles of energy dispersive X-ray spectroscopy

When high voltage electrons traverse a thin foil specimen, one of the primary inelastic interactions is that of inner-shell, ionization as shown in Fig, 2. The ejection of a K-shell electron leaves the atom in an excited state. One of the ways that it can return to ground is by an electron from an outer shell failing to the vacant inner-shell position and at the same time emitting an X-ray of characteristic energy as well as wavelength. This characteristic energy is a function of the difference in electron energy levels of the atom. Thus, it is these X-rays, which we are interested in because they provide direct information about the chemistry of the interaction between the electron and beam. In usual measurement of the EDS, a continuous X-ray is detected as well as characteristic X-rays. The continuous X-ray is emitted with Bremsstrahlung, which is caused by the interaction of the electron beam and a nucleus. In that case, the energy of the incident electrons is changed with the interaction and the extra energy is emitted as a photon. That photon is detected as the continuous X-ray. The continuous X-ray is usually seen as background in the X-ray spectrum. Fig. 3 is the typical EDS profile taken from yttria ($Y_2O_3$) doped zirconia ($ZrO_2$), which detail is shown later. Generally, for detection of the characteristic X-ray, silicon-lithium (Si(Li)) semiconductor detectors are used in almost all TEM-EDS as shown in Fig. 4. And they are sealed with different kinds of window consisting of beryllium or an ultrathin polymer to reduce a contamination with hydrocarbons and water vapor from the TEM. And a windowless detector is also used. The ultrathin window and windowless detector can detect the light elements such as B, C, N and O. However, the usual EDS detectors have a disadvantage of relative poor energy resolution, which is approximately 130-140keV in the case of the

Nanoscale Chemical Analysis in Various Interfaces with Energy Dispersive X-Ray Spectroscopy and
Transmission Electron Microscopy

247

conventional Si(Li) type detector. In order to overcome this problem, EDS with a microcalorimeter detector has been developed (Hara et al., 2010), which a superconducting transition-edge-sensor type microcalorimeter is employed as the new detector. This new EDS system will enable to detect with much higher energy resolution of 20eV compared

**(a)**                                              **(b)**

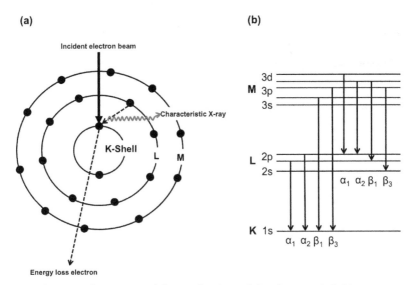

Fig. 2. (a) A schematic illustration of the mechanism of the characteristic X-rays generation. In this illustration, the generated characteristic X-ray corresponds Kα X-ray. (b) A relationship between the kind of characteristic X-ray and energy level of electrons.

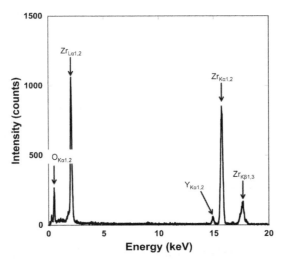

Fig. 3. Typical EDS profile obtained from $Y_2O_3$ doped $ZrO_2$.

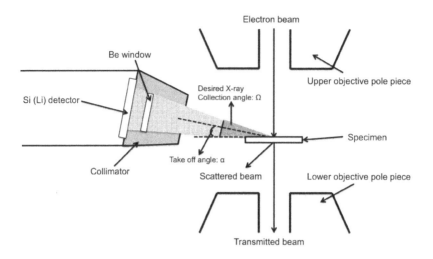

Fig. 4. A schematic drawing of a cross section of EDS detector and objective lens in TEM stage.

with the conventional Si(Li) type detector. Moreover, its detection limit will be improved because of the high sensitivity. As above mentioned, electron beam is interacted with nucleus and electrons of atoms. The behaviors of the interaction in a matter, we can predict with Monte Carlo method. Usually, the extent of the electrons depends of the mass of the matter, i.e., atomic number and is wider with increasing of the atomic number. The extent of the electron is defined as a spatial resolution, which is shown as following equation:

$$b = b_0 + 6.25 \times 10^4 \times \left(\frac{Z}{E_0}\right) \times \left(\frac{\rho}{A}\right)^{1/2} \times t^{3/2} \tag{1}$$

here, $b_0$ is incident electron probe size, Z is atomic number, $E_0$ is acceleration voltage, $\rho$ is density of specimen, A is mass number and t is thickness of the specimen, respectively.

By using above equation, spatial resolution of 0.5 to 1 nm can be obtained in TEM with FEG. Most important thing to efficiently detect the X-ray in TEM-EDS is large solid angle $\Omega$ shown in Fig. 3, which is schematic illustration of TEM-EDS system and high current density. The solid angle is explained as equation (2),

$$\Omega = \frac{S}{L^2} \tag{2}$$

here, S is cross section of the detector and L is distance between the specimen and the detector, respectively. As shown in eq. (2), the detector should be designed near the specimen. Recently, the EDS detector in TEM is designed as semi high angle side type. This type detector is well balanced between resolution of image and X-ray detection efficiency. To improve the X-ray detection efficiency, the specimen holder is also devised as well as design of the TEM-EDS instrument.

Quantitative X-ray analysis in the TEM-EDS is one of most straightforward technique. Usual quantitative analysis is used standard sample, which its composition has been already known. And both X-ray intensities in spectra obtained from standard and unknown specimen are compared to each other and quantitative values are estimated. However, since the absorption and fluorescence effect in thin film specimen is less than that in bulk specimen, most of the quantitative analyses in TEM are performed without the measurement of the standard sample and calculated from relative intensity of experimentally obtained each peaks corrected atomic number (Z), absorption (A) and fluorescence (F). Therefore, its analytical method is referred as ZAF (or standardless) method. That is, even if the X-ray correction is simplified, quantitative analysis has been done with high accuracy. In the case of thin film, which thickness is less than several 10 nm, the intensity $I_A$ of characteristic X-ray emitted from specimen A is described as,

$$I_A = \left( I\sigma_A \omega_A a_A N_0 \rho_A C_A \Omega \varepsilon t \right) / M_A \tag{3}$$

here, I is the intensity of the incident beam, $\sigma$ is ionized cross section of K (or L, M)-shell, $\omega$ is fluorescence efficiency, a is intensity ratio of K (or L, M) peak for all K (or M. L) shell, $N_0$ is Avogadro number, $\rho$ is density, C is atomic composition, $\Omega$ is solid angle, $\varepsilon$ is detection efficiency, t is thick ness of the specimen and M is atomic weight, respectively. $\sigma_a$ of K-shell is known as follows:

$$\sigma_A = 4\pi a_0 N_k b_k (R / E_0)(R / E_k) \ln(c_k E_0 / E_k) \tag{4}$$

, which $a_0$ is Bohr radius, $N_k$ is the number of electrons in K-shell, $E_0$ is the energy of electrons, $E_k$ is the ionization energy of K-shell, $b_k$ and $c_k$ are constant and R is Rydberg constant, respectively.

Additionally, the fluorescence efficiency $\omega$ is also expressed as

$$\omega = Z^4 / (b_k + Z^4) \tag{5}$$

in which, Z is the atomic number. By using above equations (3) to (5), the characteristic X-ray ratio from element A and B is represented as:

$$\frac{I_A}{I_B} = \frac{\left( \sigma_A \omega_A a_A \rho_A C_A \right) \cdot M_B}{\left( \sigma_B \omega_B a_B \rho_B C_B \right) \cdot M_A} \tag{6}$$

Therefore, between composition and intensity of the X-ray in element A and B is related as follows:

$$\frac{C_A}{C_B} = \frac{\left( \sigma_B \omega_B a_B \rho_B \right) \cdot M_A \cdot I_A}{\left( \sigma_A \omega_A a_A \rho_A \right) \cdot M_B \cdot I_B} = k_{A,B} \left( \frac{I_A}{I_B} \right) \tag{7}$$

In this equation, $k_{A,B}$ is known as k factor or Cliff-Lorimer factor and composition ratio strongly depends on the k-factor in the experimental EDS measurement. Since the k-factor is also closely related to the specimen and EDS apparatus, therefore it is quite important to clarify the k-factor with high accuracy in the quantitative EDS analysis. This k-factor is changed daily in actual EDS measurement, because of the different acquirement condition, so that it should be measured for accurate measurement. And in the case of relative thick specimen, absorption correction should be considered. Horita et al., have been proposed k-factor with considering of absorption correction, and they have shown its effectivity (Horita et al., 1998).

## 3. Experimental chemical analysis with TEM-EDS for elemental distribution

### 3.1 Chemical analysis in bonding interface of explosively welded metal/ceramic clad materials

Explosive welding technique is one of solid-state welding and clad materials fabricated by this technique have high reliability of the bonding strength (Crossland, 1982). This technique was developed in the 1950s for joining metallic plates. It has been widely used for industry due to its advantage of direct bonding without bonding medium such as braze so far. Nowadays, an application of the explosive welding technique is extended to the field of composite which ductile metals are joined onto brittle materials such as ceramics and metallic glasses as well as metal/metal clad. However, during explosive welding with high-speed deformation, many cracks are introduced and propagated into the brittle materials and it is serious problem. To overcome that problem, Hokamoto et al. have attempted the metal/brittle materials such as ceramic and metallic glass clad consisted of a metallic foil onto brittle material plate by explosive welding (Hokamoto et al., 1998, 1999, 2010). Especially, they have succeeded the fabrication of metal/ceramics clad materials of several 10 mm without cracks by regulated underwater shockwave (Hokamoto et al., 1998, 1999). In order to understand the bonding mechanism and welding process, it is a key to clarify the bonding interface in the clad. In this chapter, the observation result of the bonding interface between aluminum (Al) and silicon nitride ($Si_3N_4$) clad material with explosive welding by TEM. And based on the chemical analysis in bonding interface measured by EDS with TEM, the bonding mechanism is also discussed (Ii et al., 2010). Fig. 5 is an external view of explosively welded $Al/Si_3N_4$ clad. In this experiment, Al of a flyer plate is welded onto $Si_3N_4$ as a base material. Al is locally deformed, but Al and $Si_3N_4$ are macroscopically well bonded without macroscopic cracks in both materials. The bright field image around the bonding interface of the clad material is shown in fig. 6. There are also no microscopic cracks around the interface and fine grained layer between Al and $Si_3N_4$ indicated as intermediate layer in the center of fig. 6. The width of the intermediate layer is approximately 2.5µm. The selected area electron diffraction pattern taken from the region included this intermediate layer is mainly consisted of Debye-Scherrer ring pattern, which is often seen in fine grained polycrystalline. This pattern is also explained without inconsistency, by using the lattice parameter of Al. Therefore, it can be expected that the intermediate layer consist of nanocrystalline Al. In order to clarify the components of the intermediate layer from the chemical viewpoint, EDS measurement and qualification analysis around the interface are performed with scanning TEM (STEM). Although STEM is a type of the TEM, the electron beam of TEM in most of the observations is usually parallel without any convergence. On the other hand, that of the STEM is.

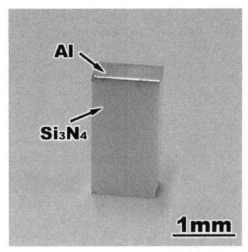

Fig. 5. An outer view of the explosively welded Al/Si₃N₄ clad. In this figure, upper Al as
flyer plate is bonded onto Si₃N₄ base material during explosive welding. And no
macroscopic cracks are observed.

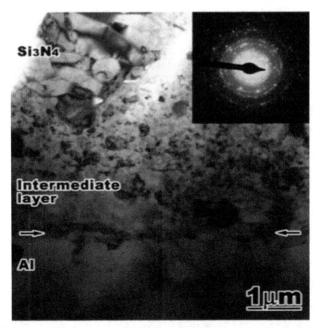

Fig. 6. A Bright field image of the interface of Al/Si₃N₄ clad. A selected area electron
diffraction pattern is inserted on the top-right of the image. At the center in this figure, the
intermediate layer consisting of a nanocrystalline grain is clearly seen.

illuminated to the specimen with large convergence angle the specimen and scanned in a desirable raster. And the STEM is imaged by the detection of the scattered electron. The detectors in STEM are set to detect the specific scattered electron. Details of STEM are described in elsewhere (Pennycook & Nellist, 2011). Fig. 7 shows the STEM bright field image and the intermediate layer was also clearly observed in fig. 7. Fig. 8 shows X-ray intensity profile across the interface in the Al/Si$_3$N$_4$ clad material obtained by EDS line analysis along the line shown in Fig. 7. Black and red lines correspond to the X-ray intensity of Al-K$\alpha$ and Si-K$\alpha$, respectively. In this figure, the X-ray intensity of the Al-K$\alpha$ peak is increasing with increase of the distance in fig. 8. Both X-ray intensities of each element are drastically changed at the 0.8$\mu$m in this figure, where approximately corresponds to the interface between Si$_3$N$_4$ and nanocrystalline region in the fig.6 and 7. The EDS line profile show that the nanocrystalline region consist of only Al. Many researchers have been reported microstructure around the interface the explosively welded clad materials (Dor-Ram et al., 1979; Kreye et al., 1976; Murdie & Blankenburgs, 1966; Nishida et al., 1993, 1995), so far. Among them, Nishida et al. have precisely investigated the interface of an explosively welded Ti/Ti clad by TEM and found the almost same structure consisting of fine grains. They concluded that the nanocrystallization is caused by the rapid solidification of a thin molten layer formed during the welding (Nishida et al., 1993). In this observation, nanocrystalline region consisting of only Al are also observed. Additionally, melting point of Al is much lower than that of Si$_3$N$_4$, and Al and Si are not solute to each other judging from Al-Si binary equilibrium phase diagram (Murray & McAlister, 1990). Therefore, it can be concluded that nanocrystalline layer is formed with rapid solidification of Al during the explosive welding.

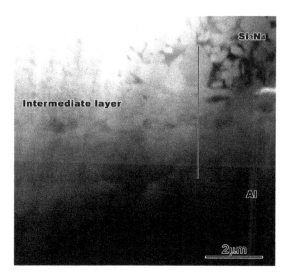

Fig. 7. A STEM image of the interface of Al/Si$_3$N$_4$ clad. EDS line profile shown in Figure 8 was measured along the black line in this figure.

Nanoscale Chemical Analysis in Various Interfaces with Energy Dispersive X-Ray Spectroscopy and
Transmission Electron Microscopy

253

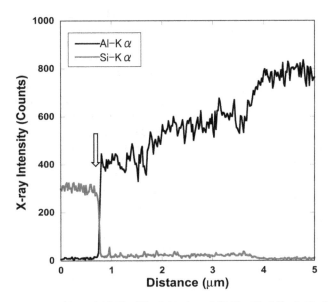

Fig. 8. X-ray intensity profiles of Al-Kα (Black line) and Si-Kα (Red line). Each X-ray intensities are drastically changed at the point shown as arrow. From our systematic analysis, the interface in this clad corresponds to the position where the X-ray intensities are steeply changed.

## 3.2 Quantitative analysis of surface segregation in $Y_2O_3$ doped $ZrO_2$ nano particles

Yttria-stabilized tetragonal zirconia polycrystal (Y-TZP) has proved to be one of important structural ceramic with excellent mechanical properties such as high fracture toughness, strength, and hardness (Gravie et al., 1975; Green et al., 1989). The microstructure and grain growth behavior in sintered bulk Y-TZP have been extensively investigated so far (Ikuhara et al., 1997; Lange, 1988; Lee & Chen, 1988; Sakuma & Yoshizawa, 1992; Yoshizawa & Sakuma, 1989) It has been known that fine grained Y-TZP with the grain size of submicrometer can be obtained because of the formation of cubic and tetragonal two-phase composites. In this case, the tetragonal–cubic phase separation is considered to take place during sintering, and therefore fine grained structure results in stable Y-TZP. In addition, raw $ZrO_2$ powders with nano-order grain size and good sinterability are commercially available. High-quality Y-TZP powders with narrow particle size distribution can be obtained by the hydrolysis process (Matsui et al., 2002). In previous paper, the microstructural development of the Y-TZP during sintering was investigated by transmission electron microscopy (TEM) observations and energy-dispersive X-ray spectroscopy (EDS) analyses (Matsui et al., 2003), and it was revealed that the yttrium cations tend to segregate in the vicinity of grain boundaries. The relationship between the microstructure of sintered body and the properties of the raw powders have been examined, and it has been pointed out that the microstructure and physical properties of the sintered body strongly depend on the properties of raw powders (Hishinuma et al., 1988). It is therefore crucial to clarify the microstructure and local composition of the starting powders. In order to clarify whole of the particles on a nano scale, the specimen preparation technique for the high resolution electron microscopy observation has been developed (Ii et al., 2006). In this section, the results of the microstructure and $Y^{3+}$ cation distribution of the $Y_2O_3$ doped $ZrO_2$ particles by high-resolution TEM (HRTEM) observations and EDS measurements are shown. Figure 9 (a) shows a typical high-resolution electron micrograph of the yttria-doped zirconia particle. First Fourier Transformed (FFT) pattern of the powder is shown as an inset at the lower-left side in the micrograph. The FFT pattern corresponds to the electron diffraction pattern taken from $[01\bar{1}]$ of the tetragonal phase and it indicates that the particle is tetragonal single phase. Fig. 9 (b) and (c) show an enlarged lattice image of the surface and internal region as indicated by B and C in (a), respectively. As shown in Figs. 9 (b) and (c), the two-dimensional lattice fringe is clearly observed inside the particle. Figure 10 shows the composition of $Y_2O_3$ analyzed from EDS spectra as a function of distance from the surface of the particle. The error bars to each measuring point are also shown. In this EDS measurements, the spectra obtained from this particle and the quantitative analysis have high reliability, because the thickness of the specimen is thin enough to obtain the lattice image into the internal region and uniform. In Fig. 10, the error is within 0.3% of each estimated value, and it is relatively small compared with the estimated $Y_2O_3$ composition. Both the $Y_2O_3$ compositions take on the maximum value at the surface of the particle, but suddenly reduce at a distance of 1 nm off from the surface. The $Y_2O_3$ composition increases again with increasing distance from the surface, but takes on a constant value over the distance of 5 nm. This fact suggests that $Y^{3+}$ cations migrate from 1 nm inside to the surface of the particle and segregate at the surface.

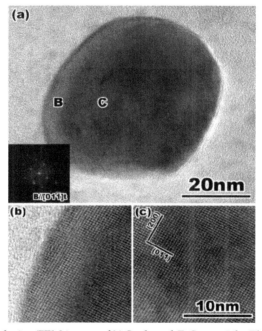

Fig. 9. (a) A high resolution TEM image of $Y_2O_3$ doped $ZrO_2$ particle. The FFT pattern in (a) indicates that the particle is tetragonal single phase. (b) and (c) Enlarged lattice image of the surface and the internal region of the particle indicated by B and C in (a), respectively.

In the case of grain boundary in polycrystalline $Y_2O_3$ doped $ZrO_2$, dopant cations have been reported to segregate at the vicinity of the grain boundary over the width of several nm to reduce excess grain boundary energy (Ikuhara et al., 1997; Shibata et al., 2004). The present results are obtained in a situation such that the surface excess energy is reduced by surface segregation of the $Y^{3+}$ cations. The observation of the other particles also showed almost same tendency. On the other hand, some of the particles, which the tetragonal phase transforms to the monoclinic phase, have a uniform distribution of $Y_2O_3$ contents, and that the formation of monoclinic phase is not attributed to $Y_2O_3$ inhomogeneous distribution. The stability of the monoclinic phase of $ZrO_2$ has been reported to depend on the grain size; the tetragonal phase of $ZrO_2$ with the grain size of less than 500 nm cannot be thermally transformed to monoclinic phase (Gupta et al., 1978; Sakamoto, 1990). Since the average size of the present zirconia particle is approximately 50 nm, the formation of monoclinic zirconia cannot be explained from a thermomechanical point of view. Moreover, the monoclinic phase was observed in the vicinity of the surface. Therefore, the monoclinic phase is probably formed by stress-induced transformation during the milling process, which was performed in the final stage of the powder fabrication.

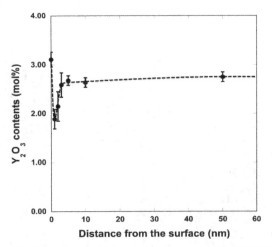

Fig. 10. A composition of $Y_2O_3$ estimated by quantitative analysis of EDS spectra in $Y_2O_3$ doped $ZrO_2$ particle as a function of the distance from the surface.

### 3.3 Yttrium cation distribution at cubic and tetragonal interphase boundary in Y-TZP

As abovementioned, $Y_2O_3$ stabilized tetragonal $ZrO_2$ polycrystal (Y-TZP) is one of the technologically important structural ceramics. For the development of the Y-TZP bulk materials, the microstructural development including phase transformation and grain growth in Y-TZP has been widely investigated by many researchers so far (Gravie et al., 1975; Lange, 1988; Lee & Chen, 1988; Yoshizawa & Sakuma, 1989). Their most important aspects of the relationship between the phase transformation and grain growth were that grain growth in the tetragonal (t) and cubic (c) dual-phase region is slower than t or c single phase region. For instance, the grain growth behavior in 4mol% $Y_2O_3$-stabilized $ZrO_2$ was explained in terms of the pinning-effect of cubic phase grains dispersed in TZP (Yoshizawa & Sakuma, 1989). However, it should be noted that their phenomenological analysis was based on the assumption that the $ZrO_2$-$Y_2O_3$ in t and c two-phase region was dual-phase composite consisting of t- and c-grains. On the other hand, Ikuhara *et al.* (1997) investigated the microstructure in TZP by high resolution transmission electron microscopy (HRTEM) and energy dispersed x-ray spectroscopy (EDS), and found that $Y^{3+}$ cations are segregated over a width of 4-6 nm across the boundaries in TZP. More recently, the detailed microstructural change during sintering process in TZP has been investigated and a formation of the cubic phase from grain boundary due to the $Y^{3+}$ cation's segregation and its growing to the grain interior are found (Matsui et al., 2003, 2006). Thus, this tetragonal to cubic phase transformation is termed as grain boundary segregation induced phase transformation (GBSIPT). Through the GBSIPT, the interphase boundary between the front of the cubic phase and the tetragonal matrix, i.e. c/t interphase boundary is likely to form inside grains. The lattice parameter ratio is slightly different between tetragonal and cubic phases, whose lattice mismatch of two phases is less than 1%. Therefore, it can be expected that the interphase boundary is coherent or semi-coherent boundary. In the case of the semi-coherent boundary, the excess strain interface energy is released by introducing misfit dislocation. In this section, the relationship between the interphase boundary structure and yttrium cation distribution in Y-TZP is focused. Fig. 11 (a) shows a typical microstructure observed in the Y-TZP specimen sintered at 1650°C. As shown in Fig. 11 (a), the dot-like contrasts between two arrows are periodically observed inside the grain.

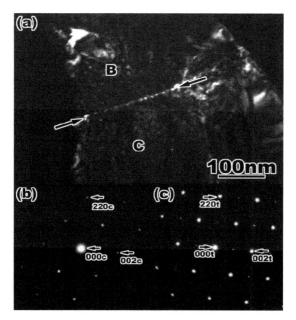

Fig. 11. (a) A Dark field image of typical microstructure of sintered $Y_2O_3$ stabilized $ZrO_2$ tetragonal zirconia polycrystal. This image were slightly tilted to keep the two beam excitation condition with $g = (110)_{c,t}$ and see misfit dislocations clearly. (b) and (c) Selected area electron diffraction patterns taken from B and C indicated in (a). The incident electron beam direction is parallel to $[1\bar{1}0]_{c,t}$. And these diffraction patterns show cubic in (b) and tetragonal (c) single phase, respectively.

Fig. 11 (b) and (c) show selected electron diffraction patterns taken from the region marked by B and C in Fig. 11 (a), which indicates the interface is formed between cubic (Fig. 11 (b)) and tetragonal (Fig. 11 (c)) phase, respectively. Therefore, we considered that their dot-like contrasts are on the interface. The average interval of these contrasts is estimated to be approximately 10 nm. An arrangement of the periodic contrast was confirmed to be close to near <111> direction from the result of the trace analysis in the obtained TEM micrograph and electron diffraction patterns. In addition, it was also confirmed that the plane normal of this interface, which is the periodic contrast along near <111> direction, is close to the <22$\bar{1}$>t and c, respectively. Consequently, it is predicted that the interphase boundary observed in this specimen is near {22$\bar{1}$}t and {22$\bar{1}$}c, respectively. In order to investigate the detailed interphase boundary structure, especially misfit dislocation structure on an atomic level, HRTEM observations for the c/t interphase boundary have been done and revealed that the misfit dislocations at the c/t interphase boundary are clearly observed (Ii et al., 2008). Fig. 12 shows the profile of the $Y_2O_3$ contents from EDS spectra as a function of distance from the c/t interphase boundary in Y-TZP. The $Y_2O_3$ contents take the uniform value both in the cubic and tetragonal regions, but steeply changes across the c/t interphase boundary. The steeply sloped change in the $Y_2O_3$ contents is in good agreement with the fact that the misfit dislocations are introduced by the difference in the lattice parameter between c and t phases with different $Y^{3+}$ compositions. It should be noted that the yttrium distribution is almost uniform inside the grain in the bulk Y-TZP sintered at 1300°C. The present results indicate that $Y^{3+}$ cation redistribution takes place in association with tetragonal-cubic phase transformation in Y-TZP.

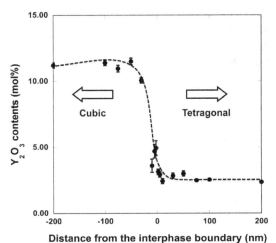

Fig. 12. A $Y_2O_3$ contents profile obtained by EDS spectra across the c/t interphase boundary as a function of the distance from the boundary.

## 4. Conclusion

In this chapter, the principles of EDS are briefly described and some results of the nanoscale chemical analysis at many interfaces measured by EDS installed to TEM with field emission type electron gun, whose probe size is less than less than 1nm. In the bonding interface in the explosively welded $Al/Si_3N_4$ clad materials, element distribution around the interface was measured by TEM and STEM-EDS as well as microscopic characterization. For this purpose, TEM and STEM observation revealed that the intermediate layer with approximately 2.5 μm between the Al and $Si_3N_4$ exists and this layer consists of fine grains whose average diameter of 100nm. And the intermediate layer consists of only Al from selected area electron diffraction pattern and EDS measurement across the interface. Consequently, we clarified that the fine grained Al is formed by rapid solidification after melting during the cladding and it plays as an important role to the welding of Al and $Si_3N_4$. On the other hand, the microstructures of $Y_2O_3$ doped $ZrO_2$ nano particles were also observed by HRTEM and the distribution of $Y_2O_3$ inside particle was measured by EDS. The experimental technique for the observation of whole particle with the size of several 100 nm was developed and the surface segregation of $Y_2O_3$ was detected within a few nm from the surface for the relaxation of the excess surface energy. The interphase boundary between the front of grown cubic and tetragonal matrix in sintered Y-TZP was investigated by CTEM and HRTEM with EDS measurement. $Y^{3+}$ cation concentration is drastically changed at the c/t interphase boundary. Since the periodic misfit dislocations are also observed at the c/t interphase boundary, therefore, the steep change in $Y^{3+}$ cation composition must cause the misfit strain at the interphase boundary, and consequently leads to introduce misfit dislocations. These results will provide the effectivity of the EDS measurement combined with TEM for the chemical analysis of advanced materials on the nano scale and the guidelines for understanding the correlation between the microstructure observation and the nano scale analysis.

## 5. Acknowledgment

The author is wish to a great thanks to Prof. R. Tomoshige (Sojo University) and Prof. K. Hokamoto (Kumamoto University) for fruitful discussion in explosive welding. Especially, the $Al/Si_3N_4$ clad materials were kindly provided from Prof. Hokamoto. The author is also grateful to Prof. Y. Ikuhara (The University of Tokyo), Dr. H. Yoshida (National Institute for Materials Science) and Dr. K, Matsui (Tosoh Corporation) for valuable comments and suggestions in the EDS measurement of the $ZrO_2$ particles and the c/t interphase boundary in the Y-TZP ceramics. And the Y-TZP particle and bulk materials were supplied from Dr. Matsui. The part of this study was financially supported by Grant-in-Aid for Scientific Research on Priority Areas "Giant Straining Process on Advanced Materials Containing Ultra-High Density Lattice Defects" (19025012) and on Innovative Area "Bulk Nanostructured Metals" (23102512) and by Grant-in-Aid for Scientific Reaserch (Kiban (S) (19106013)) and by Grant-in-Aid for Young Scientists (B) (20760508) from the Ministry of Education, Culture, Sports, Science and Technology (MEXT) and the Japan Society for Promotion of Science (JSPS), Japan.

## 6. References

Brydson, R. (2001). *Electron Energy Loss Spectroscopy*, Tayler & Francis, ISBN 1-85996-134-7, New York

Crossland, B. (1982). *Explosive Welding of Metels and its application* (Clarendon press, London 1982) ISBN 0-19-859-119-5..

Dor-Ram, B.; Weiss, Z. & Komen, Y. (1979). Explosive cladding of Cu/Cu systems: An electron microscopy study and a thermomechanical model, *Acta Metall.*, Vol. 27 Iss.9, pp. 1417-1429

Edington, J.W. (1974). *Practical Electron Microscopy in Materials Science*, MacMillan, Philips Technical Library, Eindhoven. Reprinted by Ceramic Book and Literature Service, ISBN 978-1878907356, Marietta, Ohio

Egerton, R. F. (1996). *Electron Energy-Loss Spectroscopy in the Electron Microscope*, 2nd edition, Plenum Press, ISBN 0-306-45223-5, New York

Fultz, B & Howe, J.M. (2002). *Transmission Electron Microscopy and Diffractometry of Materials*, 2nd ed., Spinger-Verlag, ISBN 978-3-540-73885-5, Berlin

Garvie, R. C.; Hannink, R. H. J. & Pascoe, R.T. (1975). CERAMIC STEEL, *Nature*, Vol.258, pp.703-704.

Green, D. G.; Hannink, R. H. J. and Swain, M. V. (1989). *Transformation Toughening of Ceramics*. CRC Press, Boca Raton, ISBN 978-0849365942, FL, 1989.

Gupta, T. K.; Bechtold, J. H.; Kuznicki, R. C.; Cadoff, L. H. & Rossing, B. R. (1977). Stabilization of Tetragonal Phase in Polycrystalline Zirconia, *J. Mater. Sci.*, Vol.12, pp.2421-2426.

Gupta, T. K.; Lange, F. F. & Bechtold, J. H. (1978). Effect of Stress-Induced Phase Transformation on the Properties of Polycrystalline Zirconia Containing Metastable Tetragonal Phase, *J. Mater. Sci.*, Vol.13, pp.1464-1470

Hammeeschmidt, M & Kreye, H. in: *Shock Waves and High-Strain-Rate Phenomena in Metals (Concepts and Applications)*, edited by M. A. Meyers and L. E. Murr, Plenum Pless, New York (1981), 961. 978-0306406331

Hara, T.; Tanaka, K.; Maehata, K.; Mitsuda, K.; Yamasaki, N. Y.; Ohsaki, M.; Watanabe, K,; Yu, X,; Ito, T, & Yamanaka, Y. (2010) Microcalorimeter-type energy dispersive X-ray spectrometer for a transmission electron microscope. *Journal of Electron Microscopy*, Vol.59, No.1, pp.17-26

Hishinuma, K.; Kumai, T.; Nakai, Z.; Yoshimura, M. & Somiya, S. Characterization of $Y_2O_3$–$ZrO_2$ Powders Synthesized Under Hydrothermal Conditions, *Adv. Ceram.*, Vol.24, pp.201-209.

Hokamoto, K.; Fujita, M. & Shimokawa, H. (1998). Explosive Welding of a Thin Metallic Plate onto a Ceramic Plate Using Underwater Shock Wave, *Rev. High Pressure Sci. Technol.* Vol. 7 (1998), pp.921-923.

Hokamoto, K.; Fujita, M.; Shimokawa, H. & Okugawa, H. (1999). A new method for explosive welding of $Al/ZrO_2$ joint using regulated underwater shock wave, *J. Mater. Proc. Tech.* Vol.85, No.1-3, pp.175-179.

Hokamoto, K.; Nakata, K.; Mori, A.; Ii, S.; Tomoshige, R.; Tsuda, S.; Tsumura, T.; Inoue, A. (2009). Microstructural characterization of explosively welded rapidly solidified foil and stainless steel plate through the acceleration employing underwater shock wave, *Journal of Alloys and Compounds*. Vol.485, pp.817-821

Horita, Z. (1998). Quantitative X-ray Microanalysis in Analytical Enectron Microscopy, *Mater. Trans. JIM*, Vol.39, No.9, pp.947-958.

Ii, S.; Yoshida, H.; Matsui, K.; Ohmichi, M. & Ikuhara, Y. (2006) Microstructure and Surface Segregation of 3mol% $Y_2O_3$ doped $ZrO_2$ particle, *J. Am. Ceram. Soc.*, Vol.89, No.7, pp.2952-2955.

Ii, S.; Yoshida, H.; Matsui, K. & Ikuhara, Y. (2008). Misfit Dislocation Formation at the c/t Interphase Boundary in Y-TZP, *J. Am. Ceram. Soc.*, Vol.91 No.11, pp.3810-3812

Ii, S.; Iwamoto, C.; Satonaka, S.; Hokamoto, K. & Fujita, M. (2010). Microstructure of Bonding Interface in Explosively Welded Metal/Ceramic Clad, *Mater. Sci. Forum.*, Vols.638-642, pp.3775-3780.

Ikuhara, Y.; Thavorniti, P. & Sakuma, T. (1997). Solute Segregation at Grain Boundaries in Superplastic SiO2-Doped TZP, *Acta Mater.*, Vol.45, pp.5275–5284.

Kreye, H.; Wittkamp, I. & Richter, U. (1976). Electron Microscope Investigation of the Bonding Mechanism in Explosive Bonding. *Zeitschrift fuer Metallkunde*, Vol.67, Iss.3, pp. 141-147. ISSN: 00443093,

Lange, F. F. (1988). Controlling Microstructures Through Phase Partitioning from Metastable Precursors The $ZrO_2$–$Y_2O_3$ System, pp. 519–532 in Ultrastructure Processing of Advanced Ceramics, Edited by J. D. Mackenzie and D. R. Ulrich. Wiley, New York, 1988.

Lee, I. G. & Chen, I. -W. (1988). Sintering and Grain Growth in Tetragonal and Cubic Zirconia, *Sintering '87*. pp. 340–345 in, Edited by S. Somiya, M. Shimada, M. Yoshimura, and R. Watanabe. Elsevier Applied Science, ISBN 978151662906, London, 1988.

Matsui, K. & Ohgai, M. (2002). Formation Mechanism of Hydrous Zirconia Par- ticles Produced by Hydrolysis of ZrOCl2 Solutions: IV, Effects of $ZrOCl_2$ Concentration and Reaction Temperature, *J. Am. Ceram. Soc.*, Vol.85, pp.545–553.

Matsui, K.; Horikoshi, H.; Ohmichi, N.; Ohgai, M.; Yoshida, H. & Ikuhara, Y. (2003). Cubic-Formation and Grain-Growth Mechanisms in Tetragonal Zirconia Poly- crystal, *J. Am. Ceram. Soc.*, Vol.86, pp.1401–1408.

Matsui, K.; Ohmichi, N.; Ohgai, M.; Yoshida, H.; & Ikuhara, Y. (2006). Effect of alumina-doping on grain boundary segregation-induced phase transformation in yttria-stabilized tetragonal zirconia polycrystal, *J. Mater. Res.*, Vol.21, pp.2278-2289

Murray, J. L. & McAlister, M. J. in: *Binary Alloy Phase Diagrams*, II Ed., edited by T.B. Massalski, Vol. 1 (1990), p 211-213

Murdie, D.C. & Blankenburgs, G. (1966). Examination of Two Explosively Welded Interfaces, *J. Inst. Met.*, Vol.94 (1966), 119-120.

Nishida, M.; Chiba, A.; Imamura, K.; Minato, H. & Shudo, J. (1993). Microstructural modifications in an explosively welded Ti/Ti clad material: I. Bonding interface, *Metall. Trans. A*, Vol. 24A, No.3, pp.735-742

Nishida, M.; Chiba, A.; Honda, Y.; Hirazumi, J. & Horikiri, K. (1995). Electron microscopy studies of bonding interface in explosively welded Ti/steel clads, *ISIJ Int.*, Vol.35, Iss.2, pp.217-219

Pennycook, S. J. & Nellist, P. D. (2011). Scanning Transmission Electron Microscopy, Springer-Verlag, ISBN 978-1441971999, New York.

Sakamoto, M. (1990). Raw Powder of $ZrO_2$ System, pp. 265–273 in Technical Hand- book of Super Fine Ceramic, Edited by S. Shirasaki and A. Makishima. Science Forum, Tokyo, 1990 (in Japanese).

Sakuma, T. & Yoshizawa, Y. (1992). The Grain Growth of Zirconia During Annealing in the Cubic/Tetragonal Two-Phase Region, *Mater. Sci. Forum.*, Vol.94–96, pp.865–870 (1992).

Shibata, N.; Oba, F.; Yamamoto, T. & Ikuhara, Y. (2004). Structure, Energy and Solute Segregation Behavior of [110] Symmetric Tilt Grain Boundaries in Yttria- Stabilized Cubic Zirconia, *Philos. Mag.*, Vol.84, pp.2381–2415.

Wiliams, D.B. & Carter, C. B. (1996). *TRANSMISSION ELECTRON MICROSCOPY: a textbook for materials science*, Plenum Press, ISBN 0-306-45247-2, New York

Yoshizawa, Y. & Sakuma, T. (1989). Evolution of Microstructure and Grain Growth in $ZrO_2$–$Y_2O_3$ Systems, *ISIJ Int.*, Vol.29, pp.746–752.

# 13

# Application of Stopped-Flow and Time-Resolved X-Ray Absorption Spectroscopy to the Study of Metalloproteins Molecular Mechanisms

Moran Grossman and Irit Sagi
*Departments of Structural Biology and Biological Regulation,*
*The Weizmann Institute of Science, Rehovot*
*Israel*

## 1. Introduction

### 1.1 Metalloproteins as mediators of life processes

Revealing detailed molecular mechanisms associated with metalloproteins is highly desired for the advancement of both basic and clinical research. Metalloproteins represent one of the most diverse classes of proteins, containing intrinsic metal ions which provide catalytic, regulatory, or structural roles critical to protein function. The metal ion is usually coordinated by nitrogen, oxygen, or sulfur atoms belonging to amino acids in a polypeptide chain and/or in a macrocyclic ligand incorporated into the protein. The presence of a metal ion rich in electrons allows metalloproteins to perform functions such as redox reactions, phosphorylation, electron transfer, and acid-base catalytic reactions that cannot easily be performed by the limited set of functional groups found in amino acids. Approximately one-third of all proteins possess a bound metal[1], and almost half of all enzymes require the presence of a metal atom to function[2]. The most abundant metal ions *in vivo* are Mg and Zn, while Fe, Ca, Co, Mn, and Ni are also frequently observed. Metalloproteins play important roles in key biological processes such as photosynthesis, signaling, metabolism, proliferation, and immune response [3-5]. For instance, the principal oxygen carrier protein in human, hemoglobin, uses iron (II) ions coordinated to porphyrin rings to bind the dioxygen molecules; calmodulin is a calcium-binding protein mediating a large variety of signal transduction processes in response to calcium binding; cytochromes mediate electron transport and facilitate a variety of redox reactions using iron, which interconverts between $Fe^{2+}$ (reduced) and $Fe^{3+}$ (oxidized) states; and many proteolytic enzymes contain zinc ions in their active sites that is highly crucial for peptide bond hydrolysis reaction.

### 1.2 Metalloenzymes architecture and function

Among metalloproteins there are many families of enzymes that utilize metal ions for catalytic activity (Figure 1). The metal ion is usually located at the active site of the enzyme, creating a highly reactive microenvironment, which is often hard to mimic by *in-vitro* synthesis. Zinc, for instance, was found to be an integral component of more than 300 biological catalysts, and it is functionally essential for enzymes belonging to the main biochemical categories: oxireductase, transferase, hydrolases, lyases, isomerases and ligases[4].

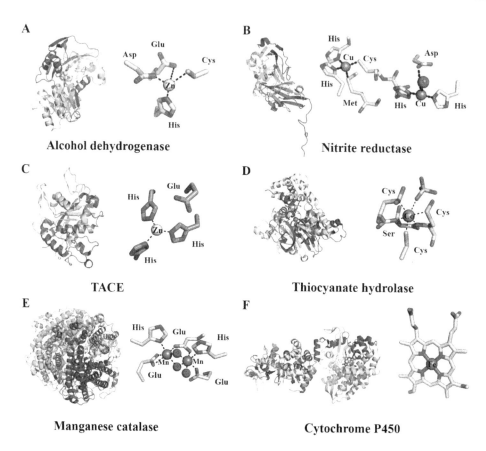

**A** Alcohol dehydrogenase

**B** Nitrite reductase

**C** TACE

**D** Thiocyanate hydrolase

**E** Manganese catalase

**F** Cytochrome P450

Fig. 1. **Metalloenzymes play key roles in biology.** A) Crystal structure of alcohol dehydrogenase (ADH) from *Thermoanaerobacter brockii* (PDB ID: 1YKF)[11]. ADH catalyzes the reversible oxidation of secondary alcohols to their corresponding ketones using NADP$^+$. The catalytic zinc ion is tetrahedrally coordinated by cysteine, histidine, aspartate, and glutamate residues. B) Crystal structure of nitrite reductase from *alcaligenes xylosoxidans* (PDB ID: 1OE1)[12]. Nitrite reductase is a key enzyme in denitrification catalyzing the reduction of nitrite (NO$_2^-$) to nitric oxide (NO). Two copper centers are involved in the electron transfer. C) Crystal structure of TNFα-converting enzyme (TACE) (PDB ID: 2OI0)[13]. TACE is the main shedding catalysis in the release of tumor necrosis factor-α (TNFα) from its pro domain. D) Crystal structure of thiocyanate hydrolase from *Thiobacillus thioparus* (PDB ID: 2DD5)[14]. Thiocyanate hydrolase is a cobalt (III)-containing enzyme that catalyzes the degradation of thiocyanate to carbonyl sulfide and ammonia. E) Crystal structure of manganese catalase from *Lactobacillus plantarum* (PDB ID: 1JKU)[15]. Manganese catalase is an important antioxidant metalloenzyme that catalyzes disproportionation of hydrogen peroxide, forming dioxygen and water. F) Cytochrome P450 from *Sulfolobus solfataricus* (PDB ID: 1IO7)[16] belongs to a family of heme containing enzymes that catalyze the oxidation of organic compounds using dioxygen.

For example, the large family of zinc-dependent matrix metalloproteinases (MMPs) is produced as cell-secreted and membrane-tethered enzymes that degrade extracellular matrix (ECM) proteins. These proteases play key roles in diverse physiological and pathological processes, including embryonic development, wound healing, inflammatory diseases, and cancer [6-9]. Zinc is also found in other important families of enzymes such as carbonic anhydrase, alcohol dehydrogenase and carboxypeptidase. This metal ion is special among first-raw transition metals due to its filled orbital configuration ($d^{10}$), providing its unique electronic characteristics, such as the ability to shuffle between four, five and six coordination geometries[10]. This flexible coordination chemistry is utilized by metalloenzymes to alter the reactivity of the metal ion, which appears to be important in the mediation of enzyme catalysis. For example, a substrate can bind to the metal ion during catalysis by substituting a bound ligand, or by binding to an empty coordination site. Alternatively, an inhibitor can bind to metal ion by the same mechanism, preventing the coordination of the substrate and thus interfering with catalysis. Therefore, there is a fundamental as well as a practical interest in quantifying the chemical and structural factors that govern mechanisms of action and contribute to catalytic efficiency in such metalloenzyme systems.

### 1.3 Enzymatic mechanisms

The reactions executed by metalloenzymes often involve a cycle of events that take place at the metal-protein's nearest environment during which several intermediate states, varied in their structure, reactivity and energetics, are formed. Therefore, structural and functional characterization of the active site's geometry and elucidation of the catalytic mechanism underlying this fascinating class of enzymes are of critical importance for assigning molecular mechanisms and for rationalizing the design of drugs. However, the identification and characterization of the short-lived metal-ligand/protein intermediates that evolve during catalysis is highly challenging. This is mainly due to the difficulty in directly investigating the detailed dynamic structural and electronic changes occurring in the active site during catalysis. In addition, several metal ions, such as the zinc ion that resides in the active site of MMPs, are spectroscopically silent and hence cannot be studied using conventional spectroscopic techniques. The majority of protein structures presently solved by X-ray crystallography actually correspond to a static state (or a state average) that poorly represents the ensemble of conformations adopted by a protein in action. Although protein structures provide information about protein folds and local structural arrangements, a single static structure is generally insufficient to enable the elucidation of enzymatic catalytic reaction mechanisms at an atomic level of detail. Typically, the catalytic cycle involves a series of intermediates and transition states that cannot be trapped by static or steady-state structural methods. Furthermore, determining the energies of the various stationary points in the cycle is not trivial from a theoretical or experimental point of view. Remarkably, time-resolved protein crystallography has been used to characterize the structure of intermediates that evolve in the course of enzymatic reactions (reviewed in [17, 18]). However, not all important functional intermediates and structural transitions can be captured in a crystalline form.

X-ray absorption spectroscopy (XAS) is one of the most sensitive method to probe the local environment of a metal ion with high spatial resolution (0.1Å). XAS was developed in the early 1970s as a method to structurally characterize the atomic environment of metal elements found in mineral, noncrystalline solid or adsorbed phases[19]. Nowadays, XAS is widely used in various fields including mineralogy, geochemistry, materials science and biology. High-quality XAS spectra can be collected on heterogeneous mixtures of gases, liquids, and/or solids with little or no sample pretreatments, making it ideally suited for many chemical and biological systems. In the field of biology, XAS is widely used in structure-function analysis of metalloproteins, as well as in the study of whole tissue or culture. In spite of the increase use of XAS, it is generally utilized as a technique for studying stable or equilibrium states, and not of intermediates. In recent years, with the advances in X-ray light sources and data analysis procedures, the time dimension was introduced into the field of XAS providing insights into the changes in the local structure of metal sites dictating metalloproteins molecular mechanisms. Here, we will describe the theoreticxal aspects of XAS as well as the development of stopped-flow and time-resolved XAS and discuss recent studies in which these techniques deciphered important molecular mechanisms driving the function of metalloproteins.

## 2. X-ray absorption spectroscopy (XAS)

### 2.1 Theoretical aspects

XAS measures the absorption of X-rays as a function of their energy. Specifically, the X-ray absorption coefficient is determined from the decay of the X-ray beam as it passes through the sample. The number of X-rays transmitted ($I_t$) through a sample is given by the intensity of X-rays impinging on the sample ($I_0$) decreased exponentially by the thickness of the sample (x) and the absorption coefficient of the sample ($\mu$):

$$I_t = I_0 e^{-\mu x} \tag{1}$$

In a typical experimental setup for XAS shown in Figure 2, the X-rays go through an ionization chamber to measure the number of incident X-rays ($I_0$), then through the sample, and then through another ionization chamber to measure the number of transmitted X-rays ($I_t$).

Fig. 2. **Schematic drawing of an EXAFS experimental setup**.

The X-ray absorption coefficient is determined by rearranging Eq. (1):

$$\mu x = \ln\left(\frac{I_0}{I_t}\right) \tag{2}$$

The X-ray absorption coefficient, $\mu$, decreases as the energy increases except for a sudden sharp rise at certain energies called edges. Each such absorption edge is related to a specific atom present in the sample, and it occurs when an incident X-ray photon has just enough energy to cause a transition that excites a particular atomic core-orbital electron to the free or unoccupied continuum level. K-edges, for example, refer to transitions that excite the innermost 1s electron. The spectral region closer to the edge, conventionally within 50 eV of the absorption edge, is called XANES (X-ray absorption near edge structure). This

Application of Stopped-Flow and Time-Resolved X-Ray Absorption Spectroscopy to the Study of Metalloproteins Molecular Mechanisms

265

region is often dominated by strong scattering processes as well as local atomic resonances in the X-ray absorption and is therefore more difficult for quantitative analysis. However, even a qualitative comparison of this region is useful in providing information about the redox state of the metal center and details about the local site symmetry, bond length and orbital occupancy.

Above the edges, there is a series of wiggles or an oscillatory structure that modulates the absorption. In these energies, the created photoelectron also possesses a kinetic energy (KE) which equals to the difference between the incident X-ray energy (E) and the electron binding energy of the photoelectron ($E_0$):

$$KE = E - E_0 \tag{3}$$

The photoelectrons can be described as spherical waves propagating outward from the absorber atoms. These photoelectron waves are scattered from the atoms surrounding the absorber. The relative phase of the outgoing photoelectron wave and the scattered wave at the absorbing atom affects the probability for X-ray absorption by the absorber atom. The relative phase is determined by the photoelectron wavelength and the interatomic distances between the absorber and scattering atoms. The outgoing and backscattered photoelectron waves interfere either constructively or destructively, giving rise to the modulations in the X-ray absorption coefficient spectra. Those oscillatory wiggles beyond ~30eV above the absorption edge contain important high-resolution information about the local atomic structure around the atom that absorbed the X-ray. This region is called extended X-ray absorption fine structure (EXAFS).

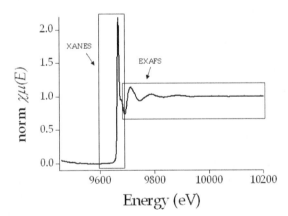

Fig. 3. **The X-ray absorption spectrum of ZnCl$_2$.** The XANES and EXAFS regions are highlighted.

The EXAFS function $\chi$ (E) is defined as:

$$\chi(E) = \frac{\mu(E) - \mu(E_0)}{\Delta\mu_0(E)} \tag{4}$$

where $\mu(E)$ is the measured absorption coefficient, $\mu_0(E)$ is a smooth background function representing the absorption of an isolated atom, and $\Delta\mu_0(E)$ is the measured edge step at the excitation energy, $E_0$. Since we can refer to the excited electron as a photoelectron, $\chi(E)$ is best understood in terms of wave behavior. Therefore, it is common to convert the X-ray absorption to $k$, the wave number of the photoelectron, which is defines as:

$$k = \sqrt{\frac{2m(E - E_0)}{\hbar^2}} \tag{5}$$

where $E_0$ is the absorption edge energy, m is the electron mass and h is the Planck constant. The EXAFS region of the spectra provides more detailed structural information and can be quantitatively analyzed to obtain highly accurate near-neighbor distances, coordination numbers and disorder parameters. Each atom at the same radial distance from the absorber contributes to the same component of the EXAFS signal. This group of atoms is called a shell. The number of atoms in the shell is called the coordination number. The oscillations corresponding to the different near-neighbor coordination shells can be described and modeled according to the EXAFS equation which is summed over all EXAFS absorber backscatterer pairs j:

$$\chi(k) = \sum_j \frac{N_j f_j(k) e^{2k^2\sigma_j}}{kR_j^2} \sin[2kR_j + \delta_j(k)] \tag{6}$$

where N is the number of neighboring atoms, $f(k)$ and $\delta(k)$ are the scattering properties of the atoms neighboring the excited atom (the amplitude and phase shift, respectively), R is the distance to the neighboring atom, and $\sigma^2$ is the thermal disorder in the metal-neighbor distance (Debye-Waller factor). Analysis of the EXAFS equation for a given experimental spectra enables determining N, R and $\sigma^2$ with high accuracy providing high resolution structural information regarding the local environment around the excited metal ion.

## 2.2 Application of XAS in biology

Applying XAS to the study of biological systems has become widely used in the last three decades[20-24]. Since XAS accurately reports the structure of metal–protein centers, early workers focused on providing additional high resolution structural and electronic information on crystallographically characterized samples. For instance, XAS enabled detailed structural investigation of metal active sites in imidazolonepropionase[25], cytochrome P450[26, 27], CO dehydrogenase/acetyl-CoA synthase[28-30], manganese catalases[31, 32], and lipoxygenase[33] by providing key insights into their electronic states and atomic structures. Moreover, insights into the enzymatic reaction mechanisms could be derived from XAS analysis, as was shown for tyrosine hydroxylase[34], molybdenum(Mo)-nitogenase[35-37], and farnesyltransferase[38]. Importantly, XAS is also a suitable method for characterizing metallodrug/protein interactions[39] and it was shown to be useful for resolving some of the apparent discrepancies among different spectroscopic and crystallographic studies of metalloproteins[30]. Also, XAS can provide structural information into metalloproteins that cannot be easily crystallized like membrane proteins.

## 2.3 Time-resolved XAS as a tool to investigate protein reaction mechanisms

One of the main advantages of XAS is that experiments can be carried in any state of the matter (e.g. solution and within membranes). This present XAS as important tools to depict structural transient changes during protein function. The first time-resolved X-ray absorption measurement was conducted at 1984 by smith and coworkers following the reaction of recombination of carbon monoxide with myoglobin after laser photolysis with 300μsec time resolution[40]. In laser photolysis (also known as 'pump-probe'), a particular bond of the reagent is cleaved by a pulse of light so that reactive intermediates are formed. EXAFS flash photolysis studies are performed by exciting the sample with a laser pulse and then exposing it to the X-ray beam for a rapid scan in order to characterize the evolving states upon excitation. Here, changes in the pre-edge structure and in the position of the iron edge of this protein as a function of time revealed several intermediates during binding of carbon monoxide to the iron metal center in myoglobin. Later experiments were used to characterize the photoproducts of Cob(II)alamin[41], the nickel site of *chromatium vinosum* hydrogenase[42] and photosystem II[43]. However, the repertoire of biological reactions to which this approach can be applied is limited because a suitable target for the flash is rarely available.

Rapid freeze quenching[44] is a different strategy to capture transient reaction intermediates to be probed by a variety of spectroscopic methods such as time-resolved electron paramagnetic resonance (EPR) and solid-state nuclear magnetic resonance (NMR)[45-49]. The freeze-quench method is based on the coupling of rapid mixing of the reactants to a freezing device that quenches the reaction mixture in a given state and creates a stable sample. By repeating this process while allowing the reactants to react for various time periods before freezing, a series of different time-points along the reaction path are produced. The first application of the rapid freezing procedure to EXAFS studies was described by George et al. in their study of the molybdenum centre of xanthine oxidase[50]. This preliminary study provided the distances of the first shell coordinating atoms of a reaction intermediate formed after 600msec, while it was known from other kinetic studies that at this time point the intermediate composed most of the reaction mixture. In later experiments, the feasibility of trapping single intermediate states during metalloenzyme catalysis by rapid freeze-quench was demonstrated for binuclear metalloenzyme systems [51, 52]. However, data analysis of multiple time-dependent samples requiring deconvolution of the data was not developed to this point.

Probing metalloenzyme reactions by time-dependent XAS procedures was suggested as a promising procedure to elucidate changes occurring in critical metal centers during the course of enzymatic reactions[53, 54]. Most challenging, however, is the correlation of structural and electronic changes in the metal ion's microenvironment with enzyme kinetics and protein structural dynamics. The principal difficulty in studying these critical questions is the inability to continuously follow the chemical and structural changes occurring at catalytic metal sites in real time. Inspired by this problem, we set out to advance the experimental approach of dynamic XAS on metalloenzymes during catalysis. Specifically, we introduced the use of quantitative structural-kinetic analyses of metalloenzymes in solution during their interactions with substrates and ligand. This experimental strategy employs stopped-flow freeze–quench XAS in conjunction with transient kinetic studies. By implementing the use of spectral convolution and non-linear curve fitting data analysis procedures, we could directly correlate structural and electronic changes as a function of distinct kinetic phases at atomic resolution[55, 56]. Owing to technical difficulties and the requirement of ample amounts of protein samples, the time-resolved XAS data collection remained challenging.

The following sections provide an overview on data collection and analyses using stopped-flow freeze-quench XAS and discuss the impact of this experimental strategy in revealing novel reaction mechanisms.

## 2.4 Stopped-flow XAS sample preparation and analysis procedures
### 2.4.1 Sample preparation
The experimental strategy utilizes stopped-flow freeze-quench XAS, which is conducted in conjunction with transient kinetic studies (Figure 4). To trap catalytic intermediates at the absorber metal ion during metalloenzyme catalysis, it is necessary to follow the off-equilibrium enzymatic rates. This allows following the formation of enzyme–substrate complexes prior to the steady-state reaction stage and improves the ability to trap and isolate evolving accumulated intermediate states resulting from their relatively high concentrations within this time frame. Transient reaction kinetics can be monitored by measuring the change in solution optical parameters, such as absorption or fluorescence, upon the formation of the reaction products, during the first catalytic cycles. Importantly, the reaction conditions of the transient kinetic experiments (concentration, buffer) should be equivalent to the conditions that will be later used during freeze-quench. Once the kinetic phases and time frame of the catalytic cycle are established, time points along the kinetic trace are selected for freeze-quench, usually 10-15 points along the different kinetic phases. In order to obtain a good XAS signal, the concentration of the metalloenzyme should be of at least hundreds of micromolars, to enable measuring the metal absorption spectra in fluorescence mode. Much higher concentrations should be used to measure the absorption in transmission mode. The samples for XAS are prepared in a stopped-flow instrument equipped with a freeze-quench device. Enzyme and substrate are mixed in the stopped-flow and freeze-quenched at selected time points into XAS sample holder by rapid freezing in a liquid nitrogen bath containing a freezing solution. These samples are kept frozen in special Dewars until taken to the synchrotron.

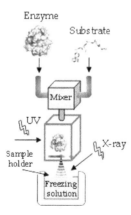

Fig. 4. **Schematic representation of a stopped-flow XAS experimental setup to monitor kinetic phases involved in the proteolytic cycle of a metalloenzyme.** This method allows one to establish a correlation between the pre-steady-state kinetic behavior of an enzyme and the structure of the evolving transient metal–protein intermediates as well as local charge transitions of the metal ion. The enzyme is rapidly mixed with substrate in a stopped-flow apparatus (mixing time = 2-5 ms). The instrument is equipped with a freeze-quench device. The pre-steady-state kinetic trace is monitored via an in-line fluorescence detector and analyzed. Enzyme/substrate mixtures are freeze-quenched at selected time point and mounted on an X-ray absorption sample holder.

## 2.4.2 Data analysis

EXAFS spectra of time-dependent measurements are particularly difficult to analyze since the spectra are composed of a disordered heterogeneous mixture of species, of which the mixing fractions change with time. Each species that contains the absorbing element may have a quite different local coordination around its nearest environment. This greatly complicates the non-linear curve fitting analysis used to extract the structural information, because the number of relevant structural parameters may be comparable to or may even exceed the number of independent data points in the experimental spectra. To account for this problem, we developed two complementary approaches for the EXAFS data analysis of mixtures, namely, **principal component analysis (PCA)** and **residual phase analysis (RPA)**[57]. Our data analysis strategy is shown in Figure 5.

PCA is initially used to estimate the number of minimal components required to convolute the EXAFS spectra. This procedure is sensitive to spectral changes that may be resolved by XAS. Using PCA, it is often possible to reveal (model-independently) the number of different species represented in the sample without a priori knowledge of their identity. However, these may not include all the reaction intermediates but instead only those that can be resolved by XAS. Each spectrum in the collection of EXAFS spectra measured at different time points may be represented by a linear combination of minimal components, which simplifies the data analysis procedure. The unique advantage of this method lies in its robustness and its model-independent determination of the number of unique species in the samples. If good experimental standards exist for representing each species, this method can also be reliably used to obtain both the identities and the mixing fractions of all the species in the sample[58].

In order to obtain a crude assessment of structural intermediates with spectral signatures above the noise level, PCA is followed by **multiple data set fitting** analysis. This step provides general trends in the dynamic changes in coordination number and metal-ligand bond distances, while employing several chemically and physically reasonable constraints between the fitting parameters. Representative time points that exhibited the most pronounced spectral changes are simultaneously fit to a theoretical model. In this analysis, the variable fitting parameters corresponding to the atomic neighbors that are not estimated to undergo dramatic changes during turnover are constrained to be the same for all time points, while fitting parameters corresponding to atomic neighbors that undergo dynamic changes in the distance or coordination number remain varied. This way, the number of fit variables is reduced, enabling crude characterization of reaction intermediates without the need to fit the entire data.

The RPA approach utilizes one of the known components as a "starting phase". The "starting phase" is then fractionated and iteratively subtracted from the total XAS signal to produce the corresponding residual spectra. In a time-dependent experiment, the reaction zero point is usually selected to be the "starting phase" since the identity of this species is known in advance. This state is usually the free enzyme state (without substrate or ligand). Thus, the known phase EXAFS data are subtracted from the experimental data at each time step. The residual phase is analyzed with the fewer number of adjustable variables providing the best fit values. It also increases the confidence that relative differences in coordination numbers of the intermediates are real. The combination of PCA and RPA in time-resolved XAS analysis of metalloenzymes has been shown to simplify the data analysis procedure[55,56].

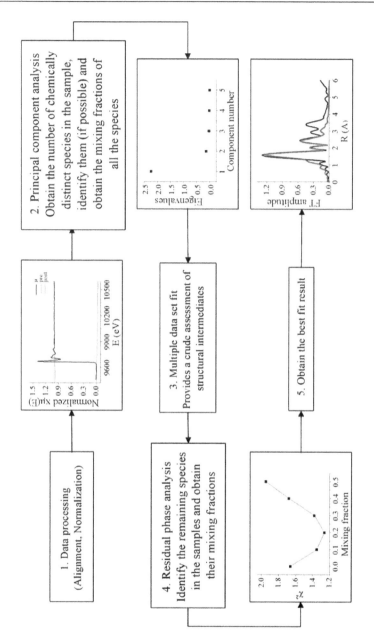

Fig. 5. **Time-resolved XAS data analysis strategy**. The processed data processing is analyzed by model-independent evaluation of the number of intermediate states by Principal Component Analysis (PCA), followed by a theoretical Multiple Data-Set (MDS) fit of the time-dependent data in order to identify possible models and, finally, the Residual Phase Analysis (RPA) study of the intermediate state, to obtain quantitative results.

Overall, these spectral analyses provide structural-kinetic models of XAS resolved reaction intermediates at a given reaction center. Importantly, the structure of the evolving intermediates and the total effective charge of the absorber ions may be directly correlated with the enzymatic kinetic trace. Thus, this experimental procedure provides the means to characterize metalloenzyme reaction pathways and hence to directly elucidate their molecular mechanisms under physiologically relevant solution conditions. Using stopped-flow XAS, we analyzed numerous bacterial and human zinc-dependent enzymes during the course of their enzymatic interactions[55, 56, 59, 60]. Remarkably, some of the derived mechanistic information was in striking contrast to the textbook mechanism in which the metal ion was proposed to remain attached to a fixed set of protein ligands. Probing the structural-kinetic reaction modes during metalloenzyme turnover indicated that there are dramatic differences in the nature of the metal ligation and the total effective charge of the catalytic metal ion. This further emphasizes the importance of dynamic fluctuations at the metal center in changing the electrostatic potential in the active site during the enzymatic reaction.

## 3. Mechanistic insights derived from stopped-flow/time-resolved XAS

### 3.1 Stopped-Flow XAS provides novel insights into enzymatic mechanisms

A functional study of zinc in enzymology is in the focus of many fields owing to the important roles that this ion plays in biological systems[61, 62]. The flexibility in coordination geometry facilitates ligand exchange processes more than for other ions and enhances the ability of zinc to promote catalysis[63]. XAS is an excellent structural tool to probe the $d^{10}$ zinc ion, which is generally spectroscopically silent. We therefore chose to begin our structural-kinetic studies on a bacterial member of the well-studied alcohol dehydrogenase (ADH) family, the zinc-dependent *Thermoanaerobacter brockii* alcohol dehydrogenase (TbADH).

This enzyme catalyzes the reversible oxidation of secondary alcohols to their corresponding ketones using NADP$^+$ as the coenzyme[64]. Crystallographic studies of various alcohol dehydrogenases and their complexes indicate that this enzyme is a tetramer comprising four identical subunits. Each subunit contains a single catalytic zinc ion tetrahedrally coordinated by a single cysteine, histidine, aspartate, and glutamate residue[11]. The reaction mechanism has been investigated by spectroscopic and kinetic tools[65]; however the details of the ADH reaction mechanism remain controversial. Specifically, the intermediate states that evolve at the zinc ion during catalysis and the nature of its coordination chemistry have been the subject of debate for many years[66-68]. Two main types of structural mechanisms have been proposed for horse liver ADH (HLADH); they differ specifically in the hypothesized coordination of the zinc ion during catalysis. Dworschack and Plapp[66] proposed that the alcohol molecule is added to the tetrahedral zinc ion to form penta-coordinated zinc during catalysis. In contrast, Dunn *et al.*[69] proposed a mechanism whereby the substrate displaces the zinc-bound water without undergoing a penta-coordinated intermediate.

Applying stopped-flow freeze-quench XAS coupled with pre-steady-state kinetics to TbADH, it was possible to quantify the structure, electronics, and lifetimes of the evolving reaction intermediates at the catalytic zinc ion[55]. Specifically, pre-steady-state kinetic analysis revealed that the reaction consists of a lag period of 25 msec followed by a kinetic burst of 35 msec. Stopped-flow freeze-quenched raw X-ray fluorescence data detected a series of alternations in the coordination number and structure of the catalytic zinc ion with concomitant changes in metal–ligand bond distances. Even though stopped-flow XAS analysis cannot determine unambiguously the metal ion coordination numbers due to the error associated with this measurement (which could be around ±1), its advantage lies in the fact that each spectrum is compared relatively to the others. Non-linear curve fitting analysis of the Fourier transform (FT) of the EXAFS signal together with principal component analysis and residual phase analysis detected relative changes in the number of first shell neighbors along the reaction coordinates. Therefore, we were able to conclude that the zinc ion does not remain stationary during the reaction, but is undergoing changes in its coordination chemistry, as proposed by Dworschack et al[66] and Makinen et al.[70] This study emphasizes the advantage of stopped-flow XAS in elucidating in real time the chemical nature of reaction intermediates that cannot be identified by any other means. Importantly, these results were recently supported by a series of high-resolution crystal structures of a relative enzyme, glucose dehydrogenase from *Haloferax mediterranei*, crystallized in the presence of substrates and products, in which dramatic differences in the nature of the zinc ligation were observed. Analysis of these structures revealed the direct consequence of linked movements of the zinc ion, a zinc-bound water molecule, and the substrate during progression through the reaction. In addition, it provided a structural explanation for multiple penta-coordinated zinc ion intermediates[71].

The high-resolution structural-kinetic analyses of ADH raised the possibility that the dynamic nature of the catalytic zinc in this metalloenzyme is directly related to its function and thus it may be shared by other analogous systems. Accordingly, by applying stopped-flow XAS on human MMPs, we could directly correlate structural and electronics dynamics with function.

## 3.2 The functional effect of active site electronics and structural-kinetic transitions evolving during metalloproteinase action

TNFα-converting enzyme (TACE) is the main shedding catalysis in the release of tumor necrosis factor-α (TNFα) from its pro domain[72]. TACE is also critical for the release of a large number of cell surface proteins from the plasma membrane, including multiple cytokines and receptors[73-75]. TACE is a member of the *a d*isintegrin *a*nd *m*etalloproteinase (ADAM) family[76]. The ADAMs are highly homologous to the MMP family and are also characterized by the presence of a zinc metalloproteinase domain. The striking homology in 3D structures of the catalytic site of both families is the main obstacle in developing selective MMP and ADAM inhibitors.

Peptide hydrolysis by the MMP and the ADAM family members is facilitated via a nucleophilic attack of the zinc-bound water hydroxyl on the peptide scissile bond carbonyl. The structural and chemical knowledge underlying the enzymatic activity of these proteases is very limited. Most structural and biochemical studies, as well as medicinal chemistry efforts carried out so far, were limited to non-dynamic structure/function characterization. Substantial efforts have been aimed at better understanding the molecular basis by which the catalytic zinc machineries of metalloproteinases execute their enzymatic reactions. Protein crystallography, proteomic, and theoretical studies have been instrumental in proposing reaction mechanisms[77-79]. Structural snapshots of protein complexes with substrate analogues and inhibitors have suggested inconsistent and controversial reaction mechanisms. Applying stopped-flow XAS methodology to TACE provided the first structural-kinetic reaction model of a metalloproteinase in real time[56]. This allows one to visualize changes in a catalytic center that researchers have, until now, been unable to investigate. Remarkably, during the kinetic lag phase, dynamic charge transitions at the active site zinc ion of TACE prior to substrate engagement with the ion were detected by time-dependent quantification of the change in edge energy at the zinc K-edge during peptide hydrolysis. This result implied communication between distal sites of the molecule and the catalytic core. The kinetic lag phase was followed by the binding of the peptide substrate to the zinc ion, forming a penta-coordinated transient complex, followed by product release and restoration of the tetrahedral zinc–protein complex. Thus, this work underscores the importance of local charge transitions critical for proteolysis as well as long-sought evidence for the proposed reaction model of peptide hydrolysis[80]. Furthermore, these results reveal novel communication pathways (yet unresolved) taking place between distal protein sites and the enzyme catalytic core (such as substrate-binding protein surfaces) and the catalytic machinery.

Interestingly, we have recently provided evidence that different MMPs and ADAMs possess variations in the polarity of the active site residues, which may be mediated by the specific enzyme/substrate surface interactions; these are reflected in typical electrostatic dynamics at the enzyme active site. Therefore, we hypothesize that these differences in active site electrostatics may be utilized in the rational design of selective inhibitors for this class of enzymes if their structural bases are also revealed[81].

### 3.3 Probing active site structural-dynamics during catalytic protein-protein interactions

The advantage of having XAS measurement in solution during physiological interactions of macromolecules was extended to the analysis of structural-kinetic behavior of the catalytic zinc ion in MMPs during zymogen activation. Here we challenged the stopped flow freeze quench technology by exploring the feasibility of rapidly mixing two proteins, namely, pro-MMP-9 and the activator serine protease enzyme while we used dynamic XAS to directly probe the nearest environment of the catalytic zinc ion in pro-MMP-9 during the physiological activation process[59].

The MMP family, as mentioned previously, comprises a large group of zinc-dependent endopeptidases involved in a variety of biological processes[82]. The MMPs are secreted or membrane tethered as zymogens as a control mechanism of MMP activity. The pro-peptide domain contains a "cysteine switch" PRCXXPD consensus sequence in which the cysteine coordinates to the zinc ion of the zymogen[83]. In this state, the catalytic zinc ion is tetrahedrally coordinated to three histidine residues and the conserved cysteine residue. Sequential proteolytic cleavages of the pro-domain by physiological proteases near the zinc-coordinated cysteine lead to replacement of the cysteine coordination shell by a water molecule and to the formation of an active enzyme. Activation of MMP zymogen is a vital homeostatic process; therefore, the structural and the kinetic bases of its molecular mechanism are of great interest[84].

The kinetic time frame for full enzyme activation was determined using gel-based analysis[85] revealing that the MMP-9 pro-domain is cleaved at two points in two sequential proteolytic points mediated by tissue kallikrein. The initial cleavage (at 1 second) resulted in a relatively inactive intermediate in which the pro-domain remained bound to the enzyme shielding the catalytic zinc site from solution. Stopped-flow X-ray spectroscopy of the active site zinc ion was used to determine the temporal sequence of pro-MMP-9 zymogen activation catalyzed by tissue kallikrein protease. The identity of three intermediates seen by X-ray spectroscopy was corroborated by molecular dynamics simulations and QM/MM calculations.

Remarkably, the cysteine-zinc interaction that maintains enzyme latency is disrupted upon MMP-9-tissue kallikrein interactions at 400 milliseconds and the fourth ligand is replaced by a water molecule. QM/MM calculations indicated active site proton transfers that mediate three transient metal-protein coordination events prior to the binding of water. It was demonstrated that these events ensue as a direct result of complex formation between pro-MMP-9 and kallikrein, and occur prior to the pro-domain proteolysis and the eventual dissociation of the pro-peptide from the catalytic site. This work revealed that synergism exists among long-range protein conformational transitions, as well as local structural rearrangements, and that fine atomic events take place during the process of zymogen activation[59]. In addition, the proposed mechanisms provide the molecular basis for physiological non-proteolytic or allosteric activation mechanisms of MMPs by free radicals and macromolecular substrates.

## 4. Integration of time-resolved XAS with structural-dynamic advanced experimental tools to probe functional protein conformational dynamics: Future prospects

Proteins are flexible entities exhibiting backbone and side chain dynamics at various timescales that seems to play a crucial role in protein stability, function, and reactivity[86]. These protein motions were found to be critical for many biological events including enzyme catalysis, signal transduction, and protein-protein interactions[87-89]. Even though the three-dimensional structures of manifold functional proteins and enzymes are available, revealing molecular dynamic details that determine protein action and enzymatic functionality still represents a scientific challenge. Moreover, following these reactions in real time represents a major challenge in structural biology. Here, we have demonstrated the advantage of time-resolved XAS in characterizing the dynamic changes in the local atomic environment of metal active sites found in metalloenzymes during turnover with atomic resolution. However, this technique cannot provide broad molecular insights into functional structural-dynamic events taking place at other regions of the protein which are essential to its function. Moreover, stopped-flow XAS can only provide information on event taking place at the slow timescales (microseconds-milliseconds) due to the stopped-flow mixing times, neglecting information on events taking place at faster timescales. Thus, combining time-resolved XAS with complimentary hard core structural-dynamic experimental tools should provide a more comprehensive description of metalloenzymes during catalysis[90].

A wealth of time-resolved structural-kinetic and spectroscopic techniques, including X-ray based techniques, have been developed in the last few decades to cope with these challenges, revealing the molecular structures of transient low-populated intermediate states[90]. Using time-resolved X-ray crystallography, it is possible to observe structural changes in crystals positioned in the X-ray beam by recording a time series of diffraction patterns using short X-ray pulses before and after the onset of a reaction[91]. By using sudden trigger events (e.g. laser-flash excitation) for initiating the reaction and by varying the delay time between the trigger and X-ray pulse, one can determine individual structures separated by time intervals of picoseconds[92-94]. Time-resolved wide-angle X-ray scattering (TR-WAXS) accurately probes tertiary and quaternary structural changes in proteins in solution with nanosecond time resolution[95]. In a TR-WAXS experiment, a laser pulse is used to trigger the protein's structural change, and transient structures are then followed by delayed X-ray pulses from an undulator in a straight section of the synchrotron. Structural changes occurring in the sample leave their 'fingerprints' represented by the differences between the signals measured before and after the laser initiates the reaction; these differences can be monitored as a function of time[96]. Attempts to probe large conformational transitions of protein side chains by small angle X-ray scattering (SAXS) in real time currently are being developed by combining stopped-flow techniques and computational analysis. Similarly, introducing time-resolved Fourier transform infrared (FT-IR) spectroscopy to biochemistry enables one to investigate structural and functional properties of soluble and membrane proteins along their reaction pathways[97-100]. Here we outline only a few methods that provide a true means to follow protein conformational dynamics in real time as well as in equilibrium. Other methods such as mass spectroscopy (MS), NMR, and circular dichroism (CD) can also provide structural information during protein/nucleic acid functional dynamic events. Each of the described methods has the potential to yield key important mechanistic information. However, combining a few experimental tools that may probe different biophysical aspects of a given biological system can be more effective in deriving structure-based mechanistic information.

This concept is illustrated in our recent work in which we have developed an integrated experimental approach combining stopped-flow XAS together with kinetic terahertz absorption spectroscopy (KITA)[101]. KITA experiments were recently introduced for investigating solvation dynamics during biological reactions such as protein folding[102]. Our newly developed approach was used to study the correlation between structural-kinetic events taking place at an enzyme's active site with protein-water coupled motions during peptide hydrolysis by a MMP. Stopped-flow XAS in conjugation with transient kinetic analysis provided the structure and life time of evolving metal-protein reaction intermediates at the metalloenzyme active site. Integration of the KITA experiment under identical reaction conditions allowed quantification of collective protein-water motion during the various kinetic phases and in correlation to the evolved structural intermediates at the enzyme catalytic site. This highly integrated experimental approach provided novel insights into the proposed role of water motions in mediating enzyme catalysis. Specifically, it was demonstrated that a "slow to fast" gradient of water motions at the enzyme surface is gradually modified upon substrate binding at the enzyme active site. This study suggest that conformational fluctuations contributed by protein conformational transitions of both enzyme and substrates effect water motion kinetics. Such synergistic effect may assist enzyme-substrate interactions via remote water retardation effects and thus impact the enzymatic function.

## 5. Conclusions

Here we have described advancements in using time-dependent XAS to probe metalloenzymes during their physiological activities. Time-resolved XAS provides atomic resolution insights into the changes in the local structure and electronics of catalytic centers found in metalloenzymes during action. This information is often critical to reveal reaction mechanisms and structure-function relationships dictating the activity of this fascinating class of enzymes. This review also highlights the need to probe functional macromolecule conformational dynamics in order to obtain critical mechanistic information at atomic level detail. Better understanding of the dynamic characteristics of these metal centers may also be crucial to drug design in structurally homologous enzymatic families. Future advancements may focus on overcoming the technical barriers encountered when combining real-time structural and spectroscopic experimental with X-ray absorption spectroscopy.

## 6. Acknowledgment

This work is supported by the Israel Science Foundation, the Kimmelman Center at the Weizmann Institute, and the Ambach family fund. I.S. is the incumbent of the Pontecorvo Professorial Chair.

## 7. References

[1] W. Shi and M. R. Chance, *Current opinion in chemical biology*, 2011, 15, 144-148.

[2] J. Finkelstein, *Nature*, 2009, 460, 813.

[3] B. L. Vallee and D. S. Auld, *Matrix (Stuttgart, Germany)*, 1992, 1, 5-19.

[4] D. S. Auld, *Biometals*, 2001, 14, 271-313.

[5] C. R. Chong and D. S. Auld, *Biochemistry*, 2000, 39, 7580-7588.

[6] C. E. Brinckerhoff and L. M. Matrisian, *Nature reviews*, 2002, 3, 207-214.

[7] A. Page-McCaw, A. J. Ewald and Z. Werb, *Nature reviews*, 2007, 8, 221-233.

[8] L. M. Matrisian, G. W. Sledge, Jr. and S. Mohla, *Cancer research*, 2003, 63, 6105-6109.

[9] C. Lopez-Otin and L. M. Matrisian, *Nature reviews cancer*, 2007, 7, 800-808.

[10] K. A. McCall, C. Huang and C. A. Fierke, *The Journal of nutrition*, 2000, 130, 1437S-1446S.

[11] Y. Korkhin, A. J. Kalb, M. Peretz, O. Bogin, Y. Burstein and F. Frolow, *Journal of molecular biology*, 1998, 278, 967-981.

[12] M. J. Ellis, F. E. Dodd, G. Sawers, R. R. Eady and S. S. Hasnain, *Journal of molecular biology*, 2003, 328, 429-438.

[13] B. Govinda Rao, U. K. Bandarage, T. Wang, J. H. Come, E. Perola, Y. Wei, S. K. Tian and J. O. Saunders, *Bioorganic & medicinal chemistry letters*, 2007, 17, 2250-2253.

[14] T. Arakawa, Y. Kawano, S. Kataoka, Y. Katayama, N. Kamiya, M. Yohda and M. Odaka, *Journal of molecular biology*, 2007, 366, 1497-1509.

[15] V. V. Barynin, M. M. Whittaker, S. V. Antonyuk, V. S. Lamzin, P. M. Harrison, P. J. Artymiuk and J. W. Whittaker, *Structure*, 2001, 9, 725-738.

[16] S. Y. Park, K. Yamane, S. Adachi, Y. Shiro, K. E. Weiss, S. A. Maves and S. G. Sligar, *Journal of inorganic biochemistry*, 2002, 91, 491-501.

[17] K. Moffat, *Chemical reviews*, 2001, 101, 1569-1581.

[18] B. L. Stoddard, *Methods*, 2001, 24, 125-138.

[19] D. E. Sayers, E. A. Stern and F. W. Lytle, *Physical review letters*, 1971, 27, 1204–1207.

[20] I. Ascone and R. Strange, *Journal of synchrotron radiation*, 2009, 16, 413-421.

[21] L. Powers, J. L. Sessler, G. L. Woolery and B. Chance, *Biochemistry*, 1984, 23, 5519-5523.

[22] E. Scheuring, S. Huang, R. G. Matthews and M. R. Chance, *Biophysical journal*, 1996, 70, Wp341-Wp341.

[23] M. R. Chance, L. M. Miller, R. F. Fischetti, E. Scheuring, W. X. Huang, B. Sclavi, Y. Hai and M. Sullivan, *Biochemistry*, 1996, 35, 9014-9023.

[24] S. K. Lee, S. D. George, W. E. Antholine, B. Hedman, K. O. Hodgson and E. I. Solomon, *Journal of the American Chemical Society*, 2002, 124, 6180-6193.

Application of Stopped-Flow and Time-Resolved X-Ray Absorption Spectroscopy to the Study of Metalloproteins
Molecular Mechanisms

277

[25] F. Yang, W. Chu, M. Yu, Y. Wang, S. Ma, Y. Dong and Z. Wu, *Journal of synchrotron radiation*, 2008, 15, 129-133.

[26] M. Newcomb, J. A. Halgrimson, J. H. Horner, E. C. Wasinger, L. X. Chen and S. G. Sligar, *Proceedings of the National Academy of Sciences of the United States of America*, 2008, 105, 8179-8184.

[27] A. Dey, Y. Jiang, P. Ortiz de Montellano, K. O. Hodgson, B. Hedman and E. I. Solomon, *Journal of the American Chemical Society*, 2009, 131, 7869-7878.

[28] W. Gu, S. Gencic, S. P. Cramer and D. A. Grahame, *Journal of the American Chemical Society*, 2003, 125, 15343-15351.

[29] J. Seravalli, W. Gu, A. Tam, E. Strauss, T. P. Begley, S. P. Cramer and S. W. Ragsdale, *Proceedings of the National Academy of Sciences of the United States of America*, 2003, 100, 3689-3694.

[30] W. Gu, J. Seravalli, S. W. Ragsdale and S. P. Cramer, *Biochemistry*, 2004, 43, 9029-9035.

[31] T. C. Weng, W. Y. Hsieh, E. S. Uffelman, S. W. Gordon-Wylie, T. J. Collins, V. L. Pecoraro and J. E. Penner-Hahn, *Journal of the American Chemical Society*, 2004, 126, 8070-8071.

[32] A. J. Wu, J. E. Penner-Hahn and V. L. Pecoraro, *Chemical reviews*, 2004, 104, 903-938.

[33] R. Sarangi, R. K. Hocking, M. L. Neidig, M. Benfatto, T. R. Holman, E. I. Solomon, K. O. Hodgson and B. Hedman, *Inorganic chemistry*, 2008, 47, 11543-11550.

[34] M. S. Chow, B. E. Eser, S. A. Wilson, K. O. Hodgson, B. Hedman, P. F. Fitzpatrick and E. I. Solomon, *Journal of the American Chemical Society*, 2009, 131, 7685-7698.

[35] C. C. Lee, M. A. Blank, A. W. Fay, J. M. Yoshizawa, Y. Hu, K. O. Hodgson, B. Hedman and M. W. Ribbe, *Proceedings of the National Academy of Sciences of the United States of America*, 2009, 106, 18474-18478.

[36] Y. Hu, M. C. Corbett, A. W. Fay, J. A. Webber, K. O. Hodgson, B. Hedman and M. W. Ribbe, *Proceedings of the National Academy of Sciences of the United States of America*, 2006, 103, 17119-17124.

[37] M. C. Corbett, Y. Hu, A. W. Fay, M. W. Ribbe, B. Hedman and K. O. Hodgson, *Proceedings of the National Academy of Sciences of the United States of America*, 2006, 103, 1238-1243.

[38] D. A. Tobin, J. S. Pickett, H. L. Hartman, C. A. Fierke and J. E. Penner-Hahn, *Journal of the American Chemical Society*, 2003, 125, 9962-9969.

[39] I. Ascone, L. Messori, A. Casini, C. Gabbiani, A. Balerna, F. Dell'Unto and A. C. Castellano, *Inorganic chemistry*, 2008, 47, 8629-8634.

[40] D. M. Mills, A. Lewis, A. Harootunian, J. Huang and B. Smith, *Science*, 1984, 223, 811-813.

[41] W. Clavin, E. Scheuring, M. Wirt, L. Miller, N. Mahoney, J. Wu, Y. Lu and M. R. Chance, *Biophysical journal*, 1994, 66, A378-A378.

[42] G. Davidson, S. B. Choudhury, Z. Gu, K. Bose, W. Roseboom, S. P. Albracht and M. J. Maroney, *Biochemistry*, 2000, 39, 7468-7479.

[43] M. Haumann, P. Liebisch, C. Muller, M. Barra, M. Grabolle and H. Dau, *Science*, 2005, 310, 1019-1021.

[44] S. de Vries, *Freeze-Quench Kinetics*, John Wiley & Sons, Ltd, 2008.

[45] R. J. Appleyard, W. A. Shuttleworth and J. N. Evans, *Biochemistry*, 1994, 33, 6812-6821.

[46] M. Boll, G. Fuchs and D. J. Lowe, *Biochemistry*, 2001, 40, 7612-7620.

[47] R. M. Jones, F. E. Inscore, R. Hille and M. L. Kirk, *Inorganic chemistry*, 1999, 38, 4963-4970.

[48] L. Skipper, W. H. Campbell, J. A. Mertens and D. J. Lowe, *The Journal of biological chemistry*, 2001, 276, 26995-27002.

[49] W. Zhang, K. K. Wong, R. S. Magliozzo and J. W. Kozarich, *Biochemistry*, 2001, 40, 4123-4130.

[50] G. N. George, R. C. Bray and S. P. Cramer, *Biochemical society transactions*, 1986, 14, 651-652.

[51] J. Hwang, C. Krebs, B. H. Huynh, D. E. Edmondson, E. C. Theil and J. E. Penner-Hahn, *Science*, 2000, 287, 122-125.

[52] P. J. Riggs-Gelasco, L. Shu, S. Chen, D. Burdi, B. H. Huynh, J. P. Willems, L. Que and J. Stubbe, *Journal of the American Chemical Society*, 1998, 120, 849–860.

[53] G. N. George, B. Hedman and K. O. Hodgson, *Nature structural biology*, 1998, 5 Suppl, 645-647.

[54] M. A. Newton, A. J. Dent and J. Evans, *Chemical Society reviews*, 2002, 31, 83-95.

[55] O. Kleifeld, A. Frenkel, J. M. Martin and I. Sagi, *Nature structural biology*, 2003, 10, 98-103.

[56] A. Solomon, B. Akabayov, A. Frenkel, M. E. Milla and I. Sagi, *Proceedings of the National Academy of Sciences of the United States of America*, 2007, 104, 4931-4936.

[57] A. I. Frenkel, O. Kleifeld, S. R. Wasserman and I. Sagi, *Journal of Chemical Physics*, 2002, 116, 9449-9456.

[58] S. R. Wasserman, P. G. Allen, D. K. Shuh, J. J. Bucher and N. M. Edelstein, *Journal of synchrotron radiation*, 1999, 6, 284-286.

[59] G. Rosenblum, S. Meroueh, M. Toth, J. F. Fisher, R. Fridman, S. Mobashery and I. Sagi, *Journal of the American Chemical Society*, 2007, 129, 13566-13574.

[60] A. Solomon, G. Rosenblum, P. E. Gonzales, J. D. Leonard, S. Mobashery, M. E. Milla and I. Sagi, *The Journal of biological chemistry*, 2004, 279, 31646-31654.

[61] D. Beyersmann and H. Haase, *Biometals*, 2001, 14, 331-341.

[62] S. Frassinetti, G. Bronzetti, L. Caltavuturo, M. Cini and C. D. Croce, *Journal of environmental pathology, toxicology and oncology*, 2006, 25, 597-610.

[63] G. Parkin, *Chemical Communications*, 2000, 1971-1985.

[64] M. Peretz, O. Bogin, E. Keinan and Y. Burstein, *International journal of peptide and protein research*, 1993, 42, 490-495.

[65] O. Kleifeld, A. Frenkel, O. Bogin, M. Eisenstein, V. Brumfeld, Y. Burstein and I. Sagi, *Biochemistry*, 2000, 39, 7702-7711.

[66] R. T. Dworschack and B. V. Plapp, *Biochemistry*, 1977, 16, 111-116.

[67] R. Meijers, R. J. Morris, H. W. Adolph, A. Merli, V. S. Lamzin and E. S. Cedergren-Zeppezauer, *The Journal of biological chemistry*, 2001, 276, 9316-9321.

[68] L. Hemmingsen, R. Bauer, M. J. Bjerrum, M. Zeppezauer, H. W. Adolph, G. Formicka and E. Cedergren-Zeppezauer, *Biochemistry*, 1995, 34, 7145-7153.

[69] M. F. Dunn, J. F. Biellmann and G. Branlant, *Biochemistry*, 1975, 14, 3176-3182.

[70] M. W. Makinen, W. Maret and M. B. Yim, *Proceedings of the National Academy of Sciences of the United States of America*, 1983, 80, 2584-2588.

[71] P. J. Baker, K. L. Britton, M. Fisher, J. Esclapez, C. Pire, M. J. Bonete, J. Ferrer and D. W. Rice, *Proceedings of the National Academy of Sciences of the United States of America*, 2009, 106, 779-784.

[72] J. J. Peschon, J. L. Slack, P. Reddy, K. L. Stocking, S. W. Sunnarborg, D. C. Lee, W. E. Russell, B. J. Castner, R. S. Johnson, J. N. Fitzner, R. W. Boyce, N. Nelson, C. J. Kozlosky, M. F. Wolfson, C. T. Rauch, D. P. Cerretti, R. J. Paxton, C. J. March and R. A. Black, *Science*, 1998, 282, 1281-1284.

[73] A. P. Huovila, A. J. Turner, M. Pelto-Huikko, I. Karkkainen and R. M. Ortiz, *Trends in biochemical sciences*, 2005, 30, 413-422.

[74] C. P. Blobel, *Nature reviews*, 2005, 6, 32-43.

[75] J. Arribas and A. Borroto, *Chemical reviews*, 2002, 102, 4627-4638.

[76] T. G. Wolfsberg, P. Primakoff, D. G. Myles and J. M. White, *The Journal of cell biology*, 1995, 131, 275-278.

[77] S. J. Benkovic and S. Hammes-Schiffer, *Science*, 2003, 301, 1196-1202.

[78] I. Bertini, V. Calderone, M. Fragai, C. Luchinat, M. Maletta and K. J. Yeo, *Angewandte Chemie (International ed)*, 2006, 45, 7952-7955.

[79] L. P. Kotra, J. B. Cross, Y. Shimura, R. Fridman, H. B. Schlegel and S. Mobashery, *Journal of the American Chemical Society*, 2001, 123, 3108-3113.

[80] D. W. Christianson and W. N. Lipscomb, *Accounts Chem. Res.*, 1989, 22, 62-69.

[81] I. Sagi and M. E. Milla, *Analytical biochemistry*, 2008, 372, 1-10.

[82] K. Maskos and W. Bode, *Molecular biotechnology*, 2003, 25, 241-266.

[83] E. B. Springman, E. L. Angleton, H. Birkedal-Hansen and H. E. Van Wart, *Proceedings of the National Academy of Sciences of the United States of America*, 1990, 87, 364-368.

[84] H. J. Ra and W. C. Parks, *Matrix biology*, 2007, 26, 587-596.

[85] P. D. Brown, A. T. Levy, I. M. Margulies, L. A. Liotta and W. G. Stetler-Stevenson, *Cancer research*, 1990, 50, 6184-6191.

[86] K. Henzler-Wildman and D. Kern, *Nature*, 2007, 450, 964-972.

[87] P. K. Agarwal, *Journal of the American Chemical Society*, 2005, 127, 15248-15256.

[88] R. G. Smock and L. M. Gierasch, *Science*, 2009, 324, 198-203.

[89] M. S. Marlow and A. J. Wand, *Biochemistry*, 2006, 45, 8732-8741.

[90] M. Grossman, N. Sela-Passwell and I. Sagi, *Current opinion in structural biology*, 21, 678-85.

[91] S. Westenhoff, E. Nazarenko, E. Malmerberg, J. Davidsson, G. Katona and R. Neutze, *Acta crystallographica*, 2010, 66, 207-219.

[92] D. Bourgeois and A. Royant, *Current opinion in structural biology*, 2005, 15, 538-547.

[93] V. Srajer, Z. Ren, T. Y. Teng, M. Schmidt, T. Ursby, D. Bourgeois, C. Pradervand, W. Schildkamp, M. Wulff and K. Moffat, *Biochemistry*, 2001, 40, 13802-13815.

[94] F. Schotte, M. Lim, T. A. Jackson, A. V. Smirnov, J. Soman, J. S. Olson, G. N. Phillips, Jr., M. Wulff and P. A. Anfinrud, *Science*, 2003, 300, 1944-1947.

[95] M. Andersson, E. Malmerberg, S. Westenhoff, G. Katona, M. Cammarata, A. B. Wohri, L. C. Johansson, F. Ewald, M. Eklund, M. Wulff, J. Davidsson and R. Neutze, *Structure*, 2009, 17, 1265-1275.

[96] M. Cammarata, M. Levantino, F. Schotte, P. A. Anfinrud, F. Ewald, J. Choi, A. Cupane, M. Wulff and H. Ihee, *Nature methods*, 2008, 5, 881-886.

[97] I. Radu, M. Schleeger, C. Bolwien and J. Heberle, *Photochemical and photobiological sciences*, 2009, 8, 1517-1528.

[98] J. A. Wright and C. J. Pickett, *Chemical communications*, 2009, 5719-5721.

[99] F. Garczarek and K. Gerwert, *Nature*, 2006, 439, 109-112.

[100] F. Garczarek, J. Wang, M. A. El-Sayed and K. Gerwert, *Biophysical journal*, 2004, 87, 2676-2682.

[101] M. Grossman, B. Born, M. Heyden, D. Tworowski, F. G. B., I. Sagi and M. Havenith, *Nature structural and molecular biology*, 2011, 18, 1102-1108.

[102] S. J. Kim, B. Born, M. Havenith and M. Gruebele, *Angewandte Chemie (International ed*, 2008, 47, 6486-6489.

# Permissions

The contributors of this book come from diverse backgrounds, making this book a truly international effort. This book will bring forth new frontiers with its revolutionizing research information and detailed analysis of the nascent developments around the world.

We would like to thank Shatendra K. Sharma, for lending his expertise to make the book truly unique. He has played a crucial role in the development of this book. Without his invaluable contribution this book wouldn't have been possible. He has made vital efforts to compile up to date information on the varied aspects of this subject to make this book a valuable addition to the collection of many professionals and students.

This book was conceptualized with the vision of imparting up-to-date information and advanced data in this field. To ensure the same, a matchless editorial board was set up. Every individual on the board went through rigorous rounds of assessment to prove their worth. After which they invested a large part of their time researching and compiling the most relevant data for our readers. Conferences and sessions were held from time to time between the editorial board and the contributing authors to present the data in the most comprehensible form. The editorial team has worked tirelessly to provide valuable and valid information to help people across the globe.

Every chapter published in this book has been scrutinized by our experts. Their significance has been extensively debated. The topics covered herein carry significant findings which will fuel the growth of the discipline. They may even be implemented as practical applications or may be referred to as a beginning point for another development. Chapters in this book were first published by InTech; hereby published with permission under the Creative Commons Attribution License or equivalent.

The editorial board has been involved in producing this book since its inception. They have spent rigorous hours researching and exploring the diverse topics which have resulted in the successful publishing of this book. They have passed on their knowledge of decades through this book. To expedite this challenging task, the publisher supported the team at every step. A small team of assistant editors was also appointed to further simplify the editing procedure and attain best results for the readers.

Our editorial team has been hand-picked from every corner of the world. Their multi-ethnicity adds dynamic inputs to the discussions which result in innovative outcomes. These outcomes are then further discussed with the researchers and contributors who give their valuable feedback and opinion regarding the same. The feedback is then collaborated with the researches and they are edited in a comprehensive manner to aid the understanding of the subject.

Apart from the editorial board, the designing team has also invested a significant amount of their time in understanding the subject and creating the most relevant covers. They scrutinized every image to scout for the most suitable representation of the subject and create an appropriate cover for the book.

The publishing team has been involved in this book since its early stages. They were actively engaged in every process, be it collecting the data, connecting with the contributors or procuring relevant information. The team has been an ardent support to the editorial, designing and production team. Their endless efforts to recruit the best for this project, has resulted in the accomplishment of this book. They are a veteran in the field of academics and their pool of knowledge is as vast as their experience in printing. Their expertise and guidance has proved useful at every step. Their uncompromising quality standards have made this book an exceptional effort. Their encouragement from time to time has been an inspiration for everyone.

The publisher and the editorial board hope that this book will prove to be a valuable piece of knowledge for researchers, students, practitioners and scholars across the globe.

# List of Contributors

**Murid Hussain, Guido Saracco and Nunzio Russo**
Politecnico di Torino, Italy

**Leonid Ipaz, Julio Caicedo and Gustavo Zambrano**
Thin Films Group, Physics Department, University of Valle, Columbia

**William Aperador**
Colombian School of Engineering Julio Garavito, Bogotá DC, Columbia
Universidad Militar Nueva Granada, Bogota DC, Columbia

**Joan Esteve**
Department of Applied Physics Optics, Universitat de Barcelona, Spain

**M. Torres Deluigi and J. Díaz-Luque**
Universidad Nacional de San Luis, Argentina

**Leonardo Abbene and Gaetano Gerardi**
Dipartimento di Fisica, Università di Palermo, Italy

**Matjaž Kavčič**
J. Stefan Institute, Slovenia

**Carlos Angeles-Chavez, Jose Antonio Toledo-Antonio and Maria Antonia Cortes-Jacome**
Instituto Mexicano del Petróleo, Programa de Ingeniería Molecular, Eje Central Lázaro, D.F. México, Mexico

**Kengo Moribayashi**
Japan Atomic Energy Institute, Japan

**Rafał Sitko and Beata Zawisza**
Department of Analytical Chemistry, Institute of Chemistry, University of Silesia, Poland

**Frédéric Christien, Edouard Ferchaud, Pawel Nowakowski and Marion Allart**
Polytech'Nantes, University of Nantes, France

**Rossana Gazia, Angelica Chiodoni and Stefano Bianco**
Center for Space Human Robotics@PoliTo, Istituto Italiano di Tecnologia, Italy

**Edvige Celasco and Pietro Mandracci**
Materials Science and Chemical Engineering Department, Politecnico di Torino, Italy

**Sergio Pilling and Diana P. P. Andrade**
Inst. de Pesquisa & Desenvolvimento, Universidade do Vale do Paraíba (UNIVAP), São Jose dos Campos, Brazil

**Seiichiro Ii**
National Institute for Materials Science, Japan

**Moran Grossman and Irit Sagi**
Departments of Structural Biology and Biological Regulation, the Weizmann Institute of Science, Rehovot, Israel

Printed in the USA
CPSIA information can be obtained
at www.ICGtesting.com
JSHW011458221024
72173JS00005B/1121

9 781632 384652